AF743977

Key Steel Manufacturing Technologies by Collaborative Innovation

Series Editors

Guodong Wang, Northeastern University, Shenyang, China

Zhouhua Jiang, Shenyang, China

Miaoyong Zhu, Northeastern University, Shenyang, China

Yuexin Han, Northeastern University, Shenyang, China

Zhenyu Liu, Chinese Academy of Engineering, Beijing, China

Zhao-Dong Wang, Northeastern University, Shenyang, China

Jianping Li, Key Laboratory of Rolling and Automation, Shenyang, China

Dianhua Zhang, Beijing, China

Shixin Liu, National Natural Science Foundation of China, Beijing, China

Yimin Zhu, Northeastern University, Shenyang, China

The series were written to address the needs of collaborative innovation and development within the iron and steel industry. They focus on the primary challenges facing the industry in China, including resources, energy, environment, and product performance. Six research directions in key common technologies throughout the entire iron and steel manufacturing process were proposed as follows: beneficiation; smelting; continuous casting; hot rolling; short-process hot rolling; cold rolling.

The series share insights into the intelligent manufacturing, the internationally original processing theory and production equipment. The publication of this series of books will be of great significance to improve the theoretical system of key common technologies in the iron and steel industry.

The series focus on expounding the new generation of ladle bottom powder spraying refining technology and equipment, high-quality continuous casting billet production process and equipment, integrated structure and performance control of smelting, casting and rolling, heat treatment technology and equipment for hot-rolled sheet and strip products with extreme specifications, thin slab endless/semi-endless rolling + non-pickling coating process technology, complete process technology and equipment for high-performance silicon steel via thin strip continuous casting, high-precision flatness and edge thinning control technology and equipment, advanced annealing and coating technology and equipment, new technology of complex refractory iron ore pre-enrichment-suspension roasting-magnetic separation (PSRM), short-process green preparation of super iron concentrate and clean steel base material, intelligent manufacturing of long profiles, intelligent flat products manufacturing and other key and common technologies urgently needed in the steel industry. Most of these key and common technologies are of original innovation of Chinese scientific and technological workers and have been successfully popularized and applied in many Chinese steel enterprises, which have greatly promoted the green transformation and development of the steel industry in China. Most of the technologies have reached the internationally leading level, helped the China's steel industry leap from "Following Suit" to "Taking the Lead", and made an important contribution to achieving the goal of "green technology, intelligent equipment, high-quality products, and service-oriented supply".

This series of books are edited by Guodong Wang, director of Northeastern University's Collaborative Innovation Center of Steel Technology, and member of China Engineering Academy, with a number of academicians in the industry working as consultants, and many authoritative experts and scholars in various fields of the steel industry as co-authors.

This series are supported by the NATIONAL PUBLICATION FOUNDATION.

Zhaozhen Cai · Miaoyong Zhu

The Control Technology for Slab Surface Crack During Continuous Casting of Micro-alloyed Steel

冶金工业出版社
Metallurgical Industry Press

Springer

Zhaozhen Cai
Northeastern University
Shenyang, China

Miaoyong Zhu
Northeastern University
Shenyang, China

ISSN 2731-9733 ISSN 2731-9741 (electronic)
Key Steel Manufacturing Technologies by Collaborative Innovation
ISBN 978-981-96-7170-0 ISBN 978-981-96-7171-7 (eBook)
https://doi.org/10.1007/978-981-96-7171-7

Jointly published with Metallurgical Industry Press
The print edition is not for sale in China (Mainland). Customers from China (Mainland) please order the print book from: Metallurgical Industry Press.
ISBN of the Co-Publisher's edition: 978-7-5240-0238-3

© Metallurgical Industry Press 2025

This work is subject to copyright. All rights are solely and exclusively licensed by the Publisher, whether the whole or part of the material is concerned, specifically the rights of reprinting, reuse of illustrations, recitation, broadcasting, reproduction on microfilms or in any other physical way, and transmission or information storage and retrieval, electronic adaptation, computer software, or by similar or dissimilar methodology now known or hereafter developed.
The use of general descriptive names, registered names, trademarks, service marks, etc. in this publication does not imply, even in the absence of a specific statement, that such names are exempt from the relevant protective laws and regulations and therefore free for general use.
The publishers, the authors, and the editors are safe to assume that the advice and information in this book are believed to be true and accurate at the date of publication. Neither the publishers nor the authors or the editors give a warranty, express or implied, with respect to the material contained herein or for any errors or omissions that may have been made. The publishers remain neutral with regard to jurisdictional claims in published maps and institutional affiliations.

This Springer imprint is published by the registered company Springer Nature Singapore Pte Ltd.
The registered company address is: 152 Beach Road, #21-01/04 Gateway East, Singapore 189721, Singapore

If disposing of this product, please recycle the paper.

Foreword

As an important raw material industry, iron and steel industry plays an important role in the "supply side." It strives to provide the society with sufficient high quality and high-performance iron and steel products with the lowest resource and energy consumption, the lowest environmental and ecological load, and the highest efficiency and labor productivity, so as to meet the needs of social development, national security and people's life.

In the early stage of reform and opening-up, China's iron and steel industry was in the "following others" stage, mainly relying on the introduction of production lines and technologies from abroad. After more than 40 years of reform, innovation and development, China has a steel production capacity of more than one billion tons, which exceeds more than half of the world's steel output, and the iron and steel industry has developed rapidly. With continuously improved technical level, in the fierce international competition, China's iron and steel industry is not in any single stage of development (i.e., "following others" to "keeping pace" and even "leading the pack"), but rather in all of the three stages concurrently. However, it still faces the following four major problems in the technological development. Firstly, the consumption of iron and steel production resources and energy is huge, pollutant emissions are serious, and the environment is overwhelmed. Therefore, there is an urgent need to realize green technology. Secondly, the stability, uniformity and consistency of production equipment are poor, and the production efficiency is low. Thus, it is an urgent task for the iron and steel industry to realize the intellectualization of equipment, in-depth perception of information, coordinated and accurate control, intelligent optimization of decision-making and improvement of autonomous learning. Thirdly, insufficient product quality, unbalanced product structure, and insufficient supply capacity of high-performance products and independent innovation products all lead to strong demand for high-quality products. Fourthly, the supply side development quality of China's iron and steel industry is not high enough and the service is not in place. Consequently, it is necessary to focus on improving the quality and efficiency of development, take supporting the supply side structural reform as the main line, take improving the quality of the supply system as the main direction, build a service-oriented iron and steel industry and realize the service-oriented supply.

Having experienced a period of rapid development, China's iron and steel industry has entered a stage of structural adjustment and transformation in recent years. Iron and steel enterprises must change the mode of development, optimize the economic structure, and foster new drives of growth. They should put quality first and give priority to performance, pursue supply-side structural reform as the main task, work hard for better quality, higher efficiency, more robust drivers of economic growth through reform, and raise total factor productivity. They should also endeavor to make China's iron and steel industry a global leader in "green technology, intelligent equipment, high-quality products and service-oriented supply," and build China's iron and steel industry into a world leading iron and steel industry cluster.

In October 2014, the Collaborative Innovation Center of Steel Technology, with the two metallurgical universities of Northeastern University and Beijing University of Science and Technology as the core, and jointly established by enterprises, research institutes and other colleges and universities, was officially put into operation after being approved by the Ministry of Education and the Ministry of Finance.

From October 2014 to the end of 2018, Collaborative Innovation Center of Steel Technology has been operating for four years. The process and equipment R&D platform focuses on the key common processes and equipment technologies in the iron and steel industry. According to the overall development plan of the platform's top-level design and the tasks and indicators formulated in various research directions, and through the in-depth integration and collaborative innovation of industry, academia and research, the platform has made great achievements in six research directions, i.e., mining and beneficiation, smelting, hot rolling, short process, cold rolling, informatization and intellectualization. These achievements are all key common technologies urgently needed by the iron and steel industry. Major achievements are as follows:

- A new generation of ladle bottom powder spraying refining process and equipment technology;
- High-quality continuous casting slab production process and equipment technology;
- Integrated structure and performance control of refining, casting and rolling;
- Heat treatment process and equipment of hot rolled strip products with extreme specifications;
- Thin slab endless/semi-endless rolling + pickling-free coating process technology;
- Complete set of process technology and equipment for preparing high-performance silicon steel by thin strip casting;
- High-precision flatness and edge thinning control technology and equipment;
- Advanced annealing and coating technology and equipment;
- New technology of complex refractory iron ore preconcentration suspension roasting magnetic separation (PSRM);

- Short process green preparation of super iron concentrate and clean steel base material;
- Intelligent manufacturing of long profile;
- Intelligent manufacturing of flat material.

Most of these key common technologies belong to the original technologies of Chinese scientific and technological workers, and have been successfully promoted and applied in China's iron and steel enterprises, greatly promoting the green transformation and development of China's iron and steel industry. Most of the technologies have reached the international leading level as a whole, played an important part in promoting the role of China's iron and steel industry from "following others" to "leading the pack," and made an important contribution to realizing the goal of "green technology, intelligent equipment, high-quality products, and service-oriented supply."

General Secretary Xi Jinping pointed out in his speech at the 2014 Academician Conference of the Chinese Academy of Sciences and the Chinese Academy of Engineering, "We must strengthen overall planning and coordination, vigorously carry out collaborative innovation, concentrate our efforts on major events, and form a strong synergy to promote independent innovation." Looking back on more than two years of refining and application, and four years of arduous research and development, we are carrying out our work under the guidance of this idea. The expectations of the leaders and the workers of steel and iron enterprises for China's original technology have struck a chord in our souls and inspired us to rectify the results of cooperation and innovation and promote them, to make them become a sharp weapon for the technicians of iron and steel enterprises to overcome difficulties and win new victories. Therefore, we have the desire to write a series of books. This series of books will be based on the series of innovative achievements of the Collaborative Innovation Center of Steel Technology, with the whole process, green technology, equipment and engineering, and industrialization as the main line. Combined with the actual engineering projects and examples of high-quality steel produced on the production line of iron and steel industry, this series of books will systematically gather the research progress of industry-university-research collaborative innovation and application, as well as key common technologies, cutting-edge leading technologies and modern engineering technology innovation, so as to provide reference for enterprise technological reconstruction, transformation and upgrading, high-quality development and planning the future development blueprint. This idea has won the positive response, and full support and close coordination of the vast number of colleagues in the enterprises. The leaders and editors of the Metallurgical Industry Press specially came to the university to give enthusiastic guidance, put forward suggestions, and discuss specific matters in publishing the series of books.

The needs of the country and the expectations of the iron and steel industry have touched our hearts and inspired us to move forward; the support and guidance of colleagues, publishers and editors have given us strong confidence. The chiefs and academic backbones of the Collaborative Innovation Center and our intimate friends in enterprises and scientific research units acted immediately, splashed ink, and made great plans. We believe that through the joint efforts of all parties in the industry, academia and research and the publishing house, we will dedicate a series of books that are visible, valuable, informative and influential to the students in the steel and iron industry, as a return for the vigorous support from colleagues in enterprises. At the same time, on the occasion of the 70th anniversary of the founding of the People's Republic of China, we would like to present a gift forged with hard work, sweat, innovation and a pure heart to the 70th birthday of our great motherland.

July 2019

Guodong Wang
Academician of Chinese Academy of Engineering
Beijing, China

Preface

Steel is an indispensable basic material for the development of human society and plays a pivotal role in human production activities. Since the mid-19th century, with the emergence of open-hearth and converter steelmaking technologies, global steel production has surged. In 2024, global crude steel production reached 1.883 billion tons, and China reached 1.005 billion tons, accounting for 53.37% of the world's crude steel production.

While steel production is growing rapidly, the increasingly demanding service environment poses higher performance requirements for steel materials, mainly reflected in the requirements for higher strength, toughness, weldability, and corrosion resistance. Against this backdrop, the technology of microalloying steel was born and developed rapidly at the beginning of the 20th century. Especially since the late 1990s, the world's major steel-producing countries have successively formulated and implemented new generation steel material research and development plans, and the production and application of micro-alloyed steel have Micro-alloyed steels, which contain trace amounts of alloy elements such as Nb, V, Ti, B, Al, etc., have reached unprecedented development. Due to their excellent performances such as high strength, high toughness, and easy welding, they are currently widely used in critical fields such as energy and chemical industry, transportation, marine engineering, and defense industry. They have become one of the main products of steel at home and abroad, accounting for more than 15% of the global steel output.

Continuous casting of steel is another revolutionary technology in the development process of the steel industry following oxygen converter steelmaking. Due to its advantages of high efficiency, high metal yield, and low energy consumption compared to traditional mold casting, it has become the main production method for modern steel blanks, and also the most important production method for micro-alloyed steel blanks. At present, the continuous casting ratio in China has exceeded 99.7%, of which about 98% of the micro-alloyed steel blanks are produced by continuous casting.

In recent years, with the significant development of the technologies of clean steel smelting and continuous casting mold flux, the conventional defects such as subcutaneous inclusions and surface longitudinal cracks in micro-alloyed steel slabs have been nearly resolved, and are no longer the key common technical problems restricting the high-quality and efficient production of micro-alloyed steel. However, micro-alloyed steels that add strong carbon and nitrogen binding elements such as Nb, V, Ti, B, and Al, are prone to significantly reduce hot ductility due to the strong precipitation of carbides, nitrides, and carbonitrides at the grain boundaries of the cast structure during the continuous casting solidification process. This can lead to the high occurrence of quality defects such as edge and corner transverse cracks, off-corner cracks, and surface hot charging cracks in the slab. The occurrence of these cracks forces steel companies to cool the high-temperature slabs offline and carry out surface flame cleaning, which not only greatly increases the labor intensity of workers and extends the manufacturing cycle of products, but also causes a great waste of energy and environmental pollution. The direct economic loss caused by this in China's steel industry can reach five billion RMB per year, and the increase in CO_2 emissions exceeds five million tons. This is a bottleneck problem restricting the high-quality, efficient, and green production of micro-alloyed steel by steel companies at home and abroad. It urgently needs to be solved.

This book consists of five chapters. Chapter 1 provides an overview of the production and surface quality control of micro-alloyed steel slab continuous casting, the characteristics of typical micro-alloyed steel slab surface cracks in steel companies, and the current state of the main control methods. Chapter 2 details the formation mechanism of slab transverse corner cracks during micro-alloyed steel continuous casting and their control, the precipitation behavior of carbonitrides during typical composition micro-alloyed steel solidification, the thermo-mechanical behaviors of the slab solidification in the mold and secondary cooling zones, and proposes a new technology for controlling the slab transverse corner cracks based on high ductility of the corner structure; Chap. 3 details the formation mechanism of thin slab corner cracks, the heat transfer and mechanical behaviors of thin slabs solidification in the mold and secondary cooling zones, and proposes a new crack control technology based on the mold with the narrow face Gaussian concave structure and rapid cooling in narrow face foot roller zone; Chap. 4 introduces the heat transfer and deformation evolution behaviors of thick slabs in the mold and secondary cooling zones, the formation mechanism of longitudinal depression in the off-corners, and proposes a control technology for the off-corner depression based on the novel convex-structure mold and its foot roller strong cooling; Chap. 5 mainly introduces the cause of hot charging cracks on the wide surface of micro-alloyed steel slabs, the current main control measures, and gives the optimal quenching process for typical micro-alloyed steel and proposes a new quenching process at the solidification end of the continuous casting slabs.

Here, I would like to thank the leaders of Bao Steel, HBIS Group Co., Ltd., and other companies, as well as the industry for their trust, support, and help. I would especially like to thank team members Zhenyu Niu, Jiazhi An, Shi Bo Wang, Qi Xin, Biao Cheng, and other students for their unremitting efforts and important contributions to the process of related theory and technology research and development. Although their names do not appear as authors of this book, without their work, the writing of this book would not have been possible.

Due to limitations in knowledge, any shortcomings or inappropriate aspects in the book are sincerely hoped to be corrected by the readers.

Shenyang, China
March 2021

Zhaozhen Cai
Miaoyong Zhu

Competing Interests The authors have no competing interests to declare that are relevant to the content of this manuscript.

Contents

Chapter 1
Overview

1.1 Development of Micro-alloyed Steel and Its Continuous Casting Technology

Micro-alloyed steel is the abbreviation for micro-alloyed high strength low alloy steel [1–3], which is structural steel based on ordinary low carbon steel and ordinary high strength low alloy steel. By adding a small amount microalloy elements of Nb, V, Ti, B, Al, et. al, controlling the rolling and cooling, as well as heat treatment to induce the precipitation of microalloy elements, the toughness, formability, and weldability of the steel will be greatly improved. The amount of the alloy addition of micro-alloyed steel generally does not exceed 0.2% [4]. Micro-alloyed technology of steel has become the most important development direction of the new generation of steel materials [5, 6].

Micro-alloyed steel first appeared in the 1930s, by adding a small amount of V to ordinary low carbon steel to improve the toughness of the steel [7, 8]. In the 1960s and 1970s, home and abroad scholars propose theories of second phase particles to prevent grain coarsening of steel structure, and the behaviors of solid solution, dissolution and precipitation of micro-alloyed carbonitrides as well as their effects on austenite recrystallization, controlled rolling and cooling. In addition, Nb, Ti, B, Al, and so on have been gradually introduced as microalloy elements, which makes the micro-alloyed technology of steel develop rapidly and be widely used [9–12]. Since the late 1990s, with the increasingly stringent requirements for the service environment of steel, the demand for high-performance steels characterized by high strength, high toughness, corrosion resistance, and easy welding has been increasing year by year. The world's major steel producers have successively formulated and implemented research and development plans for new generation steel materials based on micro-alloyed steel, which has led to unprecedented development in the production and application of micro-alloyed steel. At present, micro-alloyed steel is widely used in key and critical fields such as the energy and chemical industry, transportation, marine engineering, and defense industry and has become one of the

© Metallurgical Industry Press 2025

Z. Cai and M. Zhu, *The Control Technology for Slab Surface Crack During Continuous Casting of Micro-alloyed Steel*, Key Steel Manufacturing Technologies by Collaborative Innovation, https://doi.org/10.1007/978-981-96-7171-7_1

main products of steel at home and abroad, accounting for more than 15% of the global steel output.

Micro-alloyed technology of steel was first adopted in China in 1979 and has successively gone through three key development stages: enhancing the existing high strength low alloy steel system with micro-alloyed technology, combining micro-alloyed technology and controlled rolling and cooling, and using of micro-alloyed technology in the research and development of new generation steel materials [13, 14]. After many years of research and production practice, micro-alloyed steel has now developed into a comprehensive category that is widely produced in large and medium steel companies and is one of the core products of current steel companies in China [15–19].

Continuous casting is the main production method of slab base metal at home and abroad, and it is also the most important production method of micro-alloyed steel. Before the 1970s, the continuous casting process and device technology of steel worldwide were still in the development stage. Due to the high crack sensitivity and other solidification characteristics during the casting process of micro-alloyed steel slab, the production of micro-alloyed steel slab mainly used ingot casting. However, the production efficiency and metal yield of the ingot casting process are low, and the energy consumption is high, so the development of the continuous casting production process of micro-alloyed steel became an important direction for the development of continuous casting technology at that time [20].

The large-scale continuous casting production of micro-alloyed steel mainly emerged after the 1970s, initially focusing primarily on billet continuous casting [21]. However, as different types of micro-alloyed steel were gradually developed and the application fields of micro-alloyed steel plates gradually expanded, the market demand for micro-alloyed steel plates surged. Against this backdrop, the efficient production of high quality base material of micro-alloyed steel slab using continuous casters became an urgent requirement in the industry. During this period, scholars from North America, Europe, Japan, and other countries conducted extensive and systematic research on the behavior of high temperature solidification of micro-alloyed steel (especially the precipitation of micro-alloyed carbonitrides and their impact on the hot ductility of the steel) [22–25], laying a key theoretical foundation for the development of micro-alloyed steel continuous casting technology. Since the 1980s, based on the breakthrough in the solidification theory of micro-alloyed steel, key processes and device technologies for the efficient continuous casting production of high quality micro-alloyed steel slab were extensively developed. After nearly 20 years of rapid development, almost all types of micro-alloyed steel are produced on slab continuous casters with different cross sections.

However, since the invention of micro-alloyed steel slab continuous casting technology, due to the limitations of the cooling modes of conventional slab continuous casting process and the high temperature solidification characteristics of micro-alloyed steel, the frequent occurrence of quality defects such as slab surface cracks during continuous casting and subsequent hot working. These defects restricted the high quality, efficient, and green production of micro-alloyed steel in long time. In recent years, the country is vigorously developing the marine engineering, large bridges, ultra-high-rise buildings, large pressure vessels, oil pipelines, and other large or extra-large projects, it is required to further improve the performance and quality of micro-alloyed steel plates to cope with the harsh service environment. To meet these requirements, the types and contents of alloy components in these micro-alloyed steels have significantly increased, which significantly aggravates the crack sensitivity during slab solidification and subsequent hot working. Frequent occurrence of surface cracks of the continuous casting and subsequent hot working of micro-alloyed steel slab is the common technical challenge in the steel industry. The technology of micro-alloyed steel slab continuous casting, represented by crack-free production, is an important direction in the development of continuous casting at home and abroad.

1.2 Micro-alloyed Steel Continuous Casting and Typical Surface Quality Defects

1.2.1 Transverse Corner Cracks of Slab

Conventional and wide-thick slab continuous casters are the main continuous casting casters of production slab in steel companies at home and abroad. Among them, the typical thickness of the slab produced by conventional continuous caster is 180–230 mm, and the width is 800–1650 mm. The typical thickness of the slab produced by wide-thick slab continuous caster is 250–450 mm, and the width is 1600–2600 mm. These are the main base materials for hot rolled coils and wide-thick plates. Conventional slab and wide-thick slab continuous casters have gone through stages of vertical continuous casters, vertical bending continuous casters, and arc continuous casters. After many years of development, the main slab continuous casters have evolved into a vertical arc continuous caster composed of "mold + bender + bow segments + straight segments + horizontal segments," as shown in Fig. 1.1. To ensure slab solidification, this type of continuous caster has longer secondary cooling zones, and the height from the mold meniscus to the entrance of the straight segments is often more than 15.0 m. The corners of the slab produced are mostly right-angle structures.

Currently, influenced by slab thickness, production pace, and slab quality, the casting speed for micro-alloyed steel, such as Nb, B, Al, V, and other bearing steel with high crack sensitivity, typically ranges from 0.6 to 1.5 m/min when using vertical arc continuous caster for slab production. Under this relatively low casting speed, the corner of the slab with a right-angle structure is subjected to two-dimensional heat transfer effects from the multiple pathways such as cooling water, radiation, and roll-slab contact heat transfer. The corner temperature entering the straight segments of the continuous caster is generally in the range of 760–850 °C. Under the influence of straight and misaligned roller gaps, the corner (particularly the inner arc corner) of micro-alloyed steel slab often forms transverse cracks, as shown in Fig. 1.2. If the slab corner with the transverse crack defects is not effectively cleaned, cracks

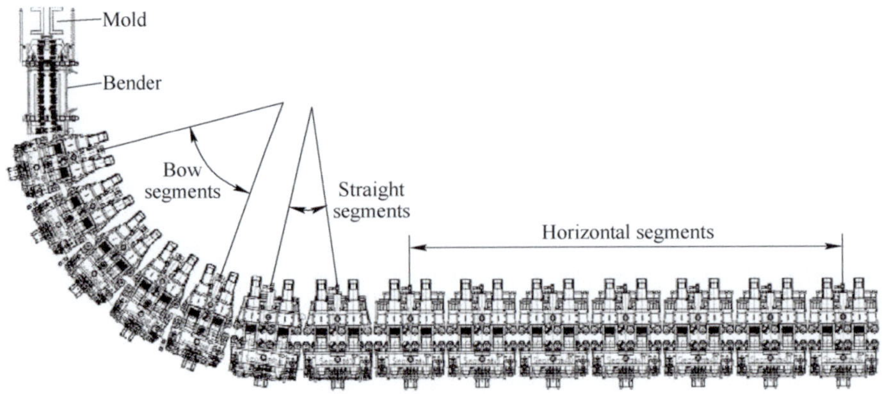

Fig. 1.1 Casting flow diagram of conventional slab continuous caster

will form on the edge of the steel plate after rolling, as shown in Fig. 1.3, which is a common technical problem in the industry.

The transverse crack forming at the corners of slab is the most common type of surface crack defect during the continuous casting production of micro-alloyed steel

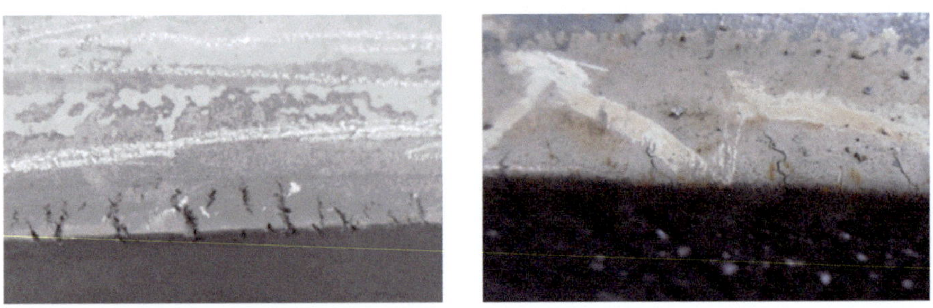

Fig. 1.2 Morphology of transverse corner cracks of micro-alloyed steel slab

Fig. 1.3 Morphology of edge cracks of steel plates. **a** Wide-thick plate; **b** hot rolled coil

slab [26–29]. They predominantly occur at the slab corner and adjacent area, with a majority found at the inner arc corner. This type of crack typically appears at the troughs of oscillation mark of slab and can propagate along the grain toward the center of the narrow and wide faces of slab, expanding to form regular or irregular cross-corner cracks. Generally, this type of slab corner crack is slight, and it is often difficult to visually detect under the cover of iron oxide scale and other covers. It often requires 1–2 mm of slab corner surface flame cleaning or pickling treatment to reveal. The general depth of the transverse corner crack of slab is generally in the range of 0.2–8.0 mm, and that of the opening width is in the range of 0.2–2.0 mm. Also, the severe cracks can reach a depth of more than 10.0 mm and the opening width can reach more than 3.0 mm.

Current steel companies worldwide have developed and implemented various technologies and devices to control corner crack in the continuous casting of micro-alloyed steel slabs, yet these measures have not been able to completely prevent the occurrence of cracks. For slab that have developed transverse corner cracks, common treatments include offline flame cleaning or corner cutting. These methods not only significantly increase labor intensity, reduce metal yield, and extend the manufacturing cycle of products, but also makes the slab unable to achieve hot charging. The process of heating slab from room temperature to the hot charging temperature in the furnace results in significant energy waste and environmental pressure.

1.2.2 Corner Cracks of Thin Slab

Thin slab continuous casting and rolling is a new short process for producing hot rolled coil, which was successfully developed in the late 1980s. It represents one of the most significant revolutionary technologies of steel industry since oxygen blown converter steelmaking and continuous casting, which has the advantages of shortened production process, increased efficiency, significant energy saving, and emission reduction effect and a high level of automation [30–33]. In recent years, with the rapid development of high-speed continuous casting and endless rolling technology, thin slab continuous casting and rolling production lines are rapidly constructed in major steel producing countries. ESP and other production lines have begun to be constructed on a large scale, especially under the considerable benefits of efficient production of thin strip steel by endless rolling technology. By the end of 2018, 68 thin slab production lines with a total of 103 strands had been constructed and put into operation worldwide, with an annual capacity exceeding 110 million tons [34]. Among these, China has constructed and put 17 thin slab production lines into operation, totaling 32 strands, with a production capacity exceeding 40 million tons, making them a vital component of Chinese hot rolled strip production lines. At the same time, multiple thin slab production lines, including Rizhao Steel, Shougang Steel, Taihang Steel, and others, are constructed, commissioned and put into operation.

Currently, there are mainly two types of thin slab continuous casters; the first is the vertical bending continuous caster, such as CSP continuous caster, as shown in Fig. 1.4a. The mold length is usually 1.1–1.2 m, and mostly funnel mold. The height of the vertical from the mold exit to the bending point is long, usually 10–12 m. After the vertical zone, the slab is bent into the continuous straight zone, then cut to size, and charged into the heating furnace. The second type of thin slab continuous caster is similar to the conventional vertical arc continuous caster, with typical production lines including FTSR, ISP, and ESP, as shown in Fig. 1.4b. The mold of this type of continuous caster is also a 1.1–1.2 m funnel mold. The slab exits the mold, enters the bow segment through a short vertical zone, and then runs out horizontally after continuous straightening. At present, the typical slab thickness produced by thin slab continuous caster at home and abroad is mainly 50–100 mm, and the width is 900–1600 mm. According to different processes and slab thicknesses, the casting speed is generally 3.8–6.0 m/min.

Micro-alloyed steel is also an important product component of the thin slab continuous casting and rolling production line. The thin slab continuous casting and rolling production line has the characteristics of short production process and low cost, and some production lines can even produce "hot rolling instead of cold rolling" high quality medium-width strip steel with the thickness of not more than 1.0 mm. In recent years, a significant number of high strength steel and medium and low automotive structural steels containing alloy elements such as Nb, V, Ti have been produced by this process [35], with the proportion growing steadily. Micro-alloyed steel now accounts for over 30% of the thin slab continuous casting and rolling production at steel companies include Han Steel and Lianyuan Steel.

However, in the actual continuous casting of thin slab production, thin slab of micro-alloyed steel, which contains alloy elements such as Nb, V, Ti, and B, has

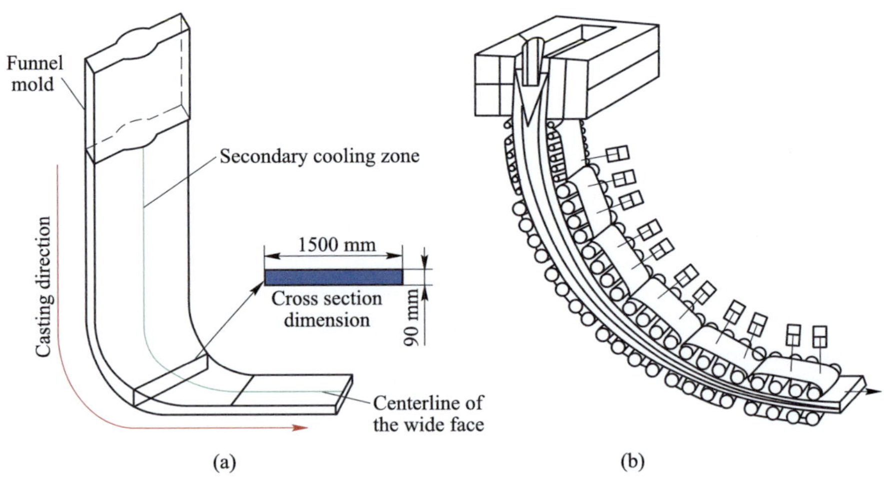

Fig. 1.4 Schematic diagrams of thin slab continuous casters. **a** Vertical bending caster; **b** arc caster

Fig. 1.5 Morphology of transverse corner crack of the micro-alloyed steel thin slab and its edge crack of hot rolled coil. **a** Transverse corner crack of slab; **b** edge crack of hot rolled coil

high crack sensitivity under the high cooling strength of mold and secondary cooling zones. The slab corner is in the third brittle temperature zone of the steel during the large deformation process under the liquid core reduction, which is prone to cause transverse corner cracks in the slab corner as shown in Fig. 1.5a. Because the thin slab continuous casting and rolling production line is very compact, the slab in the production process cannot be offline corner cleaned. The high frequency of longitudinal cracks and transverse corner cracks leads to a very high occurrence rate of crack on the surface of the micro-alloyed steel (the morphology of transverse corner crack left on the rolled products is shown in the Fig. 1.5b). The edge crack rate of micro-alloyed steel plates in some companies has surpassed 7.0% (with certain steel grades even exceeding 30%), seriously restricting the high quality and efficient production of the thin slab continuous casting and rolling production lines.

1.2.3 Off-Corner Depression and Its Crack of Wide-Thick Slab

Thick slab is the main base material for wide-thick plate rolling, and its slab thickness generally exceeds 250 mm. Currently, with the thickness of maximum continuous casting breaking through 450 mm, it is developing toward the direction of continuous casting of extra-wide and extra-thick slab with the width exceeding 2400 mm and the thickness reaching 600 mm. This type of continuous caster generally adopts a taper of 1.0–1.1% on the narrow face of the mold in the production process of micro-alloyed steel, but only a few companies such as Wuyang Steel have used a taper exceeding 1.2%. Because of the low casting speed of ultra-thick slab, their solidification shrinkage within the mold is greater compared to conventional slab casting. During actual production, the copper plate on the narrow face of the mold cannot effectively accommodate the solidification shrinkage of the slab. This causes the corner rotation to drive the wide face off-corner to form a large gap, which causes the wide face off-corner of the slab to form a "hot zone."

Similarly, limited by the low casting speed, to ensure that the micro-alloyed steel continuous casting production process has a higher temperature of slab corner through straight segment, the continuous casting production process of traditional wide-thick slab (especially micro-alloyed steel with high crack sensitivity) often adopts the overall weak cold control mode of secondary cooling including narrow face foot roller. As a result, the narrow face of shell after the mold cannot quickly form a shell with sufficient support strength, resulting in a narrow face bulge as shown in Fig. 1.6a. And then it causes a serious longitudinal depression in the "hot zone" on the wide face off-corner as shown in Fig. 1.6b. Meanwhile, it forms surface transverse crack defects and subsurface longitudinal crack defects in the depression zone during the subsequent continuous bending and straightening deformation process of the slab as shown in Fig. 1.6c, d. In the actual continuous casting production process, if the surface transverse cracks in the already formed depression zone are not timely treated with flame cleaning, the slab will form edge crack defects in the subsequent rolling process, as shown in Fig. 1.7. Thereby, it causes the steel plate to be downgraded or scrapped due to unqualified size. The large edge cutting of the steel plate also greatly reduces the metal yield of the final product.

1.2.4 Hot Charging Crack of Micro-alloyed Steel Slab

Due to the low casting speed, the temperature of the slab surface and interior at the exit of the continuous caster in the conventional and wide-thick plate continuous casting process is difficult to meet the requirements of continuous rolling. The slab hot charging technology originated in the 1960s [36], it is an important innovative technology that efficiently connects the traditional continuous casting and rolling. It cuts the slab into a specified size and delivers it directly to the heating furnace in a hot state through roller conveyor and other methods from the continuous casting workshop, thereby changing the traditional cold charging mode of the slab needing offline stacking and then charging the furnace, effectively utilizing the heat of the slab, and fully playing the role of energy saving and emission reduction. Compared with the traditional cold charging process, the slab hot charging technology has significant advantages in energy saving and emission reduction, reducing metal consumption, shortening the production cycle, and reducing the occupation of the plant area [37–39].

Most of the steel companies in China were built or rebuilt after the 1990s, and most of the conventional or thick slab continuous casting lines have the function of hot charging of high temperature continuous casting slab. According to the actual continuous casting and rolling production line layout of different steel companies, the surface temperature of the slab directly charged to the heating furnace is generally in the range of 500–700 °C. At this temperature, part of the austenite structure on the surface of the slab has undergone a $\gamma \rightarrow \alpha$ phase transition, resulting in a

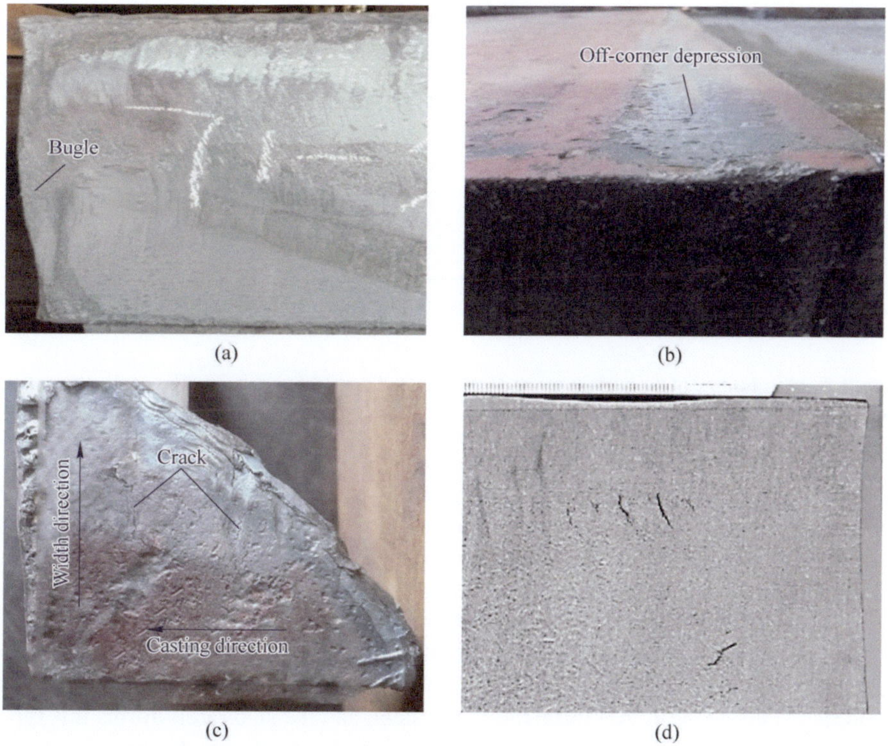

Fig. 1.6 Morphology of off-corner depression and its crack on the wide face of micro-alloyed steel thick slab. **a** Slab with bugle; **b** slab with off-corner longitudinal depression; **c** surface transverse crack on the off-corner depression zone; **d** subsurface crack below the off-corner depression zone

Fig. 1.7 Morphology of the edge crack on the steel plate caused by off-corner depression and the surface transverse crack

dual-phase structure of $\gamma + \alpha$. The untransformed austenite still exists as coarse original austenite grains, forming a mixed grain structure. Simultaneously, at this temperature, most microalloy elements have combined with the C and N in the steel, and the precipitation of microalloy carbonitride dual-phase particles occurs or completes, primarily distributed along the original austenite grain boundaries.

(a) (b)

Fig. 1.8 Morphology of the surface crack of the slab and the steel plate caused by the hot charging. **a** Surface crack after slab heating; **b** steel plate

Due to the pinning effect of the dual-phase structure consisting of coarse original austenite and grain boundary ferrite, along with the concentrated precipitation of grain boundary dual-phase particles, the hot charging slab is prone to form the hot charging cracks during the heating process, as shown in Fig. 1.8a. These defects then form surface crack defects on the steel plate after rolling, as shown in Fig. 1.6b.

In the actual production process, the slab with hot charging cracks is mostly concentrated in steel grades containing alloy elements such as Nb, V, and Al, such as Q345B/C/D/E, Q345R, AH36/DH36 series steel, and the slab rarely forms this type of crack before entering the heating furnace, which is only forms during the heating process of the slab. Therefore, it is difficult to detect or effectively prevent in the actual production process. Generally, the surface cracks of the hot charging micro-alloyed steel slab are mostly produced on its wide face, and they are continuously distributed in a network structure, and the crack opening width and depth are different. The opening size of this type of crack is mostly 0.1–1.5 mm, and the depth is 0.5–3.5 mm, but the opening size and depth of individual large cracks can reach more than 1.5 mm and 5 mm, respectively. The hot charging cracks of the micro-alloyed slab generally cannot be eliminated during the rolling process, and the surface crack defects left on the steel plate are often unable to be repaired due to their excessive depth, so they are often degraded or discarded.

1.3 Research Progress of Transverse Corner Crack Control of Micro-alloyed Steel Slab

1.3.1 Formation Position of the Crack on Slab

The continuous casting process of micro-alloyed steel goes from the mold to the exit of the continuous casting, undergoing complex solidification growth of structure and macroscopic thermo/mechanical evolution. Traditionally, the corner crack of the micro-alloyed steel continuous casting slab is formed in the mold link and expands during the secondary cooling solidification process. Cramb et al. [40] summarized the formation process of the corner crack of the slab as follows:

(1) Under the effect of unreasonable mold oscillation and mold flux physical properties, molten steel level fluctuation, and its solidification heat transfer in the meniscus, deep oscillation marks are formed on the surface of the slab corner.
(2) The structure at the bottom of the deep oscillation mark forms coarse austenite grains due to the low heat transfer rate, and there is obvious solute microsegregation at the grain boundary, thus forming a brittle structure.
(3) With the slab moving downward in the mold, the friction between the solidified shell and the mold significantly increases. When the friction force exceeds the critical strength of the shell, it will cause oscillation mark tearing.
(4) After the slab exits the mold, in the cooling zone of the mold foot roller and the bender, the forced spray of cooling water affects on the surface of the slab, causing excessive thermal stress on the surface of the slab. Thermal stress is easy to concentrate at the bottom of deep oscillation marks, further causing the bottom of the oscillation marks to crack along the austenite grain boundaries.
(5) If there are deviations in the alignment of the mold and the bender, the support rollers of the segment do not rotate and deform, and the accuracy of the bow segment and roller gap is significantly deviated, which will cause the already formed cracks to further expand.
(6) In the secondary cooling zone, the combined effects of water cooling and intermittent heat transfer from the roller and slab intensify the precipitation of carbonitrides in the surface structure of the slab corner, thereby further increasing the brittleness of the slab surface structure.
(7) During the straightening process of the slab, its upper surface is subjected to tensile stress. If the surface temperature is in the brittle zone, excessive tensile strain can easily exceed the critical strain of the slab surface structure, thereby forming surface transverse cracks.

Thus, Cramb et al. have attributed the formation of transverse corner cracks of micro-alloyed steel slabs to the entire process of mold, secondary cooling, and straightening process. They believed that the formation of deep oscillation marks on the slab surface in the mold is the initial cause, and excessive friction is the key to causing crack initiation. The accuracy of the casting roller array and cooling, straightening, and other processes are important factors in expanding cracks.

However, in the actual production process of micro-alloyed steel continuous casting, most steel companies, under the condition of ensuring the accuracy of the continuous casting, micro-alloyed steel slab corner transverse cracks often occur on the inner arc, while the outer arc is relatively less, or even no cracks occur. Therefore, most researchers currently believe that the transverse corner cracks of micro-alloyed steel slab should occur in the secondary cooling stage of continuous casting straightening or roller gap misalignment [41–45].

1.3.2 Formation Mechanism of the Slab Transverse Corner Cracks

The frequent occurrence of corner transverse crack defects in the continuous casting process of micro-alloyed steel slab is a common technical problem in the industry, which is common in major steel companies at home and abroad. Therefore, research on the causes of this type of crack has started since the 1970s. However, so far, there is no unified conclusion on its cause. Summarizing previous studies, the causes can generally be attributed to internal and external factors.

1.3.2.1 Internal Causes

The internal cause of the formation of transverse corner cracks of slab is mainly related to high temperature mechanical properties. In actual continuous casting production, the hot ductility of the steel structure is often used as a key index to measure its crack resistance. For this reason, the internal cause of transverse corner cracks in micro-alloyed steel continuous casting slab is generally attributed to the structural defect factors that reduce the hot ductility of the corner due to the solidification process of the continuous casting slab.

For the study of hot ductility of steel structure, Brimacombe et al. [46] first pointed out that there are two typical brittle temperature zones in the solidification process of steel, which are the brittle zone with temperature above 1340 °C and the range of 700–900 °C. The main reason for the formation of the second brittle zone is related to the precipitation of dual-phase particles of Ti, Nb, V, and other alloy elements. On the basis of Suzuki et al. [47, 48], the hot ductility range of steel is further divided into three brittle temperature zones. Among them, the temperature of the first brittle zone is higher than 1200 °C, the temperature of the second brittle zone is 900–1200 °C, and the temperature of the third brittle zone is 600–900 °C. Based on the theory of grain boundary embrittlement, Wolf et al. [49] further divided the three brittle temperature zones of steel and proposed that the enrichment of impurity elements such as Cu, Sn, and Sb is an important reason for the aggravation of hot ductility of steel structure. In the new century, Thomas and Mintz et al. [50–53] further proposed that the growth of proeutectoid ferrite along the austenite grain boundary is another important factor leading to the formation of the third brittle zone in steel.

Generally speaking, the internal causes that affect the hot ductility of steel mainly include the original austenite grain size [54], the precipitation size, and distribution form of microalloy carbonitride [55, 56], as well as the structural characteristics at different temperatures [57]. In the actual continuous casting production, the surface layer such as the corner of the slab solidifies to form coarse austenite grains. This will not only significantly reduce the ability of the slab surface to resist crack generation and propagation [58] but also increase the total amount and size of the grain boundary microalloy carbonitrides [59], which aggravate the grain boundary ductility. At the same time, it is also easy to form a low ductility "coarse austenite + grain boundary proeutectoid ferrite film" structure.

Under traditional continuous casting process (including thin slab liquid core reduction process), due to the lack of dynamic recrystallization behavior of large deformation reduction in the slab solidification process, the solidification growth of the original austenite grains at the corner of the slab and other surfaces is mainly related to the composition system of the steel produced by continuous casting (including C content and alloy type and content) and the heat transfer behavior of the corner of the mold. Among them, the C content in the steel affects the growth of austenite grains by changing the solidification process of the slab surface structure. Ohno et al. [60–63] used high temperature remelting solidification experiment to detect the grain growth characteristics of slab surface structure under different C content, as shown in Fig. 1.9. It is found that during the solidification process of peritectic steel, the γ phase precipitates at the δ phase grain boundary and grows from the surface to the interior along the temperature gradient when the temperature drops to the peritectic reaction temperature. During the solidification process, the residual δ phase is pinned on both sides of the γ phase, which prevents the γ phase from growing along both sides of the short axis to form fine columnar austenite, so that the short axis diameter of the γ phase is equivalent to the primary dendrite arm [64, 65]. When the temperature drops to the complete austenitizing temperature, the δ pinning phase disappears, and the austenite begins to grow rapidly into coarse columnar austenite under the action of discontinuous growth. Therefore, under the same continuous casting process conditions, when continuous casting is used to produce micro-alloyed steel with C content in the peritectic range, the grain size of the surface austenite structure such as the corner of the slab is significantly coarse, which greatly reduces the hot ductility of the surface structure such as the corner of the slab. Then it causes high surface crack defects such as the corner of the slab.

Refining the austenite grains of the slab surface solidification is one of the important ways to improve its hot ductility. As mentioned above, in the actual continuous casting production, the surface austenite grains such as the corner of the slab are difficult to be refined by means of large deformation reduction. Therefore, it should mainly rely on the refinement of the initial solidification process of the steel structure and the prevention of subsequent high temperature coarsening. Currently, the commonly used control methods of initial solidification grain refinement of steel structure are mainly realized by accelerating the cooling rate of continuous casting

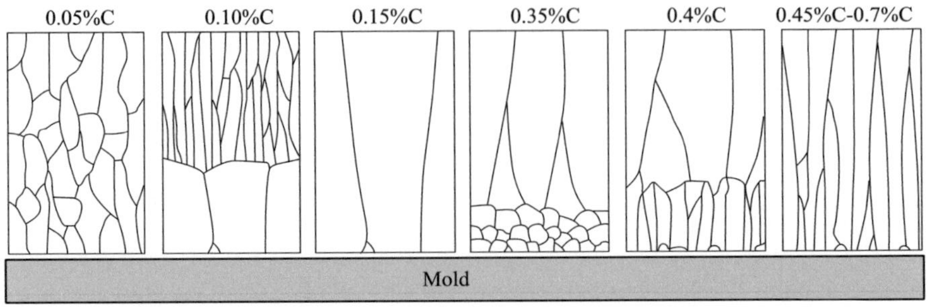

Fig. 1.9 Schematic diagram of grain growth of slab surface under different C content [60]

slab solidification process and adding microalloy components. Among them, the research on the mechanism of refining the initial solidification grain structure of steel by rapid cooling solidification has been relatively systematic [66, 67]. And practice has also proved that significantly increasing the rapid cooling of the steel solidification process can refine its structure grains to a certain extent. For the corner structure of continuous casting slab, the key is to strengthen the shell cooling in the corner area of the mold. This will refine the process of solidification of the corner structure of the slab to form austenite and prevent it from coarsening in the subsequent process. However, in the actual continuous casting production process, the solidified shell inevitably forms dynamic shrinkage under the continuous cooling of the mold and the phase transition of the structure. This causes the corner of the slab break away from the mold and hinders the efficient heat transfer of the solidified shell [68, 69]. How to improve the cooling rate of the solidified shell corner in the mold is an important connotation of the development of efficient continuous casting technology.

Adding alloy composition to refine the grain size of the micro-structure is another important mean to control the grain refinement of the solidified austenite of the continuous casting slab. At present, the most common method is to add alloys such as rare earth and Ti to steel. By introducing nucleation particles such as alloy and its oxides, carbonitrides, the austenite grains of the slab during solidification are refined. In addition, recent studies have found that the overall grain refinement of the solidification structure of the slab can be achieved by introducing a variety of alloy elements into the steel and changing the precipitation temperature, number, and distribution of the dual-phase particles. For example, the recent study [70] found that under the same solidification conditions, the as-cast austenite grains of the steel structure will be significantly refined by adding about 0.02% (mass fraction) Ti to peritectic steels with different Nb contents. This even eliminates the phenomenon of coarse columnar austenite grains, as shown in Fig. 1.10. Therefore, under the premise that the continuous casting steel meets the requirements of rolling process and mechanical properties, the solidified austenite grains can be significantly refined by adding a certain amount of Ti to the Nb-bearing steel [71], thereby improving the ductility of the steel structure. It is beneficial to control the corner crack of micro-alloyed steel continuous casting slab.

Regarding the influence of the precipitation behavior of micro-alloyed carbonitrides on the hot ductility of steel structure, Miyashita et al. [72] pointed out that Nb and V-bearing carbonitrides precipitate at austenite grain boundaries, and the generated Nb (C, N) inhibits the dynamic recrystallization of austenite grains. This leads to significant embrittlement of austenite grain boundaries, increases the N content in the steel, and promotes the precipitation of micro-alloyed carbonitrides. In turn, the transverse cracks on the surface of the continuous casting slab are aggravated, as shown in Fig. 1.11. The studies of Crowther et al. [73] also showed that micro-alloyed carbonitrides precipitated at the grain boundary of the corner structure of the slab. This results in a significant decrease in the grain boundary strength of the corner structure of the continuous casting slab. It is also the main reason for the micro-transverse cracks in the corner of the continuous casting slab. Precise control of the content of micro-alloyed elements and N in steel can effectively inhibit the occurrence of microtransverse cracks in the corner. In general, the carbonitrides containing Nb, Al, V, B, and other micro-alloyed elements precipitate and pin along the grain boundaries, which will aggravate the hot ductility of the steel structure. With the increase of alloy and nitrogen content in the steel, the ductility is aggravated. The addition of Ti alloy is beneficial to improve the hot ductility of steel.

However, as can be seen from Table 1.1, the conclusions of different researchers on the hot ductility of steel with different kinds of alloys and their contents are quite different. For example, Ohmori [75], Nakata [76], and Kang [77] considered that NbC and Nb (C, N) precipitated along the grain boundary are easy to cause stress concentration during the stress process of the steel and aggravate the hot ductility of the steel. However, Han [78] and Mejia [79] indicate that the addition of Nb elements in steel can promote the formation of Nb (C, N) within the grain, thereby improving the hot ductility of the steel. Further, Wojcik et al. [80] believe that the precipitation of fine NbC will aggravate the hot ductility of the steel, while the precipitation of large NbC has no obvious effect on the hot ductility of the steel. Similarly, the research conclusions of different scholars on the addition of V elements to steel are also different. For example, Xiaofeng et al. [81] believe that the addition of V to steel can significantly reduce the hot ductility of steel only when it exceeds 0.193%. Deng et al. [82, 83] believe that the addition of V results in a decrease of the hot ductility in the temperature range of 650–800 °C. There are a few researchers [84] believe that the addition of V in steel can reduce its pearlite content and average size, thereby improving the hot ductility of steel. For the effect of Ti on the hot ductility of steel structure, Cho [85] believe that the addition of Ti in N and B steel can reduce the precipitation of Nb (C, N) and BN along the grain boundary. Banks et al. [86] and Mejia et al. [87] also believe that the addition of Ti to steel can improve the hot ductility. However, Luo et al. [88] believe that the addition of Ti will generate a large number of fine strain-induced precipitation, which will aggravate the hot ductility of the steel. Mintz [89] also proposes that after the addition of Ti to steel, and the effect of TiN on the hot ductility of steel is more significant than that of AlN. However, due to the preferential precipitation characteristics of TiN, adding appropriate content of

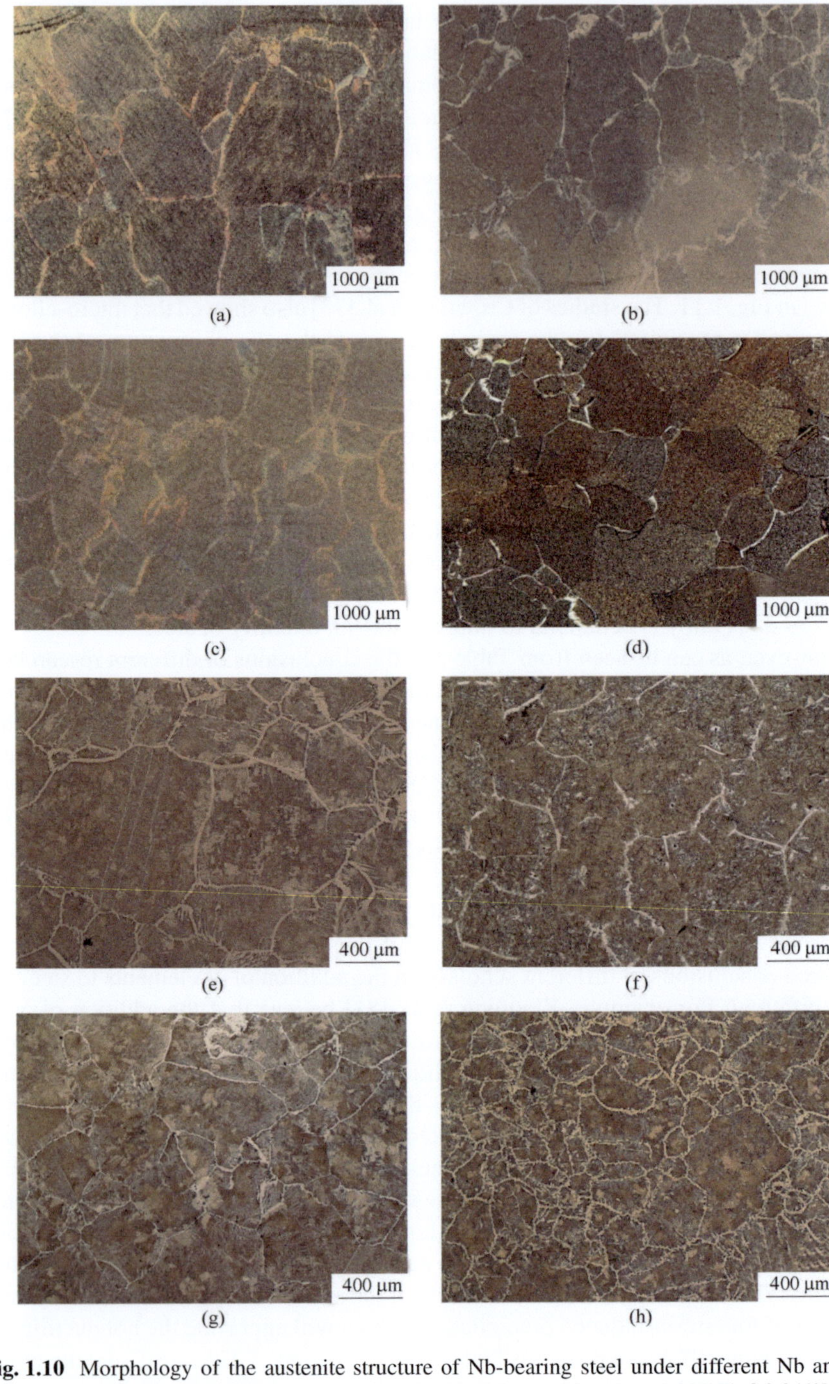

Fig. 1.10 Morphology of the austenite structure of Nb-bearing steel under different Nb and Ti content. **a** 0.02%Nb; **b** 0.04%Nb; **c** 0.06%Nb; **d** 0.08%Nb; **e** 0.02%Nb + 0.02%Ti; **f** 0.04%Nb + 0.02%Ti; **g** 0.06%Nb + 0.02%Ti **h** 0.08%Nb + 0.02%Ti

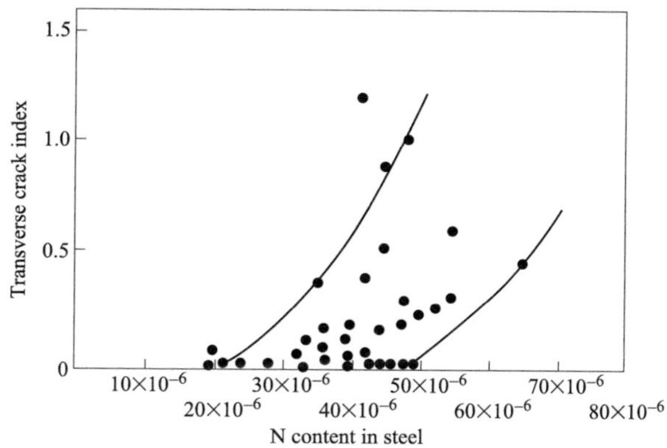

Fig. 1.11 Effect of N content on the occurrence rate of transverse cracks during Nb, V-bearing steel [74]

Ti to steel consumes a large amount of N content in steel during the actual continuous casting production of B, Nb, and Al micro-alloyed steel. This reduces the amount of Nb and V-bearing carbonitride precipitation, which has a greater impact on the ductility of the steel structure, and can significantly reduce the occurrence of corner cracks of the continuous casting slab.

In addition, different cooling processes have a significant impact on the size and distribution of carbonitrides in steel micro-structures, which have a significant effect on the ductility of steel micro-structures. In this regard, scholars at home and abroad [88, 90–94] have also extensively conducted research on the influence of different cooling rates and cooling temperatures on the ductility of micro-alloyed steel. In general, for the main Nb, Al, B, V micro-alloyed steel of steel companies at home and abroad, strengthening the cooling at precipitation temperature of carbonitrides helps to improve the hot ductility of steel micro-structures. Especially in recent years,

Table 1.1 Impact of cooling processes on the hot ductility of micro-alloyed steel

Authors	Cooling process	Conclusion
Suzuki et al. [90]	Weak cooling	Weak cooling is beneficial for reducing the brittle interval of the steel
Luo et al. [88]	Cooling rate 0.4 °C/s, 4 °C/s	Within the temperature range of 750–900 °C, the ductility of steel is high with low cooling rate
Carpenter et al. [91]	Cooling rate 100 °C/min, 200 °C/min	Increasing the cooling rate worsens the ductility of steel
Cho et al. [92]	Cooling rate 1 °C/s, 5 °C/s, 20 °C/s	The cooling rate has a small effect on the ductility of boron-free steel; increasing the cooling rate worsen the ductility of B-bearing steel because of the brittle interval move toward the high temperature zone
Lu et al. [93]	Cooling rate 3 °C/s, 8 °C/s, 15 °C/s	Increasing the cooling rate worsens the ductility of steel in the 900–700 °C temperature zone
Kang et al. [94]	Cooling rate 1 °C/s, 20 °C/s	The ductility of steel under low cooling rate is high

researchers at home and abroad, represented by Lee, Wen Guanghua, and the authors of this book, have conducted systematic research on the precipitation behavior of the dual-phase particles under different cooling rates during the solidification process of Nb-bearing steel continuous casting slab [42, 95–100]. It is also believed that high cooling rate is helpful to disperse the precipitation of microalloy carbonitrides and improve the hot ductility of steel structure.

However, it can also be seen from Table 1.1 that some scholars have pointed out that the rapid cooling process is not conducive to improving the hot ductility of steel, and even plays a harmful role through the study of the influence of different cooling rates on the hot ductility of steel structure. Compared with the production of micro-alloyed steel slab by rapid cooling process, the overall "weak cooling" secondary cooling process is beneficial to control the corner crack of slab.

From the perspective of micro-structure, the micro-structure of steel at different cooling rates and temperatures is different. However, in the main conventional perspective and thick slab continuous casting production process at home and abroad, the cooling rate of the slab is relatively high during the solidification process in the mold. In the secondary cooling bender, straight, and horizontal segments, the cooling rate of the surface structure is relatively low, and the wide face temperature is often above 850 °C. Currently, the surface structure of the slab is mostly single-phase austenite, with high overall ductility. For the corner of the slab, the temperature when it enters the straight segment is often in the range of 750–850 °C. At this temperature, its structure is mostly coarse "austenite + grain boundary ferrite film." It is generally believed that the ductility of this structure is low, and Thomas et al. [50] proposed that the formation of ferrite films at austenite grain boundaries is an important factor causing the formation of the third brittle temperature zone in steel micro-structures, and when the volume fraction of the ferrite films reaches 10%, the ductility is the worst [101].

In actual slab continuous casting production, due to constraints such as continuous casting speed, casting flow roller array length, and its cooling structure, near the straight segment, the structure of the slab corner inevitably transitions austenite to ferrite, forming the "austenite + grain boundary ferrite film" low ductility structure. Therefore, in order to avoid the low ductility structure of the slab corner pass the straight segment, steel companies at home and abroad mostly adopt the secondary cooling water process through the straight segment under high temperature and low temperature conditions, with the high temperature cooling water process when passing straight segment. However, due to the phase transition temperature of different steel grades, the heat transfer rate, and significant differences in continuous casting processes, the structure of the slab near the straight segment, especially the volume fraction of ferrite film at austenite grain boundaries, varies greatly. Using the secondary cooling water process through the straight segment under overall high or low temperature conditions is not very stable to improve the ductility of the slab corner structure. Therefore, home and abroad scholars have conducted extensive research on how to improve the structure of steel by developing suitable cooling processes and have improved the ductility of steel near the straight segment [43,

Table 1.2 Research on the impact of different cooling modes on the hot ductility of steel

Author	Cooling process	Conclusion
Yoshiki et al. [102]	Rapid cooling the structure to $\gamma \rightarrow \alpha$ transition temperature, then reheating to γ phase temperature, and then slow cooling	Greatly improve the ductility of steel
Lee et al. [103]	Cooling the structure to 600 °C at a cooling rate of 20 °C/s and then reheated to 1100 °C, then followed by cooling rate of 0.4 °C/s	Improve the ductility of the steel
Lu et al. [104]	Cooling the structure to low temperature under 3 °C/s cooling rate and then reheated by 0.3 °C/s	The ductility of the steel under cooling conditions 1 and 3 is slightly different, the ductility reduces under cooling condition 2
Ma et al. [105]	Rapidly cooled to below A_{c3} under the cooling rate of 3–6 °C/s, then reheated to above A_{r3} by the heating rate of 3.0 °C/s	Significantly refines the steel grain and greatly improves the ductility
Li [106]	Cooling the structure to ferrite transition temperature under 5 °C/s cooling rate, then reheated to austenite transition temperature by 3 °C/s	The steel grain significantly refines, and the ductility improves
Cai et al. [42]	Rapid cooling the structure to 600 °C at the rate of $\geq$ 5 °C/s, then reheated to > 900 °C with the heating rate of $\geq$ 3.5 °C/s	The steel grain greatly refines, and the ductility in the third brittle zone improves $\geq$ 30%

102–106]. It is worth mentioning that since Grange et al. [107] first used the dual-phase transition method in laboratory conditions in 1966 to achieve grain refinement in the structure of slab for the first time, home and abroad scholars have widely carried out different cooling modes of continuous casting dual-phase transition controlled cooling process to achieve significant grain refinement in the structure and improve the ductility of structure (Table 1.2).

1.3.2.2 External Causes

The external causes affecting the formation of corner cracks in micro-alloyed steel continuous casting slab are mainly the excessive stress in the slab corner caused by continuous casting processes and mechanical devices. In the actual continuous casting production process, there are many external causes that cause corner cracks in micro-alloyed steel continuous casting slab, which are mainly reflected in the following aspects:

A. Friction force between the shell and the copper plate in the mold

In the process of continuous casting production, there is a large friction stress between the mold with periodic oscillations and the slab with continuous casting speed. When the mold oscillates downward, the solidified shell is mostly in a compressed state, and the surface generally does not cause cracks. When the mold moves upward, the surface of the slab is subjected to a large tensile stress, and it is easy to form intergranular microcrack defects at the weak grain boundary at the corner of the slab. Therefore, it is an important basis for ensuring the defect-free production of the corner of the continuous casting slab of micro-alloyed steel to adopt appropriate mold taper and oscillation parameters to reduce the friction force of the slab during the solidification process in the mold.

B. Oscillation mark depth

In the actual continuous casting production process, the depth and morphology of the surface oscillation marks of the slab under different oscillation modes, oscillation parameters, and steel grades are quite different. For oscillation marks of different morphologies, their formation mechanisms are inconsistent. Emi et al. [108] believe that the "overflow" type of oscillation marks on the surface of the slab mainly form under rapid cooling conditions in the mold, and the thick shell is first formed on the meniscus to bend to the molten steel side under the action of negative strip, but it cannot bend to the mold wall again in the positive strip stage. The overflow on the upper part of the shell sweeps away the liquid flux between shell and the mold wall, thus forming a cold barrier similar to a double-layer skin. The formation of the "back bending" type oscillation marks is that under the weak cooling condition of the mold, the thin shell bends to the molten steel side during the negative strip process and is pressed back to the mold wall by the ferrostatic pressure during the positive strip stage. Tomono et al. [109] attribute the formation mechanism of the two types of mixed oscillation marks to the fact that the overflow molten steel remelts the upper part of the solidified thick shell to form an "overflow-remelting" surface oscillation mark. For the two types of oscillation marks of "overflow" and "overflow-remelting," the solute forms segregation on the upper surface of the hook-shaped oscillation marks, which makes the upper surface of the shell unable to weld with the overflow molten steel. Under the friction force with the copper plate, transverse crack defects along the growth direction of the oscillation marks are easily formed on the surface of the slab corner.

In addition, after the deep oscillation marks are formed on the surface of the slab, the trough is generally far away from the copper plate of the mold, which greatly reduces the cooling rate of the structure, resulting in the abnormal coarse austenite grains at the trough of the oscillation marks, thereby reducing the critical strain of the surface structure of the slab and aggravating the transverse corner cracks, as shown in Fig. 1.12. Studies have shown that for high strength C-Mn-Al peritectic steel containing Nb and Al, when the depth of the oscillation mark increases from 0 to 0.2 mm, the critical strain of the surface structure of the slab decreases sharply by about 3/4. In addition, deep oscillation marks will also increase the notch effect of the slab surface. Overall, the deeper the oscillation marks on the slab surface, the worse the ductility of the structure at the oscillation marks, and the easier it is to form transverse crack defects on the surface of the slab. It is one of the important tasks for steel companies to ensure the crack-free production of micro-alloyed steel slab

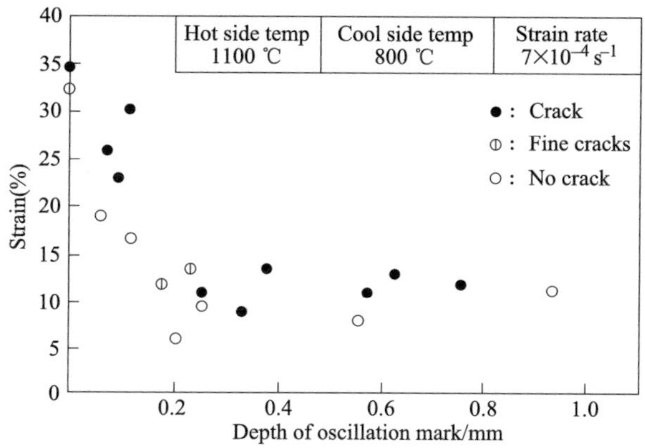

Fig. 1.12 Relationship between oscillation mark depths and critical strain of transverse cracks of the slab (strain rate 2×10^{-4} s^{-1})

corners by developing suitable mold oscillation parameters for different steel grades and minimizing the depth of surface oscillation marks on the slab.

C. Cooling process in the continuous casting process

Thermal stress is one of the main stresses subjected during the solidification process of continuous casting slab. In the actual slab continuous casting process, there is a large temperature gradient in the three-dimensional (3D) space direction along the slab casting and its transverse, slab thickness, which results in a large thermal stress in the solidification process of the casting slab. Due to the 2D heat transfer effect of the wide and narrow faces, the temperature of the slab corner decreases rapidly and fluctuates greatly, and the whole temperature is lower than that of the wide and narrow faces. At the same time, the temperature gradient of the slab corner along its direction to the center of the wide and narrow faces, and the subsurface direction is also larger, and its thermal stress is often the largest, which expands the corner crack defect of the slab. Homogenizing the secondary cooling water distribution of continuous casting to minimize the thermal stress of slab corner is another important work to control the transverse corner crack of micro-alloyed steel slab in steel companies at home and abroad.

D. Slab bending or straightening force

Currently, the main slab continuous casters at home and abroad are mostly arc continuous casters with straight molds. In the actual continuous casting production process, the slab gradually transitions from vertical to arc with specified radius in the bender, and then transitions from arc to slab in the straightening process. In these two processes, the slab surface will cause about 2% deformation at a strain rate of 10^{-3}–10^{-4} [110]. However, the difference is that in the bender, the inner arc of the slab is subjected to compressive stress, which generally does not cause corner crack defects. Although the outer arc of the slab is subjected to tensile stress, the temperature of the slab corner at the bending point often exceeds 900 °C. Under the premise that the device accuracy meets the process specifications, the slab corner generally does not form transverse crack defects. In the straightening process, the outer arc of the slab is subjected to compressive stress, and its corner generally does not form transverse crack defects. But for the inner arc corner of the slab, due to the lower temperature and ductility at this position, it is easy to form transverse corner crack defects during the tensile force process. Therefore, it is generally believed that the main link for the generation of slab corner cracks is the straight segment. For this reason, continuous casting process and device technologies such as multi-point continuous straightening, high temperature and hot ductility of slab corner should be adopted to reduce the force on the inner arc of the slab during the straightening process and reduce the crack rate of slab corner.

E. Device accuracy

The roller array accuracy of the continuous caster is the basis for the production of defect-free continuous casting slab. In the actual slab continuous casting production process, when the continuous caster runs abnormal, especially the bow segment or the roll gap is misaligned, the support roller bearing is misaligned, the support roller axis is not parallel or the parallelism between the adjacent roller bracket is inconsistent, will cause the slab subject to large deformation stress in the casting flow. For example, when the casting roller offset of the continuous caster is 0.5–1.5 mm, the strain caused by the misalignment will reach 0.2–0.4%, and the deformation of the slab is larger than that of the slab straightening stage during normal casting. Therefore, high precision device maintenance is an important content in the production of high quality micro-alloyed steel slab.

1.3.3 Typical Control Technology of the Slab Transverse Corner Cracks

The high incidence of transverse corner cracks of micro-alloyed steel slab continuous casting is a common technical problem in the steel industry. Combined with the causes of the above cracks, steel companies at home and abroad have carried out many research and development of control technology around the composition of steel, continuous casting process and device.

1.3.3.1 Steel Composition Control

As mentioned above, one of the key internal causes of corner cracks in continuous casting slab of micro-alloyed steel is that a large amount of carbonitrides in micro-alloyed steel precipitate at the grain boundary of the corner structure of the slab and brittle the grain boundary. Therefore, reducing the type and content of microalloy elements in steel, strictly controlling the N content in the process of smelting and continuous casting, and avoiding the C content of peritectic reaction as much as possible are important ways to control the corner crack of micro-alloyed steel slab from the perspective of steel composition. However, the composition of the steel is mostly subject to the mechanical properties of the product, and the type of microalloy in the steel and its content, C content, and other adjustable range are generally not large. In the actual steel production process, the N content in the steel is generally removed as a harmful element. For micro-alloyed steel, as a key component of carbonitride precipitation, N directly determines the total amount of carbonitride precipitation and is a key parameter affecting the corner crack of micro-alloyed steel continuous casting slab. Studies have shown that when the N content in the steel < 4×10^{-5}, the corner crack rate of the micro-alloyed steel slab can be significantly reduced. Therefore, how to reduce the N content of steel in the process of smelting, refining and continuous casting has become an urgent problem for most steel companies at home and abroad to solve the corner crack alloyed steel continuous casting slab in recent years [111, 112].

In addition, the temperature difference of different types of alloying elements in the precipitation of carbonitride is significant. Ti is an important microalloy element, and if an appropriate amount of Ti alloy is added to micro-alloyed steel containing Nb, Al, B, and other elements, due to its characteristic of precipitating above 1300 °C, it will consume a large amount of N in the steel first, significantly reducing the N left for carbonitrides of Nb, Al, B elements that have a greater impact on the hot ductility of the steel, which is beneficial for controlling corner cracks in micro-alloyed steel continuous casting slab. For example, in the continuous casting process of B-bearing steel, when the N content in the steel is about 4.5×10^{-5}, adding 0.025–0.030% Ti to the steel can eliminate the corner cracks of the B-bearing steel slab. At present, the "add Ti to fixate N" technology for controlling corner cracks in micro-alloyed steel slab has been widely adopted by domestic and foreign steel companies.

1.3.3.2 Continuous Casting Process Optimization

Currently, the research and development of corner crack control technology for micro-alloyed steel slab mainly focuses on the optimization of the continuous casting process. Hyundai Steel in South Korea has reduced the occurrence of corner cracks in Nb, V micro-alloyed steel thick slab by optimizing the secondary cooling water distribution process, thereby increasing the temperature of the slab entering the straight segment. Baowu Steel has improved the property of the mold flux, weakened the cooling of the mold, optimized the distribution of secondary cooling water, stabilized the taper of the mold, and adopted appropriate casting speeds and other methods [113]. To a certain extent, these methods reduced the occurrence of edge crack and warping defects in steel plates, yet have not been able to eliminate the formation of cracks. Wuhan Steel has reduced the occurrence of corner crack defects in Al, Nb micro-alloyed steel slab to a certain extent by controlling the molten steel level fluctuation, the secondary cooling system and straightening temperature, the depth of oscillation marks, and the accuracy of the continuous caster during the continuous casting process of the two continuous casters of the No.2 steelmaking plant [114].

Jigang No.3 Steelmaking has effectively reduced the occurrence of subsurface transverse cracks in slab by adopting high frequency, low amplitude oscillation of the mold, increasing the straightening temperature, and other methods [115]. Qianan Steel has studied the feasibility of rapid cooling through straightening process of continuous casting slab by using test of secondary cooling rapid cooling. The results show that the corner crack of slab of micro-alloyed steel can be reduced to a certain extent when the temperature of the corner in the straight segments is lower than 700 °C by rapid cooling. Hanye Steel has effectively controlled the occurrence of corner cracks in micro-alloyed steel slab by optimizing the oscillation parameters of the thick slab mold, strengthening the arc connection between the mold and the bender. Tianjin Steel has better controlled the occurrence of corner cracks in B, Nb micro-alloyed steel wide-thick slab by using the method of edge water reduction in straight segment of the continuous caster and adding Ti to the steel. Pan Steel has greatly reduced the occurrence rate of corner cracks in slab by optimizing the oscillation parameters of the mold, the physical and chemical properties of the mold flux, and the secondary cooling water distribution process. Mei Steel has achieved stable production of B-bearing steel slab with low crack occurrence rate and without corner cleaning by adding an appropriate amount of Ti to B-bearing steel, strictly controlling the N content of the steel, and optimizing the secondary cooling water distribution process. A steel controlled the occurrence of corner cracks in micro-alloyed steel thick slab, using the diamond mold and continuous casting secondary cooling weak cooling process.

It can be seen that at present, for the control of corner cracks in micro-alloyed steel continuous casting slab, steel companies mainly optimize steelmaking and refining, continuous casting pouring, mold cooling structure and system, secondary cooling water distribution, straightening, and other continuous casting processes. To reduce the occurrence of this type of crack, it is essential to reasonably control the steel composition and the thermo-mechanical behavior evolution during the continuous casting slab solidification process. Among the developed technologies for controlling transverse corner cracks in micro-alloyed steel slab, the optimization of the continuous casting cooling process and its technology application are the most widespread. For example, Nozaki [116], Yasunaka [117], Wang [118], and others have reduced the occurrence rate of transverse cracks by reducing the cooling intensity of the corner of the slab during the continuous casting process; Wen [119], Wang [120], and others are to reduce the occurrence rate of transverse cracks in the corner of the slab by changing the cooling conditions of the slab corner through changing the array of the secondary cooling zone nozzles. The fundamental purpose of the application of these control cooling processes is controlling the corner temperature of the continuous casting slab, to avoid the third brittle temperature zone of the corresponding steel grade when it enters the casting flow bender and straight segment, meeting the requirements of high ductility bending or straightening. However, in the actual process of the secondary cooling optimization technology for continuous casting slab, it is difficult to completely avoid the third brittle temperature zone of the corresponding steel due to restrictions such as casting speed and casting flow spray structure. For this reason, the Central Iron & Steel Research Institute has developed a large chamfered mold technology [121, 122], which has been applied in several domestic steel companies such as Shougang, Handan Steel, and Chongqing Steel, and has achieved good application. The above optimization of micro-alloyed steel continuous casting process based on slab corner temperature control and the application of new technology alleviate the occurrence of transverse corner cracks of micro-alloyed steel continuous casting slab. However, it cannot solve the problem that the microalloy carbonitride and proeutectoid ferrite film concentrated and precipitated at the grain boundary during the solidification process of the slab, and the coarse austenite grains, which leads to the weakness of the structure of the slab. The stability of crack control is poor, and it cannot fundamentally prevent occurrence of cracks.

According to the above causes of transverse corner cracks in microalloy steel slab, the key to preventing this type of crack is to form a structure with strong crack resistance on the surface of the slab, finding the essential reason for the occurrence of transverse corner cracks in the microalloy steel slab. By developing a higher level of slab solidification and control cooling process, the grain boundary chain carbonitride and the proeutectoid ferrite film of slab corner is eliminated.

Based on this idea, in recent years, researchers at home and abroad represented by Kato and others [123] have carried out research on the precipitation behavior of grain boundary carbonitrides and the evolution behavior of surface structure of micro-alloyed steel slab under different cooling rates and have developed a new technology for surface structure control cooling (SSC), which has been put into use in Kashima No.2 continuous caster [124]. The core of this technology is to reasonably control the cooling rate, cooling temperature, reheating rate, and reheating temperature of the slab in the vertical zone of the secondary cooling zone, to control the precipitation of grain boundary carbonitrides and proeutectoid ferrite film in the slab surface structure, and achieve the purpose of strengthen the crack resistance performance of the slab surface structure at high temperature. Its evolution mechanism is shown in Fig. 1.13. Based on the SSC process slab corner crack control mechanism, Guanghua Wen from Chongqing University [105] and others have also carried out application test research in Baowu Steel, Pan Steel, and other companies.

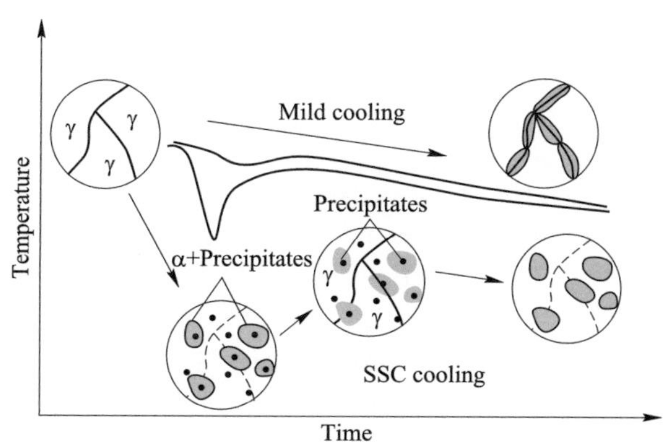

Fig. 1.13 Illustration of control mechanism of SSC technology

However, in the actual continuous casting production, the secondary cooling control of the continuous caster is very complex. Under the traditional continuous caster conditions, it is necessary to implement large-scale rapid cooling and heating control in the short area between the foot roller and the vertical zone to strengthen the slab surface structure. It is difficult to achieve the stability of the control process in the actual implementation. At the same time, the precipitation behavior of microalloy carbonitrides at the grain boundaries of the slab corner micro-structure is related to its temperature. In the actual production of micro-alloyed steel, the heat transfer of the shell corner in the mold belongs to 2D heat transfer, and the temperature of its corner after exiting the mold and entering the foot roller zone has often reduced to around 1000 °C. At this shell temperature, microalloy carbonitrides of elements such as Ti, Nb, and B in the slab micro-structure are close to complete precipitation or have precipitated significantly (the temperature range of the austenite grain boundary precipitation of TiN is about 1000–1350 °C, and that of Nb (C, N), BN is about 850–1120 °C). Using foot roller zone rapid cooling technology to control the cooling of the slab corner micro-structure of the micro-alloyed steel, the effect on improving the precipitation behavior of its grain boundary microalloy carbonitrides is limited. In addition, under the existing device conditions, it is difficult to achieve this goal well in the actual production. Therefore, it is still difficult to eliminate transverse corner cracks without other device technology guarantees to implement the above-mentioned slab SSC technology.

At the same time, Mintz et al. [125] believe that the SSC process also has its drawbacks, rapid cooling leads to large thermal stress and temperature gradient. After rapid cooling, the slab surface temperature may not be able to reheat to the range of conventional processes, leading to low temperature straightening, increasing straightening load and bender casting roller wear. Moreover, due to the evolution of the actual temperature field on the slab surface, the roller gap shrinkage control of the casting rollers and the soft reduction process at the solidification end are also affected.

1.3.3.3 Improvement of Continuous Caster Device

As mentioned above, the continuous caster device also has a significant impact on the occurrence of slab transverse corner crack, ensuring the good operation of the slab device is the basis for all corner transverse crack control technologies. In addition, the parameters of each device operation will also affect the occurrence of transverse corner crack. At present, the main research on the precision control of key continuous casting device and the optimization of its process parameters to improve the corner cracks of micro-alloyed steel continuous casting slab at home and abroad are as follows.

(1) Mold oscillation process [27]. The practice has shown that the shallower the oscillation mark the depth on the slab surface of the micro-alloyed steel, the lower the occurrence rate of transverse corner cracks. In the actual continuous

casting production process, for a specific steel grade, the depth of the slab surface oscillation mark mainly depends on the rationality of the oscillation parameters, including the oscillation frequency, amplitude, negative strip time, oscillation waveform, etc. Overall, the higher the oscillation frequency of the mold, the shallower and flatter the oscillation mark, reducing the negative strip time helps to reduce the depth of the oscillation mark. Therefore, at present, domestic and foreign steel companies, especially the mold oscillation technology provided by Pratt in recent years, tend to adopt high frequency and small amplitude oscillation mode and parameters of non-sinusoidal oscillation mold, which is conducive to reducing the depth of oscillation marks on the surface of the slab and reducing the occurrence of transverse corner cracks of micro-alloyed steel.

(2) Optimization of mold process [126]. Mold is the "heart" of the continuous caster, the process parameters such as molten steel level fluctuation and taper of the mold are closely related to the formation and expansion of the corner crack of the slab. The molten steel level fluctuation is a key index of the stabilization of continuous casting production. Severe level fluctuation will cause the uniformity of the mold flux infiltrating into the slag channel, which will lead to different depths of oscillation marks on the surface of the slab. Severe level fluctuation also easily forms flux entrapment, which aggravates the surface crack of micro-alloyed steel continuous casting slab. At present, the C content of main micro-alloyed steel products (especially wide-thick plates) is mostly in the range of peritectic and sub-peritectic. It is a common technical problem for continuous casting to produce this type of steel at high casting speed (especially adopt the weak cooling secondary cooling water process), and the molten steel level often fluctuates abnormally and severely. For above reason, in recent years, steel companies including Baowu Steel, Han Steel, and An Steel have strictly controlled the fluctuation of the molten steel level and developed level stabilization control technology combining non-uniform roller array, weak cooling of the mold, and rapid cooling of secondary cooling high temperature zone such as foot roller zone.

Reasonable mold taper is the key to eliminate the concentrated distribution of mold flux film and air gap at the corner of the slab and to improve the corner structure of solidified shell in the mold with efficient heat transfer. It is also an important parameter to prevent the bulge of the narrow face of the slab. In the actual continuous casting production of high strength micro-alloyed steel with high crack sensitivity, Baowu Steel, Lianyuan Steel, and other companies usually adopt the process of large taper mold. However, excessive mold taper, on the one hand, will not only increase the wear of the mold copper plate and shorten the lifetime of the mold; on the other hand, it will also aggravate the friction between the slab and the copper plate, which is easy to cause the micro-structure at the oscillation mark to crack and form cracks, which is not conducive to the corner crack control of the micro-alloyed steel continuous casting slab.

(3) Device maintenance. Ensuring the accuracy of the continuous caster device is a basic requirement for the production of micro-alloyed steel continuous casting

slab without defects. Regarding the accuracy of the mold roller gap, domestic and foreign steel companies usually require the deviation ≤ 0.3 mm and a high flatness. At the same time, prevent the nozzle clogging in each cooling zone of the casting flow. For this reason, the backwashing nozzle system has also been developed domestically and applied in Tianjin Steel and other companies.

(4) Optimization of the mold cavity structure. The research and development of the mold cavity structure which is beneficial to control the occurrence of transverse corner cracks is a topic in the research and application of the crack-free continuous casting technology on the surface of micro-alloyed steel slab. At present, the most widely used non-conventional mold in domestic is the large chamfered mold [122, 123]. The aim of the crack control of the mold is to greatly increase the temperature of the corner when pass straight segment of the continuous casting slab and avoid the third brittle temperature zone of the steel. The industrial test of Qianan Steel, Handan Steel, and other companies shows that the corner temperature of the continuous casting slab produced by the chamfered mold in the straight segments can be increased by more than 70 °C compared with the conventional right-angle mold, thus avoiding the third brittle temperature zone of the main continuous casting micro-alloyed steel. At present, this mold has been used in many domestic steel companies and has achieved good crack control.

In recent years, based on the study of the precipitation behavior of carbonitride and the growth behavior of solidified austenite which caused the corner crack of micro-alloyed steel slab, the author has developed a novel curved-structure mold which can effectively compensate the solidification shrinkage of the corner of the shell in the mold [43], to achieve efficient heat transfer of the slab corner during the solidification growth process in the mold, disperse and refine the precipitation of carbonitrides in the high temperature segments of microalloy and the initial solidification austenite grains, thereby improving the hot ductility of the steel micro-structure and reducing the occurrence of transverse corner cracks of the micro-alloyed steel slab.

In addition, Chen et al. [127] studied the influence of the geometric shape of the mold corner on the temperature of the slab corner. Proposing that compared to the right-angle mold, using a large chamfer, multiple chamfers, and rounded mold can significantly increase the temperature of the slab corner, avoiding the low ductility temperature range, which is beneficial for the control of transverse corner cracks, but it has not yet been seen in large-scale industrial applications.

1.4 Research Progress of Wide-Thick Slab Off-Corner Depression and Its Control

The thick slab is the base material for preparing high quality wide-thick plates and ultra-thick plates, with the characters of large width and thickness. During the continuous casting production process of the thick slab, with the slab thickness increases, the narrow face shell may lack sufficient support from the foot rollers after exiting the mold. This insufficient support weakens the shell strength in the secondary cooling zone, leading to significant bulge of the narrow face. In actual thick slab continuous casting production, once the narrow face bulge occurs, the longitudinal depression occurs in the slab wide face off-corner. Transverse crack and longitudinal crack often respectively occur in the surface and subsurface of the longitudinal depression zone in the slab wide face off-corner. Therefore, the frequent occurrence of longitudinal depression at wide face off-corner has become one of the important common technical problems in thick slab continuous casting, which needs to be solved urgently.

1.4.1 Formation Mechanism of the Depression

With the increase of thick slab continuous casting production lines in recent years, the problem of frequent occurrence of longitudinal depression at wide face off-corner during the thick slab continuous casting process has gradually emerged. So extensive research about the causes of longitudinal depression in the off-corner of thick slab has been carried out at home and abroad based on steel composition, mold flux property, solidified shell shrinkage, mold taper, and so on [128–131]. The results show that optimizing the mold process can reduce the longitudinal depression in the thick slab wide face off-corner. However, there is no unified conclusion on causes of the longitudinal depression in the wide face off-corner of the thick slab.

Regarding the position of the longitudinal depression in the thick slab wide face off-corner, there are two main perspectives at present: one proposes that the depression forms in the mold, which is referred to as the single-stage formation theory; the other proposes that the depression forms at two positions, in the mold and the foot rollers zone, evolving progressively in stages, which is known as the two-stage formation theory.

First of all, in most early studies, it was believed that the longitudinal depression in the off-corner of the thick slab was formed in the mold. Brimacombe et al. [132] through analyzing and testing of a large number of slab concluded that the cause of the depression is that non-uniform heat transfer is prone to occur between the shell and the copper plate at the meniscus of the mold, leading to non-uniform growth of the initial solidified shell. When the solidified shell moves downward to the middle and lower part of the mold, a high temperature "hot spot" will form at the thin position of the shell. If the taper of the narrow face copper plate is too small to compensate for the solidification shrinkage of the shell in the width direction at that time, the tensile stress will form at the "hot spot" of the shell. When the tensile stress reaches a certain level, it causes the shell surface at the "hot spot" sinks, or even cracks, as shown in Fig. 1.14.

However, the theory proposed by Brimacombe et al. can precisely explains the partial depression often accompanied by the formation of longitudinal cracks on the slab surface during the continuous casting process. But it cannot explain why the longitudinal depression of the thick slab wide face off-corner only stably appears

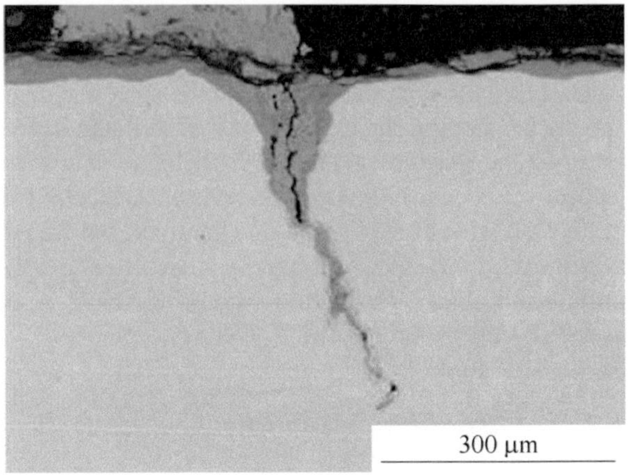

Fig. 1.14 Morphology of the surface depression and its crack of the slab

in the off-corner and through the entire slab. In addition, the internal crack occurs concurrently at the 20 mm below off-corner depression in the wide face. While the depth of subsurface cracks below the depression caused by the tensile stress from the surface less than 10 mm. It can be seen that the formation mechanism of continuous longitudinal depressions in the off-corner wide face of thick slab is not totally the same as that of concurrent partial depressions with large longitudinal cracks in the wide face.

Based on the research on the mechanism of slab surface depression formation, Thomas et al. [133] propose that in the continuous casting production process, if the taper of the narrow face copper plate of the mold is too small, the narrow face shell will not be able to get enough support force from the copper plate. Under the ferrostatic pressure of the molten steel, the narrow face solidified shell will bulge, furthermore the corner is twisted in the horizontal direction toward the wide face direction, thereby causing the shell of the wide face off-corner to detach from the wide face copper plate, and forming the continuous longitudinal depression. The process of shell twisted is shown in Fig. 1.15. Conversely, if the taper is too large (greater than 1.3%), the parts of the shell surface with higher temperatures are prone to wrinkling under pressure, also easily forming serious surface depressions, while also leading to defects such as subsurface internal cracks in the slab.

Similar to the formation mechanism of longitudinal depression under the condition of large taper, Storkman and Thomas [134] also proposed that in the linear single taper mold used by most steel companies at home and abroad, the solidification shrinkage of the shell in the middle and upper parts is faster, while the shrinkage in the middle and lower parts is slower. This will lead that the shrinkage compensation of the lower part of the copper plate is too large, which in turn extrudes the narrow face of the shell, resulting in the formation of folds at the wide face off-corner, and ultimately evolved into the longitudinal depression in the thick slab wide face off-corner, as shown in Fig. 1.16. Mahapatra et al. [135] proposed that when the heat flux density has great difference between the inner arc and outer arc of the mold, there is a noticeable

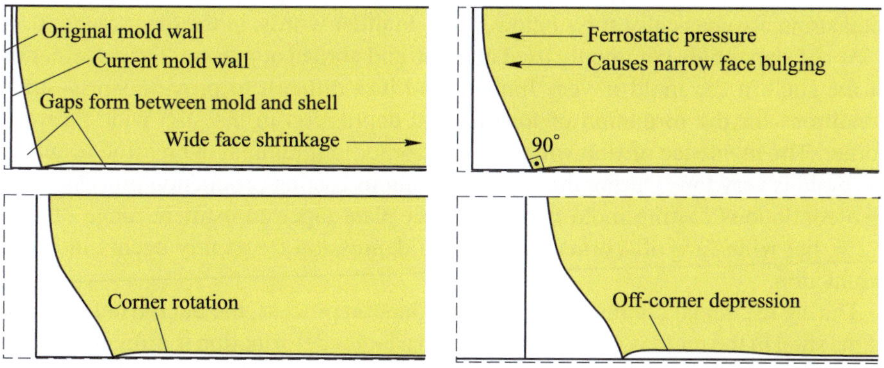

Fig. 1.15 Evolution of shell solidification deformation under the 0.3% narrow face taper and 0.1% wide face taper of the mold

difference in the shrinkage of the inner arc and outer arc when the slab solidifies to the middle and lower part of the mold. Due to the non-uniform compensation effect of the taper of the mold narrow face, the shell compensation on the outer arc side is not enough, while the shell of the inner arc side is obviously squeezed by the copper plate, which cause the wrinkle on the wide face off-corner of the inner arc side, eventually forming a longitudinal depression at the wide face off-corner. This phenomenon explains why the inner arc side depression is more severe than the outer arc side in the actual continuous casting process. On this basis, Mahapatra suggested that the concurrent cracks often occur 16 mm below the depression and determined that the depression occurs near mold exit.

Although the above mechanism for the formation of the slab wide face off-corner depression explains the formation process of depressions to a certain extent, studies on the thermo-mechanical behavior of shell solidification in the mold have shown that plastic strain is difficult to reach 0.4% during the shell solidification [136], and the

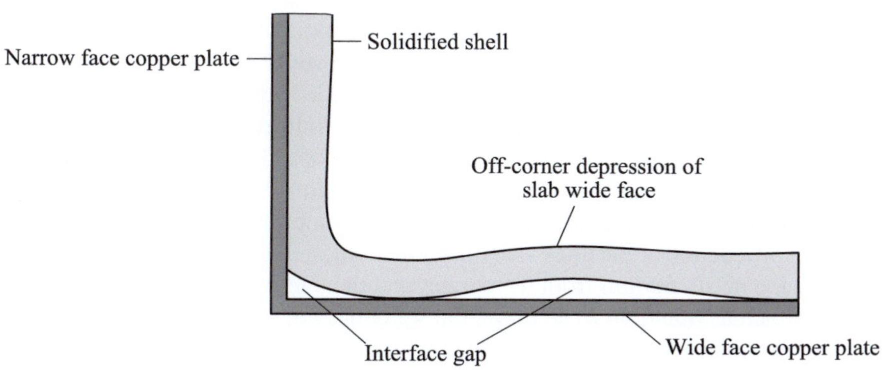

Fig. 1.16 Longitudinal depression at the wide face off-corner caused by excessive taper on the narrow face

total strain also basically stays below 0.9%. In other words, in the taper range (0.9–1.2%) of slab molds commonly used at home and abroad at present, the deformation of the shell in the mold is very limited, and it is difficult to provide stress–strain conditions for the formation of longitudinal depression in the slab wide face off-corner. The incidence of slab wide face off-corner longitudinal depression occurs in the mold is very low. During the actual continuous casting production process, thick slab continuous casting mold narrow copper plate taper adjustment range of 1.0–1.2%, but wide face off-corner longitudinal depression frequently occurs in actual production.

During the actual continuous casting production process, the ductile deformation of the shell in the mold is very limited, so that a larger deformation is difficult to occur in the shell, and it is difficult to form continuous wide face off-corner longitudinal depression. Thomas et al. [137] proposed that the longitudinal depression is only firstly occurs in the mold and develops in the foot roller zones, that is the two-stage formation mechanism of the slab wide face off-corner longitudinal depression. Through numerical simulation of the solidification behavior of the shell under typical taper and other continuous casting process conditions in the mold, the result shows that the solidified shell wide face off-corner forms a larger interface gap due to the combined effects of shell solidification shrinkage, mold taper, and copper plate deformation, which promotes the formation of a high temperature "hot spot" in this part, and this stage is defined as the incubation period for the formation of the slab wide face off-corner longitudinal depression. When the solidified shell exits the mold and enters the secondary cooling zone, the ferrostatic pressure pushes the wide face shell to form periodic bulges between the rollers, and the shell bends at the "hot spot" of the wide face off-corner, while under the action of slab bulge in the narrow face foot roller zone, the corner of the solidified shell is twisted toward the center of the wide face, until the depression is enlarged and a larger longitudinal depression is formed, and this stage is defined as the formation stage of the longitudinal depression in the slab wide face off-corner. However, some scholars also believe that there are the following shortcomings in this formation mechanism: First, the study has a direct relationship between the narrow face bulge and the formation of the wide face off-corner longitudinal depression, but the support of the side guide roller to the narrow face of the slab is not considered in the study process, and there is no detail data to show that the depression is formed in the foot roller zone; second, the study assumes that the heat transfer in the mold is uniform, but in the actual slab continuous casting production process, the existence of a "hot spot" in the slab wide face off-corner is an objective fact; Third, the depression distribution range obtained by the simulation method used in the study begins to appear around the corner, while the actual production of the thick slab off-corner longitudinal depression only begins at a distance of 30–50 mm from the corner, and the simulation results do not quite match the actual situation. In addition, the study has not yet quantitatively clarified the relationship between the slab wide face off-corner depression and the deformation of the narrow face bulge, the formation process of the off-corner depression and its evolution behavior in the casting flow, and further research is required.

1.4.2 Typical Control Technology of the Depression

To control and reduce the occurrence of longitudinal depression in the wide face off-corner during thick slab continuous casting, scholars at home and abroad have proposed a wide range of control strategies based on their understanding of the formation mechanism of longitudinal depression [138–146]. Generally speaking, the control strategies can be divided into the following aspects.

1.4.2.1 Mold Structure and Process Optimization

The heat transfer behavior of the slab in the mold directly determines the uniformity of the shell solidification. Homogenizing the heat transfer of the shell in the mold and reasonably controlling the deformation of the shell is the basis for controlling the off-corner longitudinal depression in the wide face of the thick slab. Thomas et al. [134] proposed to adopt a three-taper mold with a large compensation in the upper part, parallel in the middle part, and a small compensation in the lower part along the height direction, to ensure efficient solidification of the narrow face of the slab with the copper plate during the solidification process. During the solidification process of the slab in the mold, the upper part large taper compensation can efficiently compensate for the rapid shrinkage of the initial solidification shell and support the narrow face shell to prevent its large deformation bulge, and the tensile stress on the slab wide face caused by bulge, to prevent the wide face off-corner depression. The implementation of small taper compensation in the middle and lower part of the mold can effectively avoid large friction and squeezing between the shell and the copper plate, thereby avoiding the wrinkles or depressions of the shell in the wide face off-corner caused by large taper, and extending the lifetime of the mold copper plate. However, in actual production, optimizing the cavity structure of the mold can only alleviate the generation of wide face off-corner depression of the thick slab to a certain extent, but it cannot be stably eradicated. At present, apart from a few companies such as Wuyang Steel that use this type of mold, there is no large-scale promotion and application of this type of mold.

In addition, the practice has shown that optimizing the structure of the immersion nozzle of the mold, its immersion depth, and the amount of argon blowing can not only ensure the active molten steel level of the mold but also avoid large fluctuations in the level; it can also improve the heat transfer uniformity in the mold to a certain extent to alleviate the generation of longitudinal off-corner depression in the wide face of the thick slab.

1.4.2.2 Mold Flux Optimization

To address the frequent longitudinal off-corner depression defects on the wide face of its high aluminum steel slab in Shougang Steel, Ji Chenxi et al. [138] showed that the longitudinal depression in the wide face off-corner of this type of slab is related to the cross-sectional size of the continuous casting slab, the steel-slag reaction of the mold molten steel level, and the diffusion of the Al. For the continuous casting of large cross-sectional slab with a width of 1900 mm, it is advisable to use a mold flux with high alkalinity, low viscosity, and low melting point. For narrow cross-sectional slab with a width of 1200 mm, it is advisable to use mold flux with high viscosity and high melting point. At the same time, this study also proposes that the steel-slag reaction of the mold molten steel level will aggravate the formation of longitudinal off-corner depression in the flux face, and increasing the ratio of Al_2O_3 and SiO_2 in the mold flux can reduce the degree of steel-slag reaction and alleviate the occurrence of longitudinal off-corner depression in the wide face of the slab without affecting the viscosity and other performance indexes. For the production of general steel, the practice has shown that the application of mold flux with lower viscosity, a narrower sintering temperature range, and a higher sintering temperature can help to prevent the defect of longitudinal off-corner depression in the wide face of the slab.

1.4.2.3 Secondary Cooling Process Improvement

Thomas et al. [133] proposed that adopting tightly arranged rollers in the narrow face foot roller segment and strengthening the cooling of the narrow face of the slab can reduce the deformation of the solidified shell in this part, prevent the slab narrow face from bulging, and thus avoid the occurrence of longitudinal off-corner depression in the wide face. Lu et al. [143], based on the actual production, proposed that insufficient side cooling of the slab in the foot roller zone will lead to bulge on the narrow side, which will cause the longitudinal off-corner depression in the slab wide face of slab. By appropriately increasing the spraying on the narrow face after the slab exits of the mold, the shell thickness and strength of the narrow face of the slab can be increased, thereby preventing the occurrence of bulge on the narrow face and longitudinal off-corner depression of the wide face. Zhu et al. [140] proposed that increasing the taper of the narrow face roller to ensure its full contact with the slab, meanwhile prevent the formation of bulge on the narrow face and longitudinal off-corner depression of the wide face.

1.5 Research Progress of Slab Hot Charging Crack of Micro-alloyed Steel and Its Control

1.5.1 Introduction of Slab Hot Charging Process

The hot charging technology of continuous casting slab is an efficient green steel manufacturing technology that connects the steelmaking and rolling. It started in the 1960s and developed rapidly under the driven of two world oil crises. The basic idea of this process is to send the cut to size continuous casting slab directly from the continuous casting workshop to the heating furnace of the rolling mill under the high temperature state, thereby efficiently using the heat in the high temperature slab, and greatly shortening the production process and cycle. It has the characteristics of reducing energy consumption, improving material yield, and shortening the process, and has attracted many steel companies in recent years. The schematic diagram of the hot charging process of the continuous casting slab is shown in Fig. 1.17.

According to the different temperatures of the hot charging of the continuous casting slab, the hot charging process of the continuous casting slab is generally divided into five categories as shown in Table 1.3 [147–149].

The characteristics of each continuous casting slab hot charging process are as follows:

(1) Continuous casting-direct rolling (CC-DR), requires the hot charging temperature of the continuous casting slab to be not less than 1100 °C. The slab does not need to be heated in the heating furnace and only needs to be reheated during the slab transportation process. Then directly roll after reaching the rolling temperature. Under this process, there is no $\gamma \rightarrow \alpha \rightarrow \gamma$ transition in the slab structure,

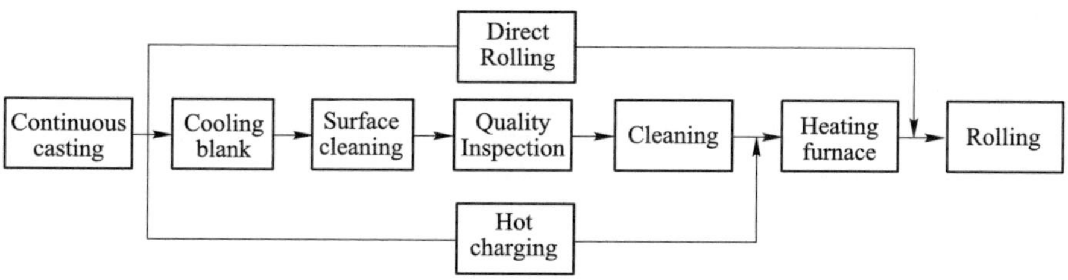

Fig. 1.17 Schematic diagram of the different charging processes of the continuous casting slab

Table 1.3 Classification of hot charging technology of continuous casting slab

Serial number	Process name	Hot charging temperature	Process features
I	Continuous casting-hot direct rolling (CC-DR)	1100 °C	Direct rolling without heating in the furnace
II	Continuous casting-hot direct rolling (CC-HDR)	A_3-1100 °C	Direct rolling heating in the furnace
III	Continuous casting-direct hot charging rolling (CC-DHCR)	A_1–A_3	Directly charge into the furnace and rolling
IV	Continuous casting-hot charging rolling (CC-HCR)	400°C-A_1	After preservation by the device, charge into the furnace and rolling
V	Cold slab	Room temperature	Offline stack cooling to room temperature, charge into the furnace and rolling

and the surface of the slab is composed of relatively coarse original austenite grains. At the same time, there is no carbonitride precipitation-redissolution behavior in the conventional cold charging slab process.

(2) Continuous casting-hot direct rolling (CC-HDR) requires the hot charging temperature of the continuous casting slab to be between 1100 °C and A_3 temperature. Similarly, it only needs to be reheated during the slab transportation process, then directly roll after reaching the rolling temperature. The metallurgical characteristics of the slab under this process are the same as those of the CC-DR process, with only some alloy elements undergoing precipitation-redissolution behavior.

(3) Continuous casting-direct hot charging rolling (CC-DHCR) requires the hot charging temperature of the slab to be between A_3 and A_1. The directly hot charging slab needs to be heated in the heating furnace until it reaches the rolling temperature before it can be rolling. Before the slab under this process is charged to the heating furnace, its surface structure is in the $\gamma + \alpha$ dual-phase zone, after the austenite grain boundary undergoes the $\gamma \rightarrow \alpha$ phase transition, ferrite film is formed. When the slab is reheated in the heating furnace, austenite grains are formed. Similarly, because the slab surface structure does not completely undergo $\gamma \rightarrow \alpha$ phase transition, most of the austenite after heating is still coarse original austenite grains, and some newly formed austenite and original austenite form a mixed grains structure with poor ductility. Under this hot charging process, most of the alloying elements in the slab surface structure have undergone precipitation-redissolution behavior, and the degree of precipitation-dissolution of each alloying element is different.

(4) Continuous casting-hot charging rolling (CC-HCR), under this process, the slab cools to between A_1 and 400 °C in the heat preservation device and then charged to the heating furnace for heating and rolling. The structure of the surface of the slab before heating has undergone complete $\gamma \rightarrow \alpha$ phase transition, and the structure is not much different from the cold charging process. Since the energy saving effect of hot charging below 400 °C is not obvious, it is assumed to 400 °C as the lower temperature limit for hot charging.

The hot charging process of continuous casting slab is an important innovative technology in the steelmaking process, which changes the mode of offline stacked cooling and charge in the furnace. It can fully use the heat inside the continuous casting slab and play the role of energy saving and consumption reduction. Compared with the conventional cold charging process of continuous casting slab, the hot charging technology of continuous casting slab has obvious advantages of cost, efficiency, and environmental protection [150].

(1) Reduce energy consumption. Because the hot charging continuous casting slab has a relatively high temperature, a large amount of heat inside the slab can be fully utilized, thus having a good effect of energy saving and consumption reduction. For every 100 °C increase in the hot charging temperature of the slab, the fuel consumption of the heating furnace heating link can be reduced by about 6%. The energy consumption is about (1.25–1.67) $\times$ 10^6 kJ/t from continuous

casting to hot rolling coil by using the traditional cold charging process of continuous casting slab. The direct rolling process can reduce energy consumption by about 75–85%. The direct hot charging process can reduce energy consumption by about 65%. The general hot charging process can reduce energy consumption by about 35% [38]; the energy saving effect is very significant.

(2) Increase the yield of metal. Due to the high temperature of the hot charging slab, it can significantly shorten the heating time of the slab in the heating furnace, thereby reducing the consumption of the iron oxide scale on the surface of the slab during the production process. The amount of iron oxide scale produced by continuous casting slab is calculated according to the total iron. When the traditional cold charging process is adopted, the amount of iron oxide scale produced by continuous casting slab in the heating furnace can be 0.6–1.0%. When adopting the hot charging process, the consumption of iron oxide scale on the surface of the slab is significantly reduced, and the metal yield can be increased by about 0.5–1.0% compared with the traditional cold charging process [151].

(3) Shorten the production cycle. When adopting the traditional cold charging process, the continuous casting slab needs to be offline stacked cooling. Depending on the thickness of the continuous casting slab and the influence of seasonal temperature, the general stacking cooling process needs to take 24–48 h, and the entire continuous casting and rolling process generally needs to take more than 40 h. When adopting the hot charging process, from continuous casting to rolling into production, it generally only takes 5–10 h, greatly shortening the production cycle of the product and greatly improving production efficiency.

(4) Reduce the plant area. Because the hot charging process eliminates the offline stack cooling link of the slab, it simplifies the production process and also reduces the plant area, saving infrastructure investment and production costs. With the compact layout adopted by modern large-scale steel companies, the hot charging technology of continuous casting slab has also become an important measure to ensure smooth production, which solves the problem of limited space for slab stacking in steel companies.

1.5.2 Introduction of Slab Hot Charging Cracks

Although the hot charging process of continuous casting slab has the advantages of reducing energy consumption, shortening production cycle, reducing metal consumption, and simplifying production process, the incidence of surface cracks on rolled products is usually greatly increased when the continuous casting slab of micro-alloyed steel is produced by hot charging process [152]. The high incidence of surface cracks during the hot charging process of micro-alloyed steel continuous casting slab has become a key technical bottleneck hindering the high quality and efficiency of steel products, and even the smooth progress of production processes.

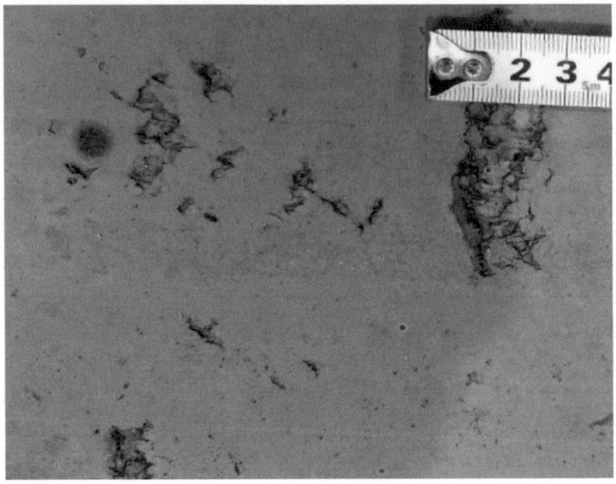

Fig. 1.18 Morphology of the surface crack of Nb-bearing steel plate caused by hot charging crack

The surface cracks caused by the hot charging process during the production of micro-alloyed steel are referred to as "hot charging cracks" [153]. During the heating process of the continuous casting slab in the heating furnace, when the sum of the thermal stress and the structural stress on the surface exceeds the high temperature strength of the slab, hot charging cracks are formed easily. Figure 1.18 shows the morphology of surface cracks caused by hot charging cracks in a Nb-bearing micro-alloyed steel plate. It can be seen that the cracks caused by the hot charging of the slab are left on the surface of the steel plate, which is dense and relatively deep, and in the actual production process, they are usually unable to be eliminated by subsequent surface grinding processes.

1.5.3 Causes and Influencing Factors of the Slab Hot Charging Cracks

The high occurrence rate of surface crack defects in the hot charging process of micro-alloyed steel slab is another common problem in the steel industry. Since the hot charging process of continuous casting slab has been widely adopted, a large amount of research has been carried out on the causes of hot charging cracks in micro-alloyed steel continuous casting slab. At present, many metallurgical workers at home and abroad proposed that the causes of hot charging cracks in micro-alloyed steel continuous casting slab are mainly influenced by two factors: (1) The evolution of the slab structure during the hot charging process. Since Nb and V have the effect of expanding the γ phase region of steel, the A_{r1} temperature of the slab is relatively low, so the hot charging temperature is mostly between A_{r3} and A_{r1} temperatures. Before heating, a part of the surface structure of the slab undergoes $\gamma \rightarrow \alpha$ phase transition. This part of the structure forms austenite grain after heating, and the other part of the structure does not undergo $\gamma \rightarrow \alpha$ phase transition. It is still the coarse original austenite grain after heating. The two form a mixed grain structure, which may cause surface cracking of the slab under thermal stress [154, 155]. (2) The behavior of microalloy element carbonitrides in the slab [156]. Since carbonitrides are prone to precipitate at austenite grain boundaries, chain large size precipitates can reduce the grain boundary strength. When the sum of thermal stress and structure stress on the slab surface exceeds the grain boundary strength, cracks are easily formed. The slab is subject to large thermal stress during the heating process of the heating furnace, and the phase transformation of the slab structure produces the structure stress. When the two stresses work together and represent as tensile stress, it is easy to cause surface cracking of the slab to form a crack source [157, 158].

Therefore, it is generally believed that the factors affecting the formation of hot charging cracks in continuous casting slab include: the surface temperature of continuous casting slab when it is charged to heating furnace, the grain size of the surface structure of the continuous casting slab, the type of microalloy and its carbonitride precipitation and distribution, and the heating process of the continuous casting slab in the heating furnace.

1.5.4 Typical Control Technology of the Slab Hot Charging Crack

Since the advent of the hot charging process of continuous casting slab, the high occurrence rate of hot charging cracks during heating process of the slab widely exists in major steel companies at home and abroad. Therefore, the following typical control methods have been developed for the hot charging crack problem of micro-alloyed steel continuous casting slab [159–163]:

(1) Alloy addition method. Adding Ti elements to micro-alloyed steel containing alloy elements such as Nb, Al, and V can preferentially consume the N content in the steel and play a role in N fixation, thereby reducing the large amount of carbonitrides such as Nb (C, N), AlN, VN that have a greater impact on the ductility of the slab surface structure from precipitating at the grain boundaries of the slab surface structure. However, some requirement of steel grades composition can only contain a lower amount of Ti or even completely avoid the addition of Ti, so this method is often restricted.
(2) Slab delayed hot charging. When the surface temperature of the continuous casting slab is below the Ar_1 temperature, the surface structure will undergo $\gamma \rightarrow \alpha$ phase transition. After the slab enters the heating furnace, the austenite formed by heating is significantly refined compared to the original austenite, and the grain boundaries of the regenerated austenite no longer coincide with the grain boundaries of the solidified austenite of the slab. The microalloy carbonitrides precipitated along the original austenite grain boundaries are not concentrated at the new austenite grain boundaries, and the crack will not be easily formed

during heating process of the slab. Although slab delayed hot charging can effectively prevent the occurrence of hot charging cracks, the continuous casting slab needs to be offline stacked cooling for a long time or heat preserved by a heat preservation device, which requires a large area for the slab, a longer production cycle, and a decrease in energy saving.

(3) Method of high temperature hot charging of slab. Using the high casting speed to increase the temperature of slab at the exit of continuous caster or adopting technology of roller table heat preservation and heat supplement, making the surface temperature of slab higher than the A_{r3} temperature when enters the heating furnace. At this temperature, the surface structure of the slab does not undergo $\gamma \rightarrow \alpha$ phase transition, and the amount of carbonitride precipitation at the grain boundary is small. The ductility of the slab is high during the heating process, which can avoid the formation of surface cracks. At present, it is mostly used for hot charging of continuous casting slab in bar and wire production lines. In plate production, due to the low casting speed of slab continuous casting and the long production line from continuous caster to heating furnace, it is difficult to realize the effective application of hot charging without adding heat supplement.

(4) Surface quenching of continuous casting slab. The surface quenching technology of the slab is a method of controlling the hot charging cracks of the continuous casting slab, which has been widely applied by steel companies at home and abroad in recent years. The process principles are as follows: the high temperature slab out of the continuous caster is subject to surface rapid spray cooling, and then the high temperature of the slab core is used to reheat, so that a high ductility fine grain structure is formed on the surface of the slab, so as to avoid the surface crack of the hot charging slab in the heating furnace. Although the surface temperature of the slab with the surface quenching process will be reduced, most of the heat in the center of the slab will be retained, which has a good effect on energy saving and consumption reduction. Since this method is more efficient and energy saving compared with the methods of adding alloys, the slab delay hot charging, and the slab high temperature hot charging, it has a wide range of application prospects.

References

1. Yong QL, Ma MT, Wu BR (1989) Micro-alloyed steel—Physical and mechanical metallurgy. Machinery Industry Press, Beijing
2. Wang ZM (1993) Development of low alloy steel and micro-alloyed steel. China Metall 3:22–26
3. Wang YK (2000) Micro alloyed steels—review and prospect. Strateg Study CAE 2:79–84
4. Xie JX, Tang H, Mao WM et al (1998) Present situation and development tendency of low-alloy steels. Spec Steel 5:1–4
5. Chen MX, Hu RZ, Li RH (2002) R & D of microalloyed steel. Steel Pipe 31(4):14–16

6. Han XY (2006) Functions of Nb, V and Ti in microalloyed steel. Wide Heavy Plate 12(1):39–41
7. Hou J, Wang F, Zhao GY et al (2007) The research situation and development tendency of micro-alloyed steel. Mater Rep 21(6):91–95
8. Xie H, Du L, Hu J et al (2014) Microstructure and mechanical properties of a novel 1000 MPa grade TMCP low carbon micro-alloyed steel with a combination of high strength and excellent toughness. Mater Sci Eng, A 612:123–130
9. Zhang CL, Shao ZH, Li J et al (2021) Application and development of Niobium microalloying technology in medium and high carbon steel. Mater Rep 35(5):5102–5106
10. Zhang YH, Zhan DP, Yang YK et al (2020) Application of Ti microalloying technology in hot rolled ribbed steel bars. J Mater Metall 19(1):51–56
11. Babenko A, Smirnov L, Upolovnikova A et al (2020) Theoretical bases and technology of steel exhaustive metal desulfurization and direct microalloying with boron beneath basic boron-containing slags. Metallurgist 63(11–12):1259–1265
12. Beatriz L, Jose R (2017) Some metallurgical issues concerning austenite conditioning in Nb-Ti and Nb-Mo micro-alloyed steels processed by near-net-shape casting and direct rolling technologies. Metall and Mater Trans A 48(6):2801–2811
13. Zhang K, Li ZD, Sun XJ et al (2015) Development of Ti-V-Mo complex microalloyed hot-rolled 900-MPa-grade high-strength steel. Acta Metall Sin 28(5):641–648
14. Yi HL, Du LX, Wang GD et al (2006) Development of hot-rolled low carbon steel with high yield strength. ISIJ Int 46(5):754–758
15. Cui J (2001) Development of microalloyed and low-alloyed steels in Baosteel. China Metall 4:32–35
16. Baoshan Iron & Steel (Group) Co., Ltd (1998) Current status and development of microalloyed and low-alloy high-strength steel production in Baosteel. China Metall 6(1):33–40
17. Lin ZQ, Zhang XG, Ao LG et al (2001) Prospect of low alloy steels of Angang in the 21st century. China Metall 3(1):30–33
18. Lin ZQ, Zhang XG, Steel Group Corporation (2007) Current status and development of low-alloy and micro-alloyed steel varieties in Angang. In: Special steel branch of Chinese society for metals
19. Dai LS, Song YH (1995) Present status of production and outlook of low alloy high strength steel, wide and heavy plate of Wuhan Iron and Steel Co. Wide Heavy Plate 3:38–41
20. Wu YS (1999) Development of world steelmaking technology and its enlightenment. Shandong Metall 2:3–6
21. He DZ, Zhou SC, Xiao HG (2013) Continuous casting of steel, 2nd edn. Metallurgical Industry Press, Beijing
22. Morrison WB (2000) Overview of microalloying in steel. In: The proceedings of the Vanitec symposium, pp 25–35
23. Bodnar RL, Shen Y, Lin M et al (1997) Accelerated cooling on Burns Harbor's 160 in. plate mill. In: Accelerated cooling/direct quenching of steels, pp 3–13
24. Funakawa Y, Shiozaki T, Tomita K et al (2004) Development of high strength hot-rolled sheet steel consisting of ferrite and nanometer-sized carbides. ISIJ Int 44(11):1945–1951
25. Shanmugam S, Ramisetti NK, Misra R et al (2008) Microstructure and high strength-toughness combination of a new 700 MPa Nb-microalloyed pipeline steel. Mater Sci Eng: A 478(1/2):26–37
26. Li SS, Zhang LF, Yang XG et al (2016) Control of transverse corner cracks on hypo peritectic steel slabs. Steelmaking 32(3):67–72
27. Yang XG (2016) Control of transverse corner cracks on low carbon micro-alloyed continuous casting slabs. University of Science and Technology Beijing, Beijing
28. Yang XF (2016) Formation reasons study on the transverse corner cracks in continuous casting slab. Baosteel Technol 2:40–42
29. El-Wazri AM, Hassani F, Yue S (1999) The effect of thermal history on the hot ductility of micro-alloyed steels. ISIJ Int 39(3):253–262

30. He AR, Jing FW, Liu C et al (2020) Development, application and prospect of process control technologies of thin slab continuous casting and rolling. Steel Roll 37(3):1–7
31. Ma YT, Li N, Yang T (2021) Application of the gene ration thin Sla continuous casting and rolling endless rolling technology. Hebei Metall 6:37–40
32. Kang YL (2015) The technology and application progress of ultra-thin hot rolled strip produced by thin slab casting and rolling. Steel Roll 32(1):7–11
33. Kromhout JK, Kawamoto M, Hanao M et al (2009) Development of mould flux for high-speed thin slab casting. Steel Res Int 80(8):575–581
34. Mao XP (2020) Nea net shape manufacturing technology of hot-rolled strip. Metallurgical Industry Press, Beijing
35. Mao XP, Chen QL, Zhu DY (2008) Recent development of microalloying technology in thin slab casting and rolling process. Iron & Steel 4:1–9
36. Yu ZX (2002) Hot delivery and hot charging technology of continuous casting slab. Metallurgical Industry Press, Beijing
37. Tian JL (2007) An evaluating method of the energy saving effect for hot delivery and hot charging of continuous casting billet. Energy Metall Ind 26(4):12–14
38. Zhuchkov SM, Kulakov LV, Lokhmatov AP et al (2004) Ways of reducing energy costs in the continuous rolling of sections. Metallurgist 48(3):174–180
39. Xi YQ (2000) Some problems of realizing the process of casting billet hot delivery and hot charging in the wire and rod production. Steel Rolling 6:38–41
40. Shuai XY (2006) Effect of CC-HCR technology on properties of pipeline steel. Electric Steel 4(4):14–18
41. Zhang PC, Wang LB, Tang D et al (2008) Effect of hot charging temperature on microstructure and precipitation behavior of X80 pipeline steel. Heat Treat Met 33(10):99–102
42. Cai ZZ, An JZ, Liu ZY et al (2019) Research and application of micro-alloyed steel slab corner transversal crack control technology. J Iron Steel Res 31(2):117–124
43. Qu SG (2021) Study on optimization of continuous casting secondary cooling process and crack control at corner of the billet. Shanxi Metall 44(1):131–132
44. Maehara Y, Yasumoto K, Tomono H et al (1990) Surface cracking mechanism of continuously cast low carbon low alloy steel slabs. Mater Sci Technol 6(9):793–806
45. Harada S, Tanaka S, Misumi H (1990) A formation mechanism of transverse cracks on CC slab surface quality. South East Asian Iron & Steel Institute 3:26–32
46. Brimacombe JK, Sorimachi K (1977) Crack formation in the continuous casting of steel. Metall Trans B 8:489–505
47. Suzuki HG, Nishimura S, Yamaguchi S (1982) Characteristics of hot ductility in steels subjected to melting and solidification. Transactions ISIJ 22:48–56
48. Hori S, Suzuki M, Unigame Y (1980) Effect of carbon on the low temperature brittleness of iron. J Jpn Inst Metal 44(2):138–143
49. Fachber WM (1982) On transverse surface cracks of CC slabs in Nb bearing high strength steels. Hiittenpraxis Metallweiterverarb 20:222–227
50. Thomas BG, Brimacombe JK (1986) The formation of panel cracks in steel ingots. ISS Trans 7:7–20
51. Suzuki HG, Nishimura S, Imamura J et al (1981) Hot ductility in steels in the temperature range between 900 and 600 deg C: related to the transverse facial cracks in continuously cast slabs. Tetsu-to-Hagane 67(8):1180–1189
52. Turkdogan E (1989) Causes and effects of nitride and carbonitride precipitation during continuous casting. Iron Steelmaker 16(5):61–75
53. Cominelia O, Abushoshaa R, Mintza B (1999) Influence of titanium and nitrogen on hot ductility of C-Mn-Nb-Al steels. Mater Sci Technol 15(9):1058–1068
54. Dippenaar R (2010) Transverse surface cracks in continuously cast steel slabs, oscillation marks and austenite grain size. Mater Sci Forum 638/639/640/641/642:3603–3609
55. Park JS, Ha YS, Lee SJ et al (2009) Dissolution and precipitation kinetics of Nb(C, N) in austenite of low-carbon Nb-microalloyed steel. Metall Mater Trans A 40(3):560–568

56. Lu ZS, Wang FM, Zhang B et al (2011) Hot ductility and carbonitride precipitation of continuous cast 36Mn2V steel slabs at different cooling rates. Trans Mater Heat Treatment 32(8):118–122
57. Bttger B, Apel M, Santillana B et al (2013) Relationship between solidification microstructure and hot cracking susceptibility for continuous casting of low-carbon and high-strength low-alloyed steels: a phase-field study. Metall Mater Trans A 44(8):3765–3777
58. Dippenaar R, Szekeres ES, Moon S (2007) Strand surface cracks—the role of abnormally large prior-austenite grains. AISE Steel Technol 4(7):105–115
59. Reiter J, Bernhard C, Presslinger H (2006) Determination and prediction of austenite grain size in relation to product quality of the continuous casting process. Mater Sci Technol Assoc Iron Steel Technol 5:805–816
60. Tsuchiya S, Ohno M, Matsuura K et al (2011) Formation mechanism of coarse columnar γ grains in as-cast hyperperitectic carbon steels. Acta Mater 59(9):3334–3342
61. Ohno M, Tsuchiya S, Matsuura K (2011) Formation conditions of coarse columnar austenite grain structure in peritectic carbon steels by the discontinuous grain growth mechanism. Acta Mater 59(14):5700–5709
62. Kencana S, Ohno M, Matsuura K et al (2010) Effects of Al and P additions on as-cast austenite grain structure in 0.2 mass% carbon steel. ISIJ Int 50(12):1965–1971
63. Ohno M, Yamaguchi T, Matsuura K et al (2011) Suppression of coarse columnar grain formation in as-cast austenite structure of a hyperperitectic carbon steel by Nb addition. ISIJ Int 51(11):1831–1837
64. Maruyama T, Kudoh M, Itoh Y (2000) Effects of carbon and ferrite-stabilizing elements on austenite grain formation for hypo-peritectic carbon steel. Tetsu-To-Hagané 86(2):86–91
65. Maruyama T, Matsuura K, Kudoh M et al (1999) Peritectic transformation and austenite grain formation for hyper-peritectic carbon steel. Tetsu-To-Hagané 85(8):585–591
66. Fu DC, Wen GH, Zhu XQ et al (2021) Modification for prediction model of austenite grain size at surface of microalloyed steel slabs based on in situ observation. J Iron Steel Res Int 28(9):1133–1140
67. Li Y, Wen G, Luo L et al (2015) Study of austenite grain size of microalloyed steel by simulating initial solidification during continuous casting. Ironmaking Steelmaking 42(1):41–48
68. Cai ZZ, Zhu MY (2013) Thermo-mechanical behavior of peritectic steel solidifying in slab continuous casting mold and a new mold taper design. ISIJ Int 53(10):1818–1827
69. Niu ZY, Cai ZZ, Zhu MY (2019) Dynamic distributions of mold flux and air gap in slab continuous casting mold. ISIJ Int 59(2):283–292
70. An JZ, Cai ZZ, Zhu MY (2022) Role of Ti on the growth of coarse columnar austenite grain during Nb-bearing peritectic steel solidification. Metall Mater Trans A
71. An JZ, Cai ZZ, Zhu MY (2022) Effect of titanium content on the refinement of coarse columnar austenite grains during solidification of peritectic steel. Int J Miner Metall Mater
72. Miyashita Y, Suzuki M, Taguchi K et al (1982) Improvement of surface quality of continuously cast slabs. Nippon Kokan Tech Rep Overseas 36:55–64
73. Crowther DN (2001) The effects of microalloying elements on cracking during continuous casting. In: Proceedings of conference on the use of vanadium in steel, Beijing, pp 99–131
74. Song JX (2018) Study on the formation mechanism and control technology of transverse corner crack of high strength microalloyed steel. Northeast University, Shenyang
75. Ohmori Y, Maehara Y (1984) Precipitation of NbC and hot ductility of austenitic stainless steels. Nippon Kinzoku Gakkai-si 48(2):158–163
76. Nakata H, Yasunaka H (1988) Influence of carbo-nitride and proeutectoid ferrite on hot ductility of Nb, V containing steel. Tetsu-To-Hagané 74(7):94–101
77. Kang SE, Tuling A, Lau I et al (2011) The hot ductility of Nb/V containing high Al, TWIP steels. Mater Sci Technol 27(5):909–915
78. Han WB, Lee JH, Kim HS et al (2012) Effect of precipitates on hot ductility behavior of steel containing Ti and Nb. J Korean Inst Metals Mater 50(4):285–292

79. Mejia I, Salas-Reyes AE, Bedolla-Jacuinde A et al (2014) Effect of Nb and Mo on the hot ductility behavior of a high-manganese austenitic Fe-21Mn-1.3Al-1.5Si-0.5C TWIP steel. Mater Sci Eng A 616:229–239
80. Wojcik T, Kozeschnik E (2016) Influence of NbC-precipitation on hot ductility in microalloyed steel—TEM study and thermokinetic modeling. Mater Sci Forum 879:2107–2112
81. Xiao F, Xu CS, Zhao KW et al (1998) Factors influencing high temperature thermoplasticity of low alloy vanadium steel at temperatures ranging from 600 to 1200 °C. Steel Vanadium Titanium 2:2–5
82. Deng S (2010) Study on the hot ductility of Nb-Ti and Nb-V-Ti steel billets in Liugang. Liugang Sci Technol 3:17–20
83. Wang JF, Deng S, Rao JP et al (2011) Study on the high temperature mechanical properties of Nb-Ti and Nb-V-Ti Micro-alloyed steels. Steelmaking 27(1):46–49
84. Kostryzhev AG, Morales-Cruz EU, Zuno-Silva J et al (2017) Vanadium microalloyed 0.25 C cast steels showing as-forged levels of strength and ductility. Steel Res Int 88(3):1600166
85. Cho KC, Mun DJ, Koo YM et al (2011) Effect of niobium and titanium addition on the hot ductility of boron containing steel. Mater Sci Eng, A 528(10/11):3556–3561
86. Banks KM, Tuling A, Mintz B (2011) Influence of V and Ti on hot ductility of Nb containing steels of peritectic C contents. Mater Sci Technol 27(8):1309–1314
87. Mejia I, Salas-Reyes AE, Calvo J et al (2015) Effect of Ti and B microadditions on the hot ductility behavior of a high-Mn austenitic Fe-23Mn-1.5Al-1.3Si-0.5C TWIP steel. Mater Sci Eng A 648:311–329
88. Luo H, Pentti KL, Porter DA et al (2002) The influence of Ti on the hot ductility of Nb-Bearing steels in simulated continuous casting process. ISIJ Int 42(3):273–282
89. Spradbery C, Mintz B (2005) Influence of undercooling thermal cycle on hot ductility of C-Mn-Al-Ti and C-Mn-Al-Nb-Ti Steels. Ironmaking Steelmaking 32(4):319–324
90. Suzuki HG, Nishimura S, Nakamura Y (1984) Improvement of hot ductility of continuously cast carbon steels. Trans Iron Steel Inst Jpn 24(1):54–59
91. Carpenter KR, Dippenaar R, Killmore CR (2009) Hot ductility of Nb-and Ti-Bearing microalloyed steels and the influence of thermal history. Metall Mater Trans A 40(3):573–580
92. Cho KC, Mun DJ, Kim JY et al (2010) Effect of boron precipitation behavior on the hot ductility of boron containing steel. Metall Mater Trans A 41(6):1421–1428
93. Lu Y, Wang Q, Li J et al (2012) Effect of the charging temperature on the hot ductility of Nb-bearing steel in the simulated hot charge process. Steel Res Int 83(7):671–677
94. Kang M, Lee J, Koo Y et al (2014) Correlation between MnS precipitation, sulfur segregation kinetics, and hot ductility in C-Mn steel. Metall Mater Trans A 45(12):5295–5299
95. Yue EB, Qiu ST, Gan Y (2007) Ther Modynamics of nitride and carbide precipitation in LSLA Steel. J Iron Steel Res 19(1):35–38
96. Liu WJ, Wang ZJ, Li SJ et al (2017) The influence of strong second surface cooling and temperature-reverting process on the slab surface corner crack of micro-alloyed steel. Steelmaking 33(5):72–78
97. Liu ZY, Wang ZJ, Cai ZZ et al (2018) New secondary cooling process for transverse corner crack control of Nb micro-alloyed steel slab. China Metall 28(3):22–29
98. Zhang XH, Liu K, Li Y et al (2011) Application of carbo-nitride precipitation model in concasting process of Nb-bearing micro-alloyed steel for ship plate. Spec Steel 32(3):40–43
99. Ma FJ, Wen GH, Tang P et al (2011) Effect of cooling rate on the precipitation behavior of carbonitride in micro-alloyed steel slab. Metall Mater Trans B 42(2):81–86
100. Wang HB, Zhang JM, Zhao XY et al (2013) Effect of cooling pattern on surface microstructure of slab. Iron Steel 48(4):35–39
101. Weisgerber B, Hecht M, Harste K et al (2002) Improvement of surface quality on peritectic steel slabs. Steel Res 73(1):15–19
102. Kato T, Ito Y, Kawamoto M et al (2003) Prevention of slab surface transverse cracking by microstructure control. ISIJ Int 43(11):1742–1750
103. Lee HU, Park ET, Son SK et al (2010) Assessment of hot ductility with various thermal histories as an alternative method of in situ solidification. ISIJ Int 50(4):540–545

104. Lu ZS, Wang FM, Zhang B et al (2011) Hot ductility and carbonitride precipitation of continuous cast 36Mn2V steel slabs at different cooling rates. Cailiao Rechuli Xuebao/Trans Mater Heat Treat 32(8):118–122
105. Ma FJ (2010) Precipitation behavior of the second phase and microstructural evolution of the surface layer of micro-alloyed slabs in continuous casting. Chongqing University, Chongqing
106. Li YF (2014) Study on refinement of prior austenite grain of slab subsurface for microalloyed steel casting. Chongqing University, Chongqing
107. Grange RA (1966) Strengthening steel by austenite grain refinement. ASM Trans Quarter 59(1)
108. Emi T, Nakato H, Iida Y et al (1978) Influence of physical and chemical properties of mold powders on the solidification and occurrence of surface defect of strand cast slabs. Steelmak Proces 61:350–360
109. Tomono H, Kurz W, Heineman W (1981) The liquid steel meniscus in molds and its relevance to the surface quality of castings. Metall Trans B 12(2):409–411
110. Suzuki M, Hayashi H, Shibata H et al (1999) Simulation of transverse crack formation on continuously cast peritectic medium-carbon steel slabs. Steel Res Int 70:412–419
111. Chen B, Yu H (2012) Hot ductility behavior of VN and V-Nb micro-alloyed steels. Int J Miner Metall Mater 19(6):525–529
112. Papillon Y, Jäeger W, Konig M et al (2003) Determination of high temperature surface crack formation criteria in continuous casting and thin slab casting. EUR 20897
113. Zhao P, Wang XH, Wu SP et al (1996) Cause of transverse corner cracks of slab in BaoSteel (GR) SS41 Continuous Casting Slabs. Iron Steel 31(2):21–24
114. Yin BQ, Yang JH (1991) Surface cracks of Nb-V-Ti micro-alloyed steel continuous casting slabs. Iron Steel Vanadium Titanium 1:39–45
115. Liu H, Wen WX (2014) Genetic analysis and control corner transverse on heavy slabs. Contin Cast 2:34–38
116. Nozaki T, Matsuno J, Murata K et al (1976) Secondary cooling pattern for the prevention of surface cracks in continuous-cast steel slabs. Tetsu-To-Hagané 62(12):1503–1512
117. Yasunaka H, Nakayama K, Ebina K et al (1995) Improvement of transverse corner cracks in continuously cast hypoperitectic slabs. Tetsu-To-Hagané 81(9):894–894
118. Wang XH, Wang WJ, Liu XY et al (1998) Study on transverse corner cracking occurrence prevention of the Nb, V and Ti microalloying steel CC slabs. Iron Steel 33(1):22–25, 72
119. Wen GH, Tang P, Han ZW et al (2003) Influence of nozzle arrangement in secondary cooling zone on slab quality cast on 1930 on slab the caster at Baosteel. Iron Steel 38(1):22–24
120. Wang XY, Liu Q, Hu ZG et al (2010) Influence of nozzle layouts on the secondary cooling effect of medium and thickness slabs in continuous casting. J Beijing Univ Sci Technol 32(8):1064–1070
121. Hou JL, Liu HQ, Gao XJ (2019) Reason analysis and preventive measures on breakout for chamfered-corner mold in Angang Steel. Henan Metall 27(2):44–46
122. Liu QL, Tao HB, Zhang H et al (2018) Effect of chamfer structure on temperature of slab corner and industrial test for controlling transverse corner crack. Steelmaking 34(2):31–37
123. Ma FJ, Wen GH, Tang P et al (2010) Causes of transverse corner cracks in micro-alloyed steel in vertical bending continuous slab casters. Ironmaking Steelmaking 37(1):73–79
124. Baba N, Ohta K, Ito Y et al (2005) Prevention of slab surface transverse cracking at Kashima n°2 caster with surface structure control (SSC) cooling. In: 5th European continuous casting conference
125. Mintz B, Crowther DN (2010) Hot ductility of steels and its relationship to the problem of transverse cracking in continuous casting. Int Mater Rev 55(3):175–176
126. Cai KK (2010) Quality control of continuous casting slab. Metallurgical Industry Press, Beijing
127. Chen DF, Yang XD, Gao XJ et al (2009) Simulation of stress on bending bow of slab caster. J Process Eng 9(S1):333–336
128. Zappulla MLS, Thomas BG (2018) Surface defect formation in steel continuous casting. Mater Sci Forum 941:112–117

129. Zappulla MLS (2020) Mechanisms of longitudinal depression formation in steel continuous casting. Colorado School of Mines, Ann Arbor, Washtenaw
130. Ji C, Cui Y, Zeng Z et al (2015) Continuous casting of high-Al steel in Shougang Jingtang steel works. J Iron Steel Res Int 22(S1):53–56
131. Cui H, Zhang K, Wang Z et al (2019) Formation of surface depression during continuous casting of high-Al trip steel. Metals 9(2):204–212
132. Brimacombe JK, Weinberg F, Hawbolt EB (1979) Formation of longitudinal, midface cracks in continuously-cast slabs. Metall Trans B 10(2):279–292
133. Thomas BG, Storkman WR, Moitra A (1990) Optimizing taper in continuous slab casting molds using mathematical models. In: Proceedings of 6th international iron and steel congress, pp 348–355
134. Stockman WR, Thomas BG (1988) Mathematical models of continuous slab casting to optimize mold taper. In: Modeling of casting and welding processes, Pittsburgh, pp 287–297
135. Mahapatra RB, Brimacombe JK, Samarasekera IV (1991) Mold behavior and its influence on quality in the continuous casting of steel slabs: Part II. Mold heat transfer, mold flux behavior, formation of oscillation marks, longitudinal off-corner depressions, and subsurface cracks. Metallurg Mater Trans B 22(6):875–888
136. Li C, Thomas BG (2004) Thermomechanical finite-element model of shell behavior in continuous casting of steel. Metall Mater Trans B 35(6):1151–1172
137. Thomas BG, Moitra A, Zhu H (1995) Coupled thermo-mechanical model of solidifying steel shell applied to depression defects in continuous-cast slabs. In: Proceedings of the 1995 7th conference on modeling of casting, welding and advanced solidification processes, pp 241–248
138. He S, Li Z, Chen Z et al (2018) Review of mold fluxes for continuous casting of high-alloy (Al, Mn, Ti) steels. Steel Res Int 90(1)
139. Srinivasan R, Weiss I (1990) Formation of surface depressions during Hot Isostatic Pressing (HIP). Scr Metall Mater 24(12):2413–2418
140. Zhu ZY, Zhang YL, Wang KZ (2019) Analysis and control of the causes of Slab bulging and depression. In: National seminar on high efficiency continuous casting application technology and billet quality control, pp 241–244
141. Thomas BG, Moitra A, McDavid R (1996) Simulation of longitudinal off-corner depressions in continuously cast steel slabs. ISS Trans 23(4):57–70
142. Matsumiya T (2006) Recent topics of research and development in continuous casting. ISIJ Int 46(12):1800–1804
143. Lu YJ, Wang Q, Li YG et al (2011) Prevention of longitudinal off-corner depression in continuously cast steel slabs. Contin Cast 27(6):33–37
144. Li Y, Zhang L (2017) Causes and measures of surface depressions in continuous casting. Baosteel Technol 35(6):53–59
145. Hu H (2020) Analysis and control of surface depression in continuous casting billet. Strait Sci Technol Ind 31(6):40–44
146. Zhao JP, Wang S, Feng S et al (2020) Reason and control of surface depression defects in slab of low carbon low-silicon Aluminum killed steel. Contin Cast 45(2):31–35
147. Gao ZL, Wen Z, Liu ML (1997) Current status and prospects of energy-saving technology of heating furnace for steel rolling. Rolling 6:48–52
148. Zhao LC, Zhou JZ, Sun BR (1990) Analysis of the connection between continuous casting and rolling. Steelmaking 1:29–37
149. Zhang ST (1999) Systematic optimizing Technology of CC-DR & CC-HCR etc. Contin Cast 3(1):12–14
150. Wang DW (2004) Development and prospects of hot delivery and hot charging technology for continuous casting billets. Metallurg Manag 12:50–52
151. Shi DR (1997) Analysis of the CC-HCR and CC-DHCR technology plans for wire and bar mill. Rolling Steel 1:54–57
152. Chen C, Ding CJ (2016) Research on surface cracks of sensitive materials during CC-DHCR process. Ind Heat 45(3):58–60

153. Wan YT (2008) Mechanism analysis of surface crack in medium plate during hot slab sending and charging. Steel Research 3:14–16
154. Wada T, Tsukamoto H, Suga M (1988) Effect of hot charge rolling conditions from austenite region on microstructure and mechanical properties of Nb and Ti bearing steel plates. ISIJ Int 74(7):1438–1445
155. Gao Y, Sun JL (2011) Effect of microstructure characteristic of Q460C steel on surface cracks. Mater Sci Technol 19(5):139–143
156. Herman JC, Donnay B, Leroy V (1992) Precipitation kinetics of microalloying additions during hot rolling of HSLA steels. ISIJ Int 32(6):779–785
157. Liu ZM, Zhang JM, Luo YZ (2012) Investigation of surface crack on micro-alloyed steel plate by DHCR. Steel 47(2):67–71
158. Wang G, Zhao LP, Liu HZ (2001) Formation mechanism on hot charging process crack of low alloy steel ingot. Steel Res Rep 3:16–17
159. Gao XJ, San Z, Wang HS (2005) Hot brittleness of slab and quenching treatment. Contin Cast 6:28–29
160. Lu YJ (2013) The Formation and prevention mechanism of surface cracks for low alloy steel slabs during hot charging process. Chongqing University, Chongqing
161. Teshima T, Kitagawa T, Miyahara S et al (1988) The secondary cooling technology of continuous casting for the production of high temperature and high quality slab. Tetsu-to-Hagane 74(7):1282–1289
162. Cai CS (2007) Surface quenching technology for cast billets used for hot delivery. In: Proceedings of the 8th national continuous casting conference, continuous casting branch of China Metal Society: China Metal Society, vol 6, pp 71–76
163. Carboni A, Ruzza DW, Feldbauer SL (1999) Quenching for improved direct hot charging quality. Iron Steelmaker 26(8):3

Chapter 2
Transverse Corner Crack of Micro-alloyed Steel Slab and Its Control

As the main source of slab for hot rolled coil, wide-thick slab, and ultra-thick plate (strip) products, continuous casting slab are mostly produced by vertical arc and arc continuous casters. Micro-alloyed steel containing Ti, Nb, B, Al, V, and other microalloy elements is the main component of steel in companies at home and abroad. It adopts the production process of vertical arc or arc continuous casters. The slab (especially the inner arc) often forms transverse corner crack, resulting in serious quality defects such as edge crack, warping and broken edge in the subsequent rolling process, which restricts the high quality, high efficiency, green and low-cost preparation of micro-alloyed steel plate (strip). This is a common technical problem in the steel industry. This chapter will focus on the formation mechanism of transverse corner crack in conventional and wide-thick slab of micro-alloyed steel, the thermo-mechanical and kinetics of dual-phase particle precipitation in the solidification process of typical microalloy components, the evolution of high temperature structure, the solidification thermo-mechanical evolution of slab in the mold and secondary cooling casting flow and its control. On this basis, the development and application of new crack control technology and device technology based on novel convex-structure mold and corner grain refinement of slab in secondary cooling high temperature zone are introduced.

2.1 Formation Mechanism and New Control Method of Transverse Corner Crack of Micro-alloyed Steel Slab

Currently, the main micro-alloyed steel produced by steel companies at home and abroad mainly use microalloy elements such as Ti, Nb, V, B, Al. By adding a certain amount of one or several microalloy elements to the steel, and using controlled rolling and cooling to improve the strength and ductility of steel products by precipitation hardening and grain refinement. So micro-alloyed steel becomes the main product of

© Metallurgical Industry Press 2025

Z. Cai and M. Zhu, *The Control Technology for Slab Surface Crack During Continuous Casting of Micro-alloyed Steel*, Key Steel Manufacturing Technologies by Collaborative Innovation, https://doi.org/10.1007/978-981-96-7171-7_2

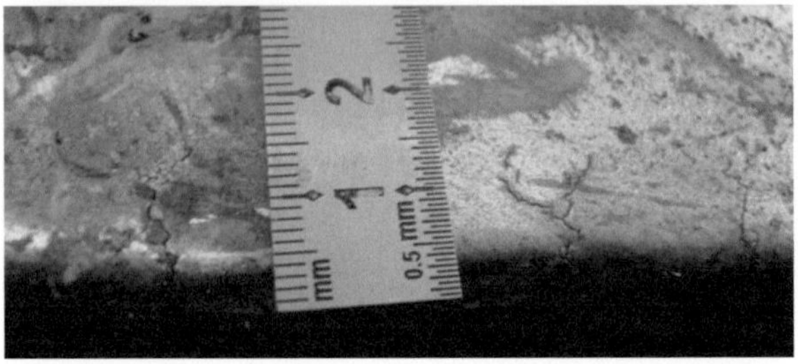

Fig. 2.1 Morphology of transverse corner cracks on micro-alloyed steel slab

steel companies and its production accounts for about 15% of global steel production. However, this type of steel has a high sensitivity to surface crack in the continuous casting process, which forms transverse corner cracks as shown in Fig. 2.1.

The continuous casting production process of steel is a complex process which couples the chemical reactions, the solidification growth and evolution of high temperature structure, as well as the macroscopic thermo-mechanical evolution of solidification of the slab, so many factors affect the formation of cracks in the slab. For the continuous casting production of micro-alloyed steel, the microalloy elements such as Ti, Nb, V, B, Al which are widely used by steel companies at home and abroad, all have strong capacities of combining C and N during the solidification process of the slab. So that they are very likely to combine with free elements such as C, N in the steel which forms carbides, nitrides, or carbonitrides at high temperatures. Therefore, the high temperature precipitation of microalloy carbonitrides is one of the key characteristics distinguishing the continuous casting process of micro-alloyed steel from the production of other ordinary steels.

In addition, selective crystallization is another natural characteristic of high temperature solidification process of steel structure. In the actual continuous casting process of micro-alloyed steel slab, alloys of different types and their contents have different solubilities in the solidification process of the steel structure. For microalloy elements such as Ti, Nb, V, B, Al, because of their low solid–liquid interface distribution coefficients, the slab is prone to significant alloy segregation or clustering during the solidification process, greatly increasing the micro-alloy content at the grain boundaries of the steel structure. At the lower cooling rate of slab continuous casting, when the slab structure is cooled to the precipitation temperature zone of carbonitrides of the microalloy elements, a large amount of corresponding microalloy carbonitrides will preferentially precipitate at the austenite grain boundaries of the slab corner.

Figure 2.2 is a local transmission and energy spectrum analysis of the austenite grain boundary at 5 mm below the surface of the low carbon high strength steel and Ti and Nb micro-alloyed steel. It can be seen from the fig. that coarse dual-phase particles with sizes of 200–450 nm are concentrated on the grain boundary. According to its morphology and energy spectrum analysis, these dual-phase particles are mainly composite precipitated carbonitrides containing elements such as Ti and Nb.

Figures 2.3 is the metallographic structure at 5 mm below the surface of corresponding high-strength steel slab. As can be seen from fig., the original austenite

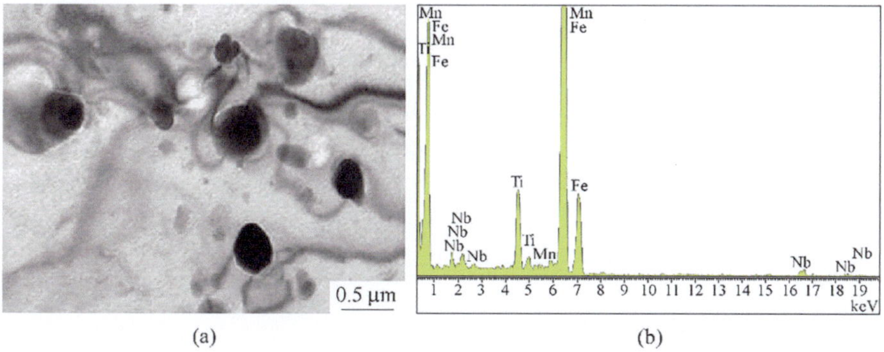

Fig. 2.2 Carbonitride precipitation at 5 mm below the corner of a high-strength steel slab and its component. **a** Morphology of carbonitride precipitation; **b** Energy spectrum analysis

grain boundaries of the slab corner structure have microvoids of about 1–3 μm in size. These microvoids grow in the direction of the ferrite film along the grain boundary and are distributed in a chain shape.

Maehara et al. [1] studied the effect of the microalloy carbonitride precipitation on the hot ductility of steel structure. The results show that the concentrated precipitation of microalloy carbonitride at the austenite grain boundaries will reduce the combining capacity of the grain boundaries and hinder the slip of grain boundaries by pinning, thereby significantly reducing the hot ductility of the steel. At the same time, the structure of dual-phase particles which precipitated at the austenite grain boundaries will gradually separate from each other under the effect of stress and strain to form microvoids, and then gradually aggregate and grow to form large and continuously distributed chain grain boundary microvoids (as shown in the metallography of Fig. 2.3 and the schematic of Fig. 2.4). According to the precipitation temperature of microalloy carbonitrides precipitated at high temperatures such as Nb,

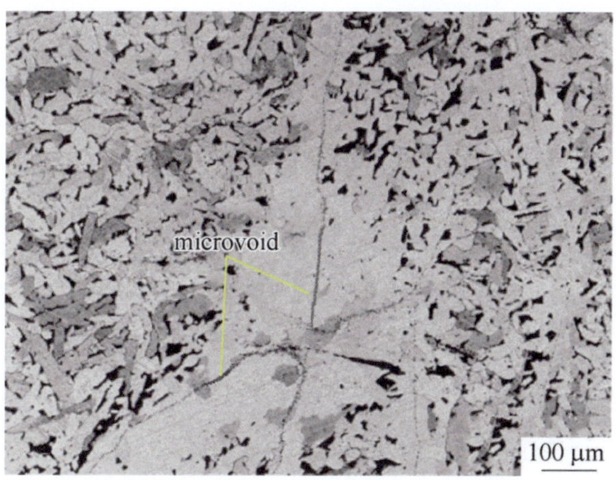

Fig. 2.3 Metallographic structure of a high strength steel slab corner

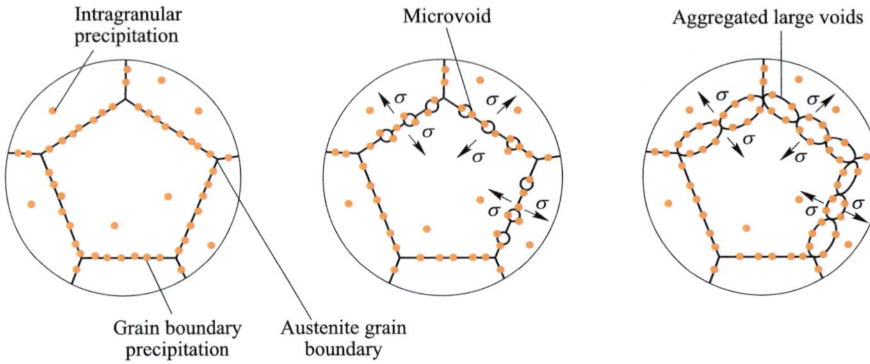

Fig. 2.4 Schematic diagram of the formation and growth evolution of grain boundary microvoids under the action of micro-alloyed carbonitride precipitated at austenite grain boundaries

the complete precipitation temperature of carbonitrides at the austenite grain boundaries often drops below 900 °C. At this temperature, the austenite grain boundaries of the steel begin to gradually form ferrite film. Under the effects of straightening and thermal stresses, microvoids are generated around the micro-structure where microalloy carbonitrides precipitate at the austenite grain boundaries, coinciding with the formation and growth of ferrite film. These microvoids will aggregate and grow during the grain boundary deformation process, and form "microvoids chains" along the growth direction of the ferrite film, thereby significantly reducing the hot ductility of the steel structure. Due to the poor arc accuracy of the bow segments roller gap and the large deformation stress such as straightening, the low ductility slab corner structure is prone to intergranular crack and expansion. Therefore, in the slab continuous casting process of micro-alloyed steel, the microalloy carbonitrides are concentratedly precipitated at the austenite grain boundaries of the structure of the slab corner, resulting in a significant reduction in the hot ductility of the corner structure, which is one of the key factors causing the transverse corner crack of the slab.

The structure morphology of a high strength steel slab with peritectic composition system at a depth of 5 mm below the corner surface is shown in Fig. 2.5. As can be seen from the fig., the casting structure of the slab corner of this steel at room temperature is mainly composed of ferrite and pearlite structures. Generally speaking, the austenite grain morphology during the slab solidification process at room temperature is mainly determined by the growth direction of the grain boundary ferrite film. It can be seen from Fig. 2.5a that the structure of the slab corner at the dual-phase zone temperature is mainly the low ductility structure of "austenite + grain boundary ferrite film", and the grain size of the austenite structure is large, about 1.2–1.5 mm. The thickness of the grain boundary ferrite film is about 20–25 μm, which is wrapped with austenite distribution as a whole. The corner crack of the slab cracks and expands along the growth direction of ferrite in the austenite grain boundary.

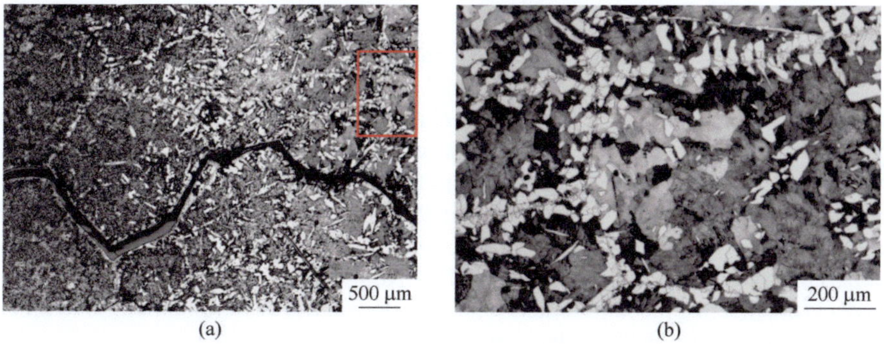

Fig. 2.5 Micro-structure of the slab corner of a Nb-bearing steel. **a** 5 mm below the corner of the slab; **b** Partial enlarged of Fig. 2.5a

Research shows [2–5] the sensitivity of cracks is closely related to the size of austenite grains. When the size of the austenite grains in the slab exceeds 1.0 mm, the area of its grain boundaries will significantly decrease. So, it causes the ability to consume cracking energy through dislocation and other structures to significantly weaken. When the slab is subject to large stress and strain, it is very likely to cause grain boundary cracking and crack expanding.

At the same time, according to the composition system of micro-alloyed steel slab produced at home and abroad, most of them are hypo eutectoid steel. In the actual continuous casting production process of Hypo eutectoid steel, when the high temperature steel structure cools to the temperature of austenite to ferrite transition, the grain boundaries will preferentially generate ferrite structure, and then it will form a film-like structure surrounding the austenite grain, thereby breaks the continuity of the austenite distribution at high temperature. At the same time, the strength of the ferrite is only about 1/4 of the austenite, and the stress and strain generated during the deformation of the slab will concentrate on the ferrite film at the grain boundary. According to Fig. 2.3, the "chain microvoids" formed with the growth of the grain boundary ferrite film will become the concentration point of stress and strain. When the stress and strain subjected by the grain boundaries of the slab structure exceed critical fracture stress and strain of the "chain microvoids" structure, cracks will form on this chain microvoids, crack and expand along its growth direction, thereby forming macroscopic crack defects that grow along the austenite grain boundaries.

It can be seen that the current main slab of micro-alloyed steel during the continuous casting process forms a coarse austenite at the slab corner, and generates a lower strength proeutectoid ferrite film at the grain boundaries. It causes the hot ductility of the slab corner structure to significantly decrease, which is another important factor causing transverse corner crack at the micro-alloyed steel slab.

As can be seen from Fig. 2.5b, there are obvious fine cracks that extend along the growth direction of ferrite film at the austenite grain boundary, passing through the "chain-like microvoids" and the ferrite film. From this phenomenon, the following conclusion can be drawn: the transverse corner crack at the micro-alloyed steel slab will form after the formation of a stable ferrite film at the austenite grain boundaries. According to the structure of the current main slab continuous caster at home and abroad and the production of micro-alloyed steel continuous casting, only when the slab moves downward to the No. 4 segment and below, the temperature of the slab corner drops to below 900 °C preferentially. At this temperature, on the one hand, microalloy carbonitrides containing Ti, Nb, B, Al, etc. have precipitated significantly at the grain boundaries, on the other hand, as the temperature of the slab corner continues to increase, its austenite grain boundaries will gradually generate a ferrite film. If the roller gap assembly accuracy of the No.4–No.6 segments of the continuous caster is not meeting the requirements, and the slab undergoes straightening and other processes afterward, the low ductility structure of the slab corner is very likely to crack along the "microvoid chain" in the direction of the ferrite film development at the austenite grain boundaries due to excessive stress or strain and expand to form macroscopic transverse corner cracks. Therefore, the corner cracks of micro-alloyed steel slab mainly occur in the middle and later parts of the bow segments of the continuous caster and during the straightening process.

Based on the above links and structure and precipitation characteristics of the transverse corner crack at the micro-alloyed steel slab, combined with the actual production of micro-alloyed steel slab continuous casting, the mechanism of transverse corner cracks can be summarized as follows.

During the continuous casting process of micro-alloyed steel slab, due to the poor heat transfer conditions at the mold corner, coarse austenite grains are formed at the corner of the slab. At the same time, due to the selective crystallization in solidification of the steel structure, microalloy elements segregate at the grain boundaries during the solidification process of the slab corner. When the temperature of the slab corner drops to the precipitation temperature zone of the microalloy carbonitrides, as well as the weak cooling conditions and the large amount of alloy segregation at the grain boundaries cause the microalloy elements to preferentially combine with the free C and N elements in the steel, forming carbides, nitrides or carbonitrides and then concentrated precipitation. When the slab pass through the middle of the bow segments and straight segments, the austenite grain boundaries at the slab corner form a ferrite film with lower strength, and under the roller gap assembly accuracy of the bow segments is unqualified, and the large stress and strain of the slab pass through the straight segments, the grain boundary ferrite film structure of the precipitated carbonitrides gradually forms microvoids and dynamically aggregates to form a "chain microvoids" with a size of about 1–3 μm, which significantly reduces the hot ductility of the slab corner structure. Under unqualified accuracy of the bow segments roller gap, and the large stress and strain of the slab pass through the straight segments, the stress concentrates around the ferrite film and "chain microvoids" in the corner structure, and finally cracks along the chain holes in the direction of ferrite film growth due to insufficient ductility, and then expands to form a macroscopic transverse corner crack.

According to the crack formation mechanism of the micro-alloyed steel slab corner, the key to fundamentally controlling the transverse corner crack generation is to accelerate the cooling rate of the slab corner in the mold and other links, to solidify and form smaller austenite grains, and to disperse the precipitation of microalloy carbonitrides in the high temperature zone of the continuous caster and inhibit the formation of the proeutectoid ferrite film at coarse austenite grain boundaries, thereby improving the overall ductility of the slab corner structure.

2.2 Thermodynamics and Kinetics of Micro-alloyed Carbonitride Precipitation

As can be seen from Sect. 2.1, improving the ductility of the slab corner structure is key to eliminating its transverse crack. The dispersion of microalloy carbonitride precipitation during the solidification process is one of the keys to improving the ductility of the slab corner structure. Therefore, this chapter will combine the composition system of typical micro-alloyed steel in China and its actual slab continuous casting production, analyze the thermodynamic and kinetic behaviors of the carbonitride precipitation, and then determine the temperature and cooling control process of its dispersion precipitation.

2.2.1 Thermodynamic of Micro-alloyed Carbonitride Precipitation

2.2.1.1 Precipitation Thermodynamic Model

The precipitation of carbonitrides during the solidification process of micro-alloyed steel is a very complex process, which is mainly described by developing corresponding thermodynamic and kinetic models of carbonitride precipitation. At present, many mathematical models can describe the thermodynamic behavior of carbonitride precipitation during the solidification process of micro-alloyed steel, including the ideal solution model, regular solution model, sub-regular solution model, and double sublattice model. Among them, the ideal solution model has strict requirements for the macro thermal effect and micro structure of the components, and there are few actual materials that meet its conditions [6]; Although the regular solution model of the binary system requires fewer parameters and is easy to calculate, it only considers the interaction between nearest neighbor atoms and assumes that atoms are randomly arranged. However, in the actual solidification process of micro-alloyed steel, it is difficult to follow this regular solution system; The subregular solution model is an improved model on the regular solution model, but this model does not consider excess entropy, so it also cannot accurately describe the thermodynamic behavior of carbonitride precipitation during the solidification process of micro-alloyed steel.

The double sublattice model was proposed and widely used in the 1970s. It is a thermodynamic calculation model for carbonitride precipitation, which has obvious advantages in phase equilibrium calculation of intermittent solid solutions. In recent years, the double sublattice model has been widely used in the study of the thermodynamic behavior of austenite carbonitride precipitation during the solidification process of micro-alloyed steel [7, 8].

For microalloy elements such as Ti and Nb, the carbonitrides precipitating during their solidification process have the same crystal structure and similar lattice constants, and can usually dissolve to form carbonitrides M (C, N). At the same time, in the micro-alloyed steel produced by current companies, the content of C, N and microalloy elements M is few, so it satisfies Henry law, and the atomic vacancies of carbonitride can be ignored, the lattice positions occupied by microalloy elements are the same as the interstitial positions occupied by C, N, and they conform to the ideal chemical coordination ratio. Therefore, the molar free energy of M (C, N) can be expressed as:

$$G_{\mathrm{M(C,N)}} = G_{MC_xN_{1-x}} = xG^{\ominus}_{\mathrm{MC}} + (1-x)G^{\ominus \mathrm{L}S^{\mathrm{m}} \mathrm{E}G^{\mathrm{m}}}_{\mathrm{MN}} \tag{2.1}$$

In the formula:

$G_{\mathrm{M\,(C,\,N)}}$	Molar free energy of $MC_xN_{1\text{-}x}$, J/mol;
M	Microalloy elements Nb, Ti, V, etc.;
x	The molar fraction of MC in $MC_xN_{1\text{-}x}$;
$G^{\ominus}_{\mathrm{MC}}$	The standard Gibbs free energy of MC, J/mol;
$G^{\ominus}_{\mathrm{MC}}$	The standard Gibbs free energy of MN, J/mol;
T	Thermodynamic temperature, K;
${}^{\mathrm{L}}S^{\mathrm{m}}$	Ideal mixing entropy, J/(mol K);
${}^{\mathrm{E}}G^{\mathrm{m}}$	Excess free energy of mixing, J/mol.

where

$$-\frac{{}^{\mathrm{L}}S^{\mathrm{m}}}{R} = x\ln x + (1-x)\ln(1-x) \tag{2.2}$$

$${}^{\mathrm{E}}G^{\mathrm{m}} = x(1-x)L^{\mathrm{M}}_{\mathrm{CN}} \tag{2.3}$$

In the formula:

R	Gas constant, 8.314 J/(mol K);
$L^{\mathrm{M}}_{\mathrm{CN}}$	The interaction free energy of C and N in $MC_xN_{1\text{-}x}$, J/mol.

It is generally believed that during the precipitation process when the partial molar free energy of the solute in the two phases is equal, the carbonitride no longer precipitates from the austenite, that is:

$$\Delta G_{MC} = \overline{G}_{MC} - \overline{G}_{M} - \overline{G}_{C} \tag{2.4}$$

$$\Delta G_{MN} = \overline{G}_{MN} - \overline{G}_{M} - \overline{G}_{N} \tag{2.5}$$

The partial molar free energy of the binary compound in the austenite is:

$$\overline{G}_{MC} = G^{\ominus}_{MC} + RT\ln x + x^2 L^{M}_{CN} \tag{2.6}$$

$$\overline{G}_{MN} = G^{\ominus}_{MN} + RT\ln(1-x) + (1-x)^2 L^{M}_{CN} \tag{2.7}$$

The partial molar free energy of solute elements M, C and N in austenite is:

$$\overline{G}_{C} = RT\ln\alpha_C = RT\ln f_C x_C \tag{2.8}$$

$$\overline{G}_{N} = RT\ln\alpha_N = RT\ln f_N x_N \tag{2.9}$$

$$\overline{G}_{M} = RT\ln\alpha_M = RT\ln f_M x_M \tag{2.10}$$

where

α_C The activity of solute element C;
α_N The activity of solute element N;
α_M The activity of solute element M;
f_C The activity coefficient of solute element C;
f_N The activity coefficient of solute element N;
f_M The activity coefficient of solute element M;
x_C The molar concentration of solute element C;
x_N The molar concentration of solute element N;
x_M The molar concentration of solute element M.

Since in actual micro-alloyed steel, the dissolution amount of M, C, and N solute elements is very few, take $f_C = 1, f_N = 1, f_M = 1$, the mutual interaction free energy of binary compounds in austenite is 0. Therefore, the above equation is simplified to:

$$x\Delta G_{MC} + (1-x)\Delta G_{MC} = 0 \tag{2.11}$$

Substitute the variables ΔG_{MC} and ΔG_{MN}

$$x\ln\frac{xK_{MC}}{x_M x_C} + (1-x)\ln\frac{(1-x)K_{MC}}{x_M x_C} = 0 \tag{2.12}$$

$$x_M x_C = xK_{MC} \tag{2.13}$$

$$x_M x_N = (1 - x)K_{MN} \tag{2.14}$$

In the formula:

K_{MC} The solubility product of MC in austenite;
K_{MN} The solubility product of MN in austenite.

Converting mass percentage concentration to mole percentage concentration, the values for binary compounds are as shown below [9, 10]:

$$\lg([Nb][C])_\gamma = 3.42 - 7900/T \tag{2.15}$$

$$\lg([Nb][N])_\gamma = 2.80 - 8500/T \tag{2.16}$$

$$\lg([Ti][N])_\gamma = 0.32 - 8000/T \tag{2.17}$$

$$\lg([Ti][C])_\gamma = 2.75 - 7000/T \tag{2.18}$$

$$\lg([V][C])_\gamma = 6.72 - 9500/T \tag{2.19}$$

$$\lg([V][N])_\gamma = 3.60 - 8700/T \tag{2.20}$$

$$\lg([B][N])_\gamma = 5.24 - 13970/T \tag{2.21}$$

$$\lg([Al][N])_\gamma = 1.79 - 7180/T \tag{2.22}$$

For Ti, Nb-bearing steel, assume that at temperature T, the $f_{1(TiCxN1-x)}$ and $f_{2(NbCxN1-x)}$ precipitated from austenite are substituted into formulas (2.13) and (2.14):

$$\begin{cases} x_{Ti}x_C = xK_{TiC} \\ x_{Ti}x_N = (1 - x)K_{TiN} \\ x_{Nb}x_C = yK_{NbC} \\ x_{Nb}x_N = (1 - y)K_{NbN} \end{cases} \tag{2.23}$$

The following relationship is obtained from the mass balance equation:

$$\begin{cases} x^0_{Ti} = (1 - f_1 - f_2)x_{Ti} + 0.5f_1 \\ x^0_{Nb} = (1 - f_1 - f_2)x_{Nb} + 0.5f_2 \\ x^0_C = (1 - f_1 - f_2)x_C + 0.5xf_1 + 0.5yf_2 \\ x^0_N = (1 - f_1 - f_2)x_N + 0.5(1 - x)f_1 + 0.5(1 - y)f_2 \end{cases} \tag{2.24}$$

In the formula:

$x^0_{Ti}, x^0_{Nb}, x^0_C, x^0_N$ Initial concentration of solute elements Ti, Nb, C, N in austenite;
$x^0_{Ti}, x^0_{Nb}, x^0_C, x^0_N$ Concentration of solute elements Ti, Nb, C, N in austenite when precipitation reaches equilibrium.

For the precipitation of V (C, N), it can be assumed that f_2 is 0, then the solubility of V, C, N elements and the content of V (C, N) at different temperatures can be obtained. When the above formula x is 0, the precipitation of BN and AlN at different temperatures can be obtained.

2.2.1.2 Nb, Ti Carbonitride Precipitation

Figure 2.6 shows the solubility of various elements in austenite at different temperatures during the solidification process of the typical low carbon Nb, Ti micro-alloyed steel (C, Ti, Nb, and N contents are 0.075, 0.01, 0.025, and 0.003%, respectively). As can be seen from the fig., in the solidification process of this low carbon micro-alloyed steel, Ti (C, N) begins to precipitate at 1386 °C and completes precipitation around 800 °C. During this Ti (C, N) precipitation process, because the solubility of TiN in austenite is more than three orders of magnitude lower than that of TiC, the precipitation process of Ti (C, N) is close to the binary phase TiN precipitation. Therefore, during the precipitation process of Ti carbonitride, it will preferentially consume a large amount of free nitrogen in the steel, which greatly reduces the amount of Nb (C, N) precipitated by combining with the Nb element.

As can be seen from Fig. 2.6, under the composition system of this micro-alloyed steel, the precipitation temperature of Nb carbonitride starts at about 1130 °C, and the precipitation completion temperature is about 750 °C, significantly lower than the precipitation temperature of Ti (C, N). In actual continuous casting production, the precipitation of Nb (C, N) has a stronger detrimental effect on the hot ductility of steel than Ti (C, N). If the characteristic of Nb (C, N) precipitation significantly lagging

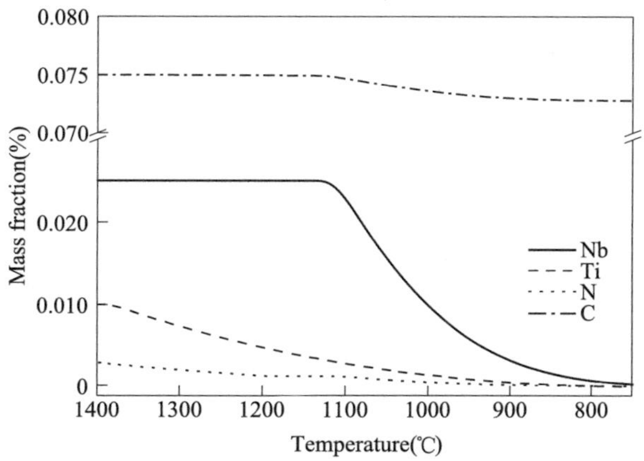

Fig. 2.6 Solubility evolution of various elements in austenite with temperature 1—Nb; 2—Ti; 3—N; 4—C

behind Ti (C, N) precipitation can be fully utilized, under the premise of satisfying the performance of the steel, it is possible to consider adding an appropriate amount of Ti to the steel to achieve the purpose of "add Ti to fixate N", thereby reducing the precipitation of Nb (C, N) at the austenite grain boundaries, and enhancing the crack resistance ability during the production process of slab.

In the actual production process of micro-alloyed steel in steel companies, due to the influence of steel performance requirements and smelting process fluctuations, the content of microalloy and N in micro-alloyed steel often fluctuates to a certain extent. Generally speaking, steel companies pay more attention to the control of N content in steel during the production of micro-alloyed steel. Mainstream steel companies such as An Steel and Baowu Steel generally control the N content in micro-alloyed steel within the range of 0.003–0.006%, and most domestic steel companies currently can control the N content in steel to 0.0045% and below. Therefore, the following text will discuss the precipitation thermodynamics in combination with the typical fluctuation range of Ti, Nb, V, B, Al, and N element contents in current steel companies.

Figure 2.7 shows the evolution in the precipitation amount of Ti (C, N) with temperature in micro-alloyed steel with a Ti content of 0.01% under different N contents. As can be seen from the fig., when the N content in the steel is higher than 0.0045%, the starting precipitation temperature and the amount of precipitation of Ti (C, N) do not change significantly with temperature. However, when the N content in the steel is less than 0.0045%, due to the reduction of free nitrogen in the steel, the precipitation amount of Ti (C, N) at the same temperature during the solidification process is significantly reduced, and the starting precipitation temperature also decreases significantly.

However, when the Ti content in the steel increases, taking the N content of 0.0045% under the mainstream N control level of domestic steel companies as an example, when the Ti addition in the steel increases from 0.01 to 0.02% and 0.03%, respectively, the final precipitation amount of Ti (C, N) increases from 0.013 to 0.025% and 0.038%, respectively, and the starting precipitation temperature overall significantly increases, as shown in Fig. 2.8. As can be seen from fig., when the

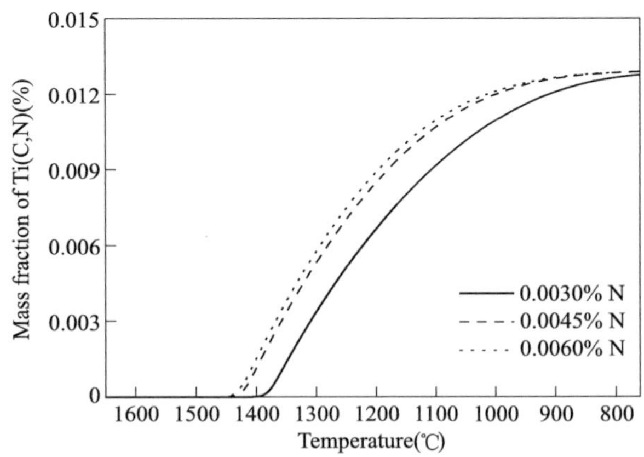

Fig. 2.7 Precipitation of Ti (C, N) with temperature at the Ti content of 0.01% under different N content

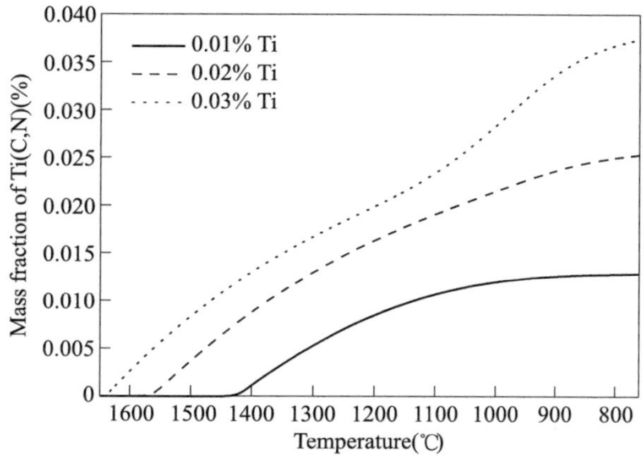

Fig. 2.8 Precipitation of Ti (C, N) with temperature under different Ti content

Ti content in the steel reaches the level of 0.02%, Ti (C, N) can precipitate under the condition of molten steel, which is beneficial to the nucleation of the structure during the subsequent solidification process of the slab. In the actual continuous casting production process, it is often found that under the same continuous casting process, the solidification structure of the Ti-bearing steel slab is more uniform, and the overall internal quality is better. The precipitation of TiN in the molten steel promotes the nucleation of the slab solidification.

Nb is the most widely used microalloy component in steel companies at home and foreign, widely used in high strength ship plate steel, bridge steel, container steel, automotive structural steel, high-rise construction steel, etc. According to the purpose and mechanical properties of different steel grades, the addition of Nb is often in the range of 0.02–0.06%, and the range of 0.02–0.03% is the most widely used.

Figure 2.9 shows the change in the amount of Nb (C, N) precipitated in austenite during the solidification process of Nb-bearing micro-alloyed steel under different Nb and N content conditions. As can be seen from the figure, within the range of Nb and N content of 0.025–0.045% and 0.003–0.006%, respectively, the starting precipitation temperature of Nb (C, N) is approximately in the range of 1120–1160 °C, and it generally increases with the increase of Nb and N contents, but the change is not significant. At the same time, it can be seen that the main precipitation temperature range of Nb (C, N) under different Nb and N contents are above 950 °C. According to the evolution of the corner temperature in the production process of slab, the slab corresponding to this precipitation temperature range is in the mold (middle and lower parts of the mold).

In addition, it can also be seen from Fig. 2.9 that when the N content in the steel is not higher than 0.0045%, the precipitation amount of Nb (C, N) does not change significantly with the change of N content, and the final precipitation amount mainly depends on the Nb content in the steel. This phenomenon indicates that in the production process of Nb-bearing micro-alloyed steel, the N content in the steel should be controlled as low as possible to below 0.0045%.

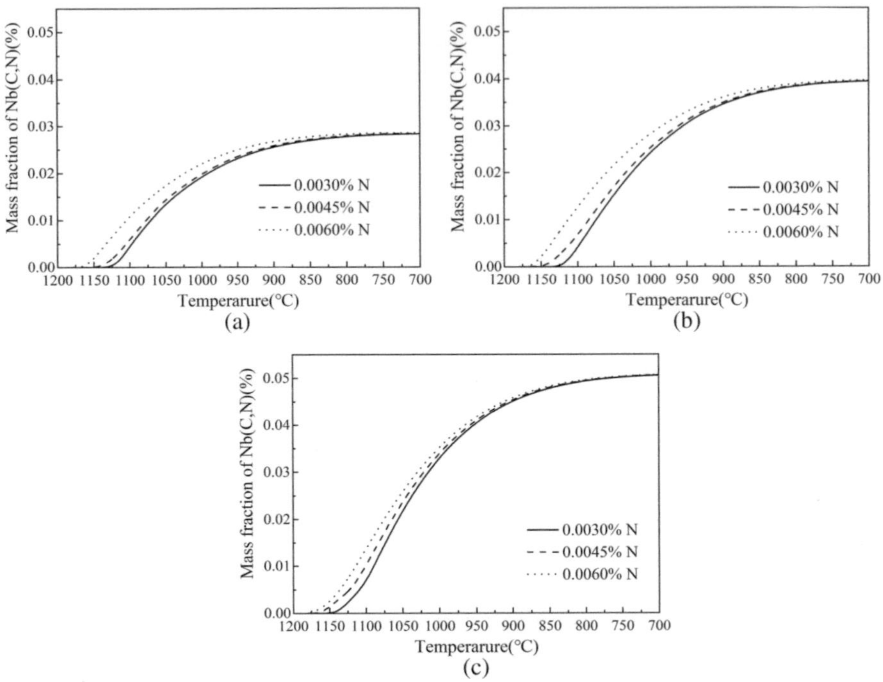

Fig. 2.9 Nb (C, N) precipitation with temperature under different N content. **a** 0.025% Nb; **b** 0.035% Nb; **c** 0.045% Nb

Further, for Nb-bearing micro-alloyed steel with a N content of 0.0045%, the change in the solid solution amount of Nb in austenite with temperature is shown in Fig. 2.10. As can be seen from the fig., when the Nb content in the steel is 0.025, 0.035, and 0.045%, respectively, the temperature at which Nb (C, N) begins to precipitate from austenite during the solidification process is approximately 1120, 1133 and 1160 °C, respectively, and the precipitation is completed around 700 °C. Figure 2.11 shows the relationship between the precipitation amount of Nb (C, N) in austenite and temperature at different Nb contents. It can be seen from the diagram that in the early stage of precipitation, the precipitation rate of Nb (C, N) is faster, and the precipitation amounts under different Nb contents are not much different. When the temperature drops to 850 °C and below, the change of Nb (C, N) precipitation in austenite with different Nb content gradually slows down and is close to the completion of precipitation. Combined with the actual slab continuous casting process, when the slab enters the straight segment, its corner temperature often drops to 850 °C and below. This means that most of the Nb (C, N) in the corner structure of the slab has approached the maximum precipitation amount when entering the continuous casting straight segment. During the straightening deformation process of the slab, the concentrated precipitation of Nb (C, N) will form chain distributed microvoids at the grain boundaries, thereby significantly reducing the ductility of the corner structure of the slab, causing the corner of the slab to crack along the austenite grain boundaries during straightening and other processes, forming transverse crack defects.

Combining with Fig. 2.8, when producing Nb-bearing steel with a N content of 0.0045%, due to the initial precipitation temperature of Nb (C, N) being around 1120 -1160 °C, when the steel is added with more than 0.02% Ti, Ti (C, N) begins to precipitate above 1550 °C. When the corner temperature of the slab drops to 1160 °C, a large amount of free nitrogen in the steel will be consumed in the form of TiN precipitation, significantly reducing the free nitrogen left for Nb, thereby reducing the precipitation of Nb (C, N) and helping to alleviate the hot ductility deterioration of Nb-bearing steel. Therefore, domestic and foreign steel companies often add a certain amount of Ti to the steel during the actual production of Nb-bearing steel, which not only guarantees the mechanical properties of the steel but also helps to prevent the transverse corner cracks. The practice has shown that in order to prevent the transverse corner cracks at the slab of "add Ti to fixate N", the amount of Ti added to the steel requires to reach more than 0.025%.

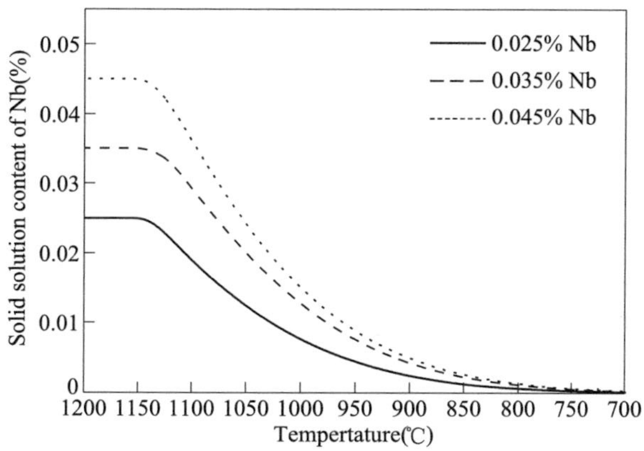

Fig. 2.10 Solid solution content of Nb in austenite with temperature under different Nb content

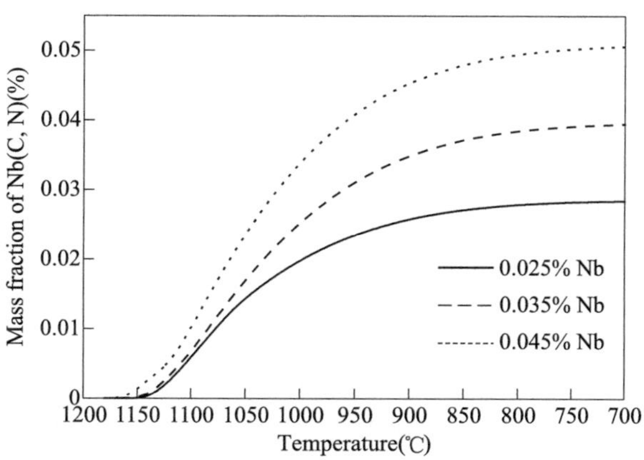

Fig. 2.11 Nb (C, N) precipitation in austenite with temperature under different Nb content

2.2.1.3 V (C, N) Precipitation

V is an important microalloy element in modern steel production. Adding an appropriate amount of V to the steel can significantly enhance its hardenability and has a secondary hardening effect. In recent years, studies have shown that V micro-alloyed steel can use its VN precipitation in austenite to promote intragranular ferrite nucleation, thereby achieving the effect of refining ferrite grains. Domestic and foreign experts and steel companies, based on this precipitation characteristic of VN in steel, have combined it with recrystallization control rolling (RCR) technology to develop the third-generation TMCP process, which fully utilizes the precipitation strengthening advantage of V in steel. And it uses VN austenite precipitation to promote intragranular ferrite nucleation and refines steel structure grains. The TMCP process has been widely used in the development of micro-alloyed steel such as tool steel, automobile steel, advanced bridge steel, high strength rebar, etc. In particular, it has been well applied in some steel products that are difficult to achieve low temperature control rolling, such as thick wall steel and high strength thick plate.

Due to the similar precipitation characteristics of V, Ti, Nb and other elements, it is also a strong C and N binding element. During the solidification process, V (C, N) dual-phase particles are easily formed and precipitated in the micro-structure of high temperature steel. Currently, the amount of V added in main V micro-alloyed steel in domestic and foreign steel companies is mainly concentrated in the range of 0.04–0.08%. Figure 2.12 shows the relationship between the precipitation of V (C, N) in austenite of different V content micro-alloyed steel under different N contents with temperature. As can be seen from the fig., with the increase of V and N content in steel, the precipitation temperature of V (C, N) in austenite begins to gradually increase. When the N content in the steel is 0.003–0.006%, the starting precipitation temperature of V (C, N) in austenite is 900–980 °C, and the precipitation completion temperature is about 600 °C during the solidification process of V-bearing micro-alloyed steel with 0.04–0.08% main addition. According to this precipitation temperature range, when the V content in the steel does not exceed 0.04%, the V (C, N) in the corner structure of the slab during the solidification process mainly precipitates in the secondary cooling zone. However, when the V and N content in the steel is high (especially in the actual solidification process, due to the grain boundary segregation effect of the V, the actual V content in the austenite grain boundary is significantly higher than the average V content), the V (C, N) in the corner structure of the slab begins to precipitate in the lower part of the mold until the slab exits the continuous caster. In addition, it can also be seen that when the N content in the steel is in the range of 0.003–0.006%, the total precipitation amount of V (C, N) in the steel structure is almost the same, indicating that the V content in the steel is the main factor limiting the precipitation amount of V (C, N) during solidification process.

To further analyze the precipitation behavior of V (C, N) during the solidification process of the slab, the relationship between the precipitation amount of V (C, N) and the temperature during the solidification process of the micro-alloyed steel with different V contents when the N content is 0.0045% is compared and analyzed, as

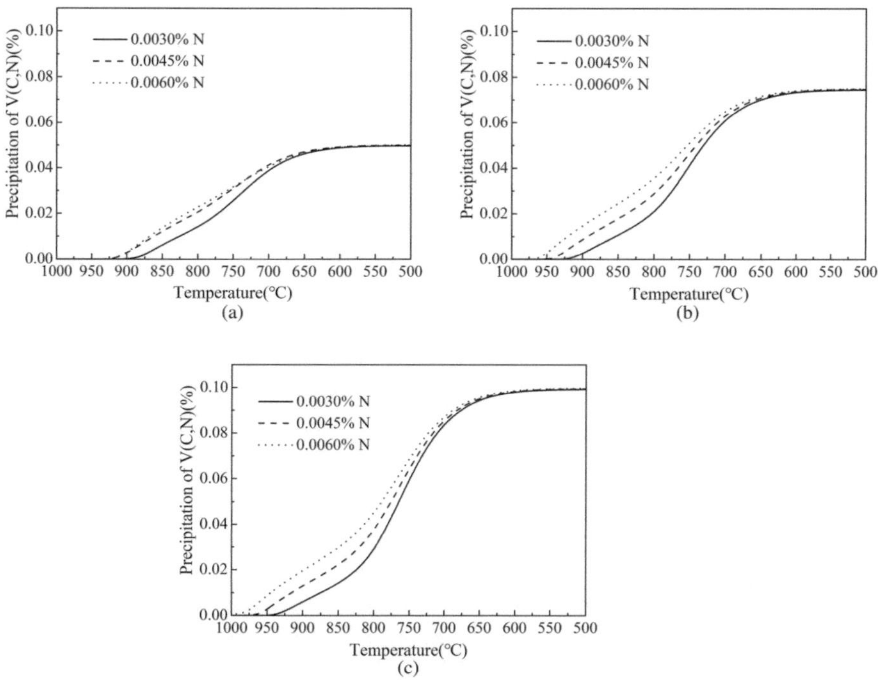

Fig. 2.12 V(C, N) precipitation with temperature under different V content in the micro-alloyed steels with different N content. **a** 0.04%V; **b** 0.06%V; **c** 0.08%V

shown in Fig. 2.13. It can be seen that when the V content in the steel is 0.04, 0.06 and 0.08%, the starting precipitation temperatures of V (C, N) are about 915, 935 and 960 °C, respectively, and the precipitation completion temperatures are basically the same, which are close to 640 °C. However, the precipitation rate of V (C, N) in the high temperature zone is slower compared to the precipitation rate of Ti and Nb carbonitrides. When the temperature drops to 840 °C, the precipitation amount of V (C, N) is only about 30% of the final precipitation amount. In the slab continuous casting process, the corner temperature of the slab in the straight segment of various steel companies is mostly in the range of 750–840 °C. If the process of high temperature weak cooling of secondary cooling water distribution can be adopted to make the corner temperature of the slab enter the straight segment above 840 °C, or if the V addition in the steel can be minimized as much as possible, it will help to reduce the precipitation of the grain boundary V (C, N) at the corner of the slab, so as to achieve the high ductility through straight segment of the slab corner and reduce the crack.

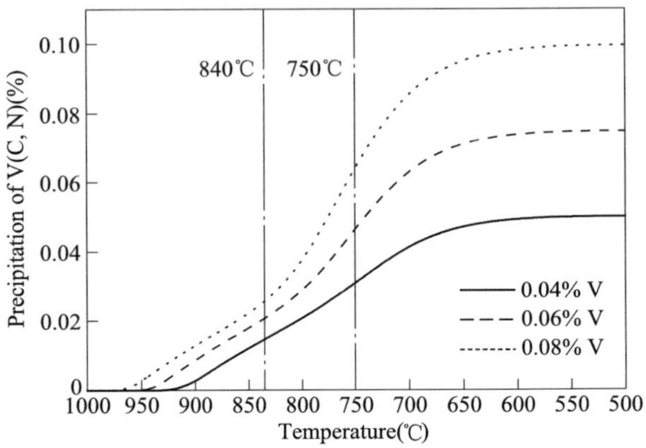

Fig. 2.13 V (C, N) precipitation with temperature under different V content

2.2.1.4 BN Precipitation

B is an important microalloy element. Adding an appropriate amount of B to steel can not only significantly improve the hardenability of the steel, but also enhance its high temperature strength and creep strength for refractory steel. It can also improve the hot hardness of high-speed steel and the cutting ability of tools. B-bearing steel as an important micro-alloyed steel, which is mainly used as structural steel. Especially in the manufacture of bolts and other grades of fasteners, it can be used to replace some medium carbon steel, medium carbon Cr-Ni-Mo steel and other steel grades. In addition, high quality B-bearing steel can also be used as spring steel, low alloy high strength steel, cold rolled steel, wear resistant steel, heat resistant steel, atomic energy steel, etc. At present, most of the B-bearing steel produced on a large scale in China is exported, and its B content is generally controlled in the range of 0.0008–0.0020%. Therefore, this book mainly discusses the BN precipitation behavior of B-bearing steel in this B content range.

Figure 2.14 shows the relationship between the amount of BN precipitated during the solidification process of B-bearing steel with B contents of 0.0008, 0.0014 and 0.0020% at different N contents and temperatures. It can be seen that when the B content is in the range of 0.0008–0.0020% and the N content is in the range of 0.003–0.006%, the temperature range at which BN begins to precipitate is mainly concentrated in the range of 1000–1100 °C, and the precipitation temperature is completed at about 850 °C. The width of the precipitation temperature zone is significantly reduced compared to the precipitation temperature zone of Nb (C, N) and V (C, N). At the same time, it can be seen from the fig. that the precipitation process of BN is relatively fast, and most of the BN has been precipitated at temperatures above 950 °C. It is particularly worth mentioning that the B element has a significant micro segregation effect during the solidification process of steel. Studies have shown that [11], the grain boundary segregation of B in B-bearing steel during solidification can reach more than 30 times. Under the significant segregation amount of the grain boundary, the precipitation temperature of BN at the grain boundary will be higher, and the precipitation amount will also be significantly increased. Therefore, according to the temperature evolution of slab corner, it can be considered that in the continuous casting process of B-bearing steel slab with 0.0008–0.0020% B content in domestic steel companies, most of the BN in the grain boundary of the corner austenite structure has been precipitated in the mold, and the precipitation is completed before the slab enters the secondary cooling straight segment of the continuous caster, which severely aggregates the hot ductility of the slab corner structure, leading to transverse corner cracks during the slab straightening process.

The relationship between BN precipitation amount and temperature of B-bearing steel with different B content under the condition of N content of 0.0045% is further analyzed, as shown in Fig. 2.15. It can be seen that under the same N content, a slight change of B content in the steel can cause a significant change in the final precipitation amount of BN. Under the control condition of 0.0045% N level in most domestic steel companies, when the B content in the steel increases from 0.0008 to 0.0014% and 0.0020%, the starting precipitation temperature of BN in the austenite during the solidification process increases from about 1015 to 1045 °C and 1090 °C, respectively, and the total precipitation amount also increases from 0.0014 to 0.0032% and 0.0046%, respectively. Particularly, it should be pointed out that when the solidification temperature drops to about 950 °C, the change of BN precipitation in the steel

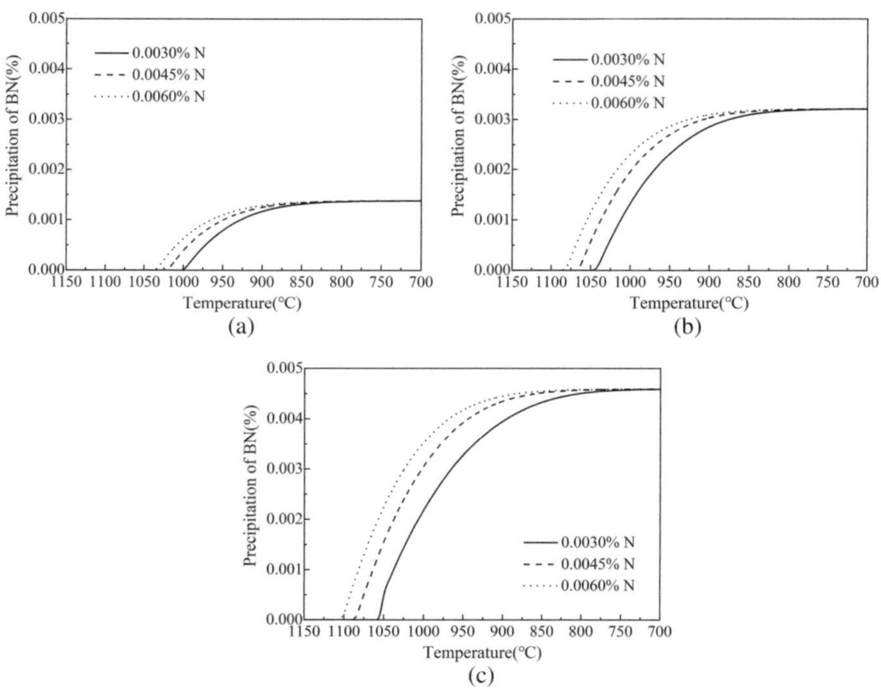

Fig. 2.14 BN precipitation with temperature under different B content in the micro-alloyed steel with different N content. **a** 0.0008% B; **b** 0.0014% B; **c** 0.002% B

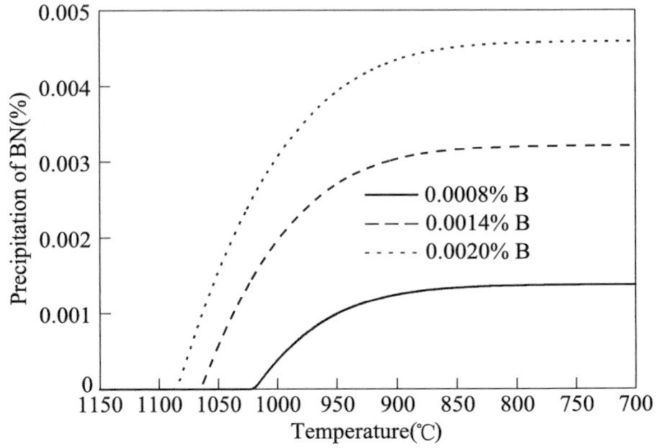

Fig. 2.15 BN precipitation with temperature under different B content

structure with different B content begins to slow down, which means that most of the BN has been precipitated in the steel. Due to the actual slab continuous casting production process, the slab is still in the mold when the slab corner temperature is above 950 °C. Therefore, the key link of BN precipitation control should be in the mold.

However, under the solidification conditions of the traditional slab continuous casting mold, due to the dynamic shrinkage deformation of the shell during the solidification process in the mold, air gap and flux film is often thick in the corner and its adjacent zone of the mold, causing the low heat transfer rate at the corner of the shell, leading to concentrated precipitation of BN at the austenite grain boundaries. The significant grain boundary segregation effect of the B element will aggravate the precipitation of BN at the austenite grain boundaries, thereby greatly aggravating the hot ductility of the slab corner structure. Therefore, the incidence of transverse corner cracks is often high in the continuous casting of B-bearing steel in steel companies, and it is difficult to control the transverse corner cracks of this kind of steel slab by the traditionally developed corner crack control technology.

However, it can also be seen from the above precipitation characteristics of BN that the changes in B and N contents in steel have a very sensitive effect on the precipitation of BN. On the premise of satisfying the performance of the steel grade and improving the smelting level, if the content of B and N in the steel can be appropriately reduced, it can effectively reduce the precipitation of BN at the austenite grain boundaries during the solidification process, thereby alleviating the reduction of its hot ductility. The practice of producing low carbon and peritectic series B steel by continuous casting in Mei Steel shows that the crack rate of slab corner can be controlled to be less than 0.03% by strictly controlling the N content in steel to be less than 0.003% and strengthening the cooling strength of the mold corner and the high temperature zone of the secondary cooling of the slab.

2.2.1.5 AlN Precipitation

Al is an important deoxidizer in steelmaking and also an important micro-alloyed component of steel. Adding an appropriate amount of Al to steel can refine the grain of steel, inhibit the aging of low carbon steel, and improve the toughness and oxidation resistance of steel at low temperature. It is widely used in steel production. However, Al also easily combines with free N to form AlN precipitation during the solidification process of steel, and under unreasonable cooling conditions, it precipitates on a large scale at the austenite grain boundaries of the slab, thereby reducing the hot ductility of the steel structure.

Figure 2.16 shows the relationship between the amount of AlN precipitated during the solidification process of steel under different N content conditions when the Al content is 0.02, 0.03 and 0.04%, respectively. As can be seen from the fig., when the Al and N contents are in the range of 0.02–0.04% and 0.003–0.006%, respectively, the starting precipitation temperature of AlN in austenite structure during solidification of slab is concentrated at 950–1055 °C, and the end temperature of precipitation is about 650 °C. According to the distribution of the precipitation temperature zone and the characteristics of the temperature evolution of the slab corner. If the contents of Al and N in the steel are high, the AlN precipitation zone of the corner structure of the slab will cover the lower part of the mold and the entire secondary cooling casting flow. However, if the contents of Al and N in the steel are low, the precipitation zone of AlN in the corner structure of the slab is mainly in the secondary cooling casting flow. Due to the structure of the traditional slab mold and its controlled cooling process, it is difficult to control the solidification of the corner of the slab inside the mold, so we hope that the AlN precipitation zone of the corner structure of the slab can exit the mold as much as possible. Under the premise of satisfying the performance of the steel and the smooth production of continuous casting, the content of Al and N in the steel should be reduced as much as possible. At present, domestic companies such as Mei Steel strictly control the Al and N accumulation in the steel to no more than 4.5×10^{-5} during the production process of Al-bearing steel grades such as GR4160Al, which can significantly reduce the occurrence of transverse corner cracks in the slab.

Figure 2.17 shows the evolution in the amount of AlN precipitated during the solidification process of Al-bearing steel with different Al content when the N content in the steel is 0.0045%. Combined with Fig. 2.16, it can be seen that the total amount of AlN completely precipitated during the solidification process is the same with different Al content in the steel, indicating that when the N content is 0.0045%, the precipitation of AlN during the solidification process of Al-bearing steel with an Al content of 0.02–0.04% is mainly limited by the N content in the steel. According to this precipitation characteristic of AlN, reducing the N content in the steel can significantly reduce the amount of AlN precipitated at the austenite grain boundaries during the solidification process of the corner of the slab, thereby alleviating the hot ductility loss of the corner structure and helping to prevent the occurrence of transverse corner cracks.

It can be seen from the thermodynamics of chemical precipitation in the solidification process of carbonitride micro-alloyed steel slab containing Ti, Nb, V, B,

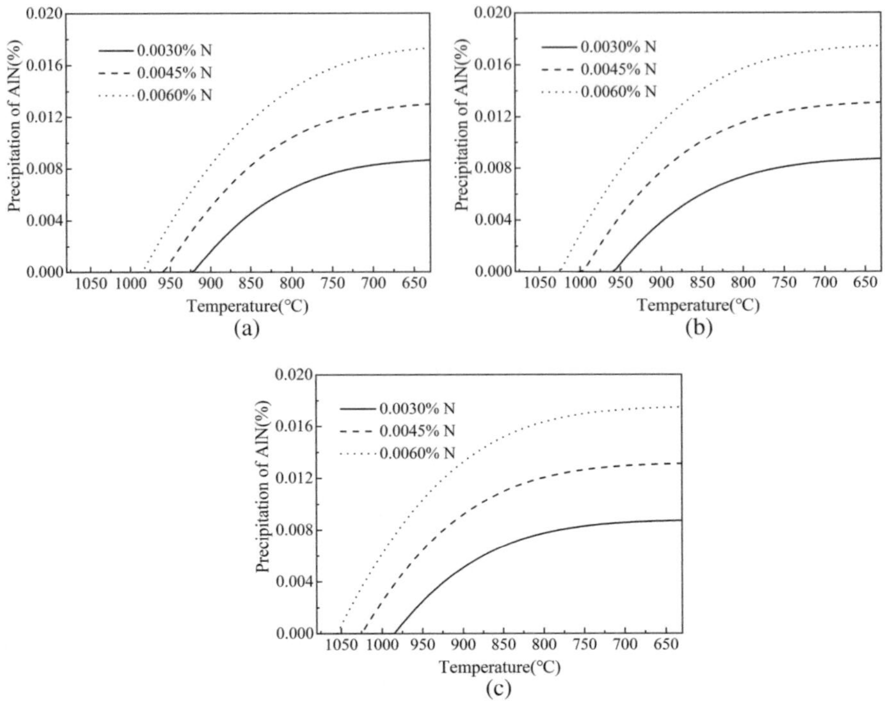

Fig. 2.16 AlN precipitated with temperature under different Al content in the micro-alloyed steel with different N content. **a** 0.02% Al; **b** 0.03% Al; **c** 0.04% Al

Fig. 2.17 AlN precipitation with temperature under different Al content

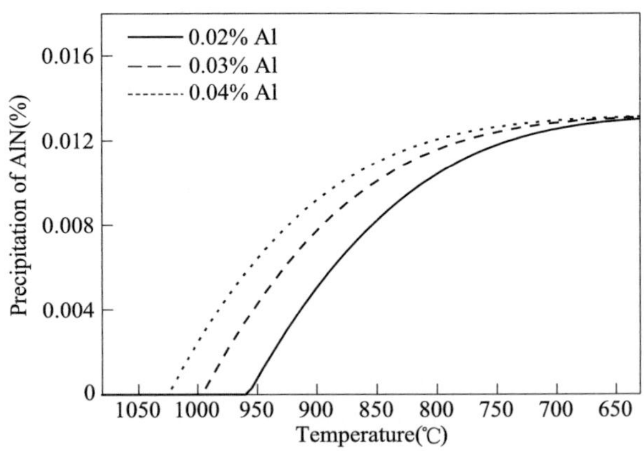

Al and other components produced by domestic and foreign steel companies. The precipitation of microalloy carbonitrides in the corner of the slab mostly covers the mold and secondary cooling zone (especially the secondary cooling high temperature zone).In the slab continuous casting production process of domestic and foreign steel companies, it is often necessary to meet the production requirements of multi-component steel grades and multi-sections of slab after the mold is adopted, and it is often difficult to effectively control the precipitation of carbonitrides during the solidification process of the slab in the mold. Therefore, a novel convex-structure mold and control cooling technology of secondary cooling with high adaptability and dispersion of different types of carbonitride precipitation should be comprehensively developed to eliminate the phenomenon of carbonitride precipitation along the grain boundary during the solidification process of the corner of the slab, so as to achieve high ductility of the corner structure of the slab to eliminate transverse cracks.

2.2.2 Kinetics of Micro-alloyed Carbonitride Precipitation

According to the thermodynamic behavior analysis of carbonitride precipitation during the solidification process of different types of micro-alloyed steel in the previous chapter. It is known that the carbonitride in the corner structure of the micro-alloyed steel slab main production at home and abroad steel companies can precipitate on a large scale in the mold and secondary cooling zone. It is necessary to carry out the research and development of its control cooling structure and process technology in key links such as the mold and secondary cooling, to disperse its carbonitride precipitation. However, the carbonitride precipitation characteristics and key temperature control requirements of micro-alloyed steel with different alloy types and contents are quite different. The determination of the optimal precipitation temperature control zone often requires to be determined according to the precipitation kinetics of the corresponding carbonitrides. Therefore, this chapter will focus on the solidification process of micro-alloyed steel with the above mentioned main chemical component system and content at home and abroad, and analyze the kinetics behavior of carbonitride precipitation.

The precipitation of microalloy carbonitrides belongs to the typical diffusion phase transition. For the diffusion phase transition, when the microalloy carbonitrides precipitate at a higher temperature, the diffusion rate of each solute element is fast but the undercooling is low, resulting in a lower phase transition driving force and a slower transition rate. When the temperature decreases, the undercooling increases, and the transition rate speeds up. As the temperature further decreases, the diffusion ability of microalloy elements decreases and gradually dominates, causing the phase transition rate to decrease. Under the combined action of the high temperature phase transition free energy and low temperature atomic diffusion, the precipitation-temperature–time curve (PTT curve) often shows typical C-curve characteristics. This book will find out the temperature control interval to avoid rapid precipitation at grain boundaries through the theoretical calculation of the PTT curve of microalloy carbonitrides under different nucleation mechanisms, providing a theoretical basis for controlling the dispersion of carbonitrides.

2.2.2.1 Precipitation Kinetics Model

The nucleation mechanism of microalloy carbonitrides usually includes grain boundary nucleation, dislocation nucleation and homogeneous nucleation. Assuming that the precipitates under these three different nucleation mechanisms are all spherical, the chemical driving force during nucleation and precipitation can be expressed by the following formula [12]:

$$\Delta G_{\mathrm{n}}^{\mathrm{b}} \cong \frac{RT}{2}\left[x\ln\frac{x_{\mathrm{M}}^{0}x_{\mathrm{C}}^{0}}{x_{\mathrm{M}}^{\mathrm{b}}x_{\mathrm{C}}^{\mathrm{b}}} + (1-x)\ln\frac{x_{\mathrm{M}}^{0}x_{\mathrm{N}}^{0}}{x_{\mathrm{M}}^{\mathrm{b}}x_{\mathrm{N}}^{\mathrm{b}}}\right], \tag{2.25}$$

where

R	gas constant, 8.314 J/mol K;
T	thermodynamic temperature, K;
$x_{\mathrm{M}}^{0}, x_{\mathrm{C}}^{0}, x_{\mathrm{N}}^{0}$	Solute elements M (microalloy elements), C, N in initial concentration of austenite;
$x_{\mathrm{M}}, x_{\mathrm{C}}, x_{\mathrm{N}}$	Concentration of solute elements M, C, N in austenite when precipitation reaches equilibrium;
x	Equilibrium MC component fraction.

The phase transition free energy of per unit volume can be expressed as follows:

$$\Delta G_{\mathrm{V}} = \frac{\Delta G_{\mathrm{n}}^{\mathrm{b}}}{V_{\mathrm{M(C,\,N)}}}, \tag{2.26}$$

where $V_{\mathrm{M\,(C,\,N)}}$—Molar volume fraction of M (C, N), $\mathrm{m^3/mol}$, can be obtained by linear interpolation of the corresponding MC, MN lattice constants and linear expansion coefficients, specifically:

$$\begin{aligned} V_{\mathrm{M(C,N)}} = {} & \left[\{a_{\mathrm{MC}} \times [1 + \alpha_{\mathrm{MC}} \times (T - 293)]\}^{3}x \right. \\ & \left. + \{a_{\mathrm{MN}} \times [1 + \alpha_{\mathrm{MN}} \times (T - 293)]\}^{3}(1-x)\right] \\ & \times \frac{6.02 \times 10^{-4}}{4}, \end{aligned} \tag{2.27}$$

where a_{MC}, a_{MN} are the lattice constants of MC, MN; α_{MC}, α_{MN} are the linear expansion coefficients of MC, MN.

$a_{\mathrm{MC}}, a_{\mathrm{MN}}, \alpha_{\mathrm{MC}}, \alpha_{\mathrm{MN}}$, the numerical values are shown in Table 2.1 [13]:

The interfacial energy ratio of M (C, N) can be obtained by linear interpolation of the corresponding interfacial energy ratios of MC and MN at different temperatures:

$$\delta_{\mathrm{M(C,N)}} = \delta_{\mathrm{MC}}x + \delta_{\mathrm{MN}}(1-x), \tag{2.28}$$

where δ_{MC} and δ_{MN} are the interfacial energies of MC and MN, respectively, the corresponding formulas for each carbide and nitride are shown as formulas (2.29) to

Table 2.1 Lattice constants and linear expansion coefficients of typical microalloy carbonitrides

Carbonitrides	NbC	NbN	TiC	TiN	VC	VN
Lattice constant *a*/nm	0.4469	0.4394	0.43167	0.4239	0.4182	0.4136
Linear expansion coefficient *α*/K^{-1}	7.02×10^{-6}	10.1×10^{-6}	7.86×10^{-6}	9.35×10^{-6}	8.29×10^{-6}	8.1×10^{-6}

(2.34) [13]:

$$\delta_{NbC-\gamma} = 1.3435 - 0.6054 \times 10^{-3}T \tag{2.29}$$

$$\delta_{NbN-\gamma} = 1.2999 - 0.5858 \times 10^{-3}T \tag{2.30}$$

$$\delta_{TiC-\gamma} = 1.2360 - 0.5570 \times 10^{-3}T \tag{2.31}$$

$$\delta_{TiN-\gamma} = 1.1803 - 0.5318 \times 10^{-3}T \tag{2.32}$$

$$\delta_{VC-\gamma} = 1.1292 - 0.5088 \times 10^{-3}T \tag{2.33}$$

$$\delta_{VN-\gamma} = 1.0879 - 0.4902 \times 10^{-3}T \tag{2.34}$$

The critical nucleation size of homogeneous nucleation carbonitrides can be determined by the formula (2.35):

$$d^* = -\frac{4\delta}{\Delta G_V + \Delta G_{EV}}, \tag{2.35}$$

where

ΔG_{EV} the elastic strain energy when carbonitrides nucleate.

For a highly stable second phase, its elastic strain energy and the free energy of phase transition per unit volume differ greatly, and can usually be ignored. Therefore, formula (2.35) can be simplified to $d^* = \frac{4\delta}{\Delta G_V}$. Its critical nucleation work ΔG^* is:

$$\Delta G^* = \frac{16\pi\delta^3}{3\Delta G_V^2} \tag{2.36}$$

For grain boundary nucleation, the research results of Christian et al. [14] show that the critical nucleation size under the grain boundary nucleation mechanism can still be described by formula (2.35). The critical nucleation work under the grain boundary nucleation mechanism is significantly reduced. Compared with the critical nucleation work under the homogeneous nucleation mechanism, the following relationship is:

$$\Delta G_{\mathrm{g}}^{*} = A_1 \Delta G^{*} \tag{2.37}$$

$$A_1 = \frac{1}{2}\left(2 - 3\cos\theta + \cos^3\theta\right) \tag{2.38}$$

$$\cos\theta = \frac{1}{2}\frac{\sigma_{\mathrm{B}}}{\sigma}, \tag{2.39}$$

where

$\Delta G_{\mathrm{g}}^{*}$ the critical nucleation work under the grain boundary nucleation mechanism;
σ_B the specific interface energy of the austenite phase, at 1100, the value of σ_B is 0.756 J/mol;
σ the specific interface energy of the second phase and the austenite phase.

For dislocation nucleation, because the energy on the unit dislocation line of the edge dislocation is higher compared to the screw dislocation, so that the following mainly considers nucleation on the edge dislocation. Therefore, the critical nucleation work and critical nucleation size under the dislocation nucleation mechanism can be described by formulas (2.40) and (2.41), respectively:

$$d_{\mathrm{d}}^{*} = -\frac{2\sigma}{\Delta G_{\mathrm{V}}}\left[1 + (1+\beta)^{\frac{1}{2}}\right] \tag{2.40}$$

$$\Delta G_{\mathrm{d}}^{*} = (1+\beta)^{\frac{3}{2}} \Delta G^{*}, \tag{2.41}$$

where the parameter β can be obtained from formula (2.42):

$$\beta = \frac{Gb^2 \Delta G_{\mathrm{V}}}{8\pi^2(1-\nu)\sigma^2}, \tag{2.42}$$

where

ν Poisson ratio, generally taken as 0.35;
b Burgers vector of dislocation, taken as 0.2 nm;
G Shear modulus of the matrix.

For the austenitic matrix, the shear modulus of the matrix can be shown by formula (2.43):

$$G = \frac{0.5E}{1+\nu} \tag{2.43}$$

$$E_{\gamma-\mathrm{Fe}} = 254680 - 114.76T \quad (1184 - 1665\mathrm{K}), \tag{2.44}$$

where E—the elastic modulus of the austenitic matrix, MPa.

For the current main micro-alloyed steel composition systems at home and abroad, because the content of their microalloy is relatively low, when carbonitride are formed in a local area, the supersaturation of solute elements in this area will decrease, significantly increasing the critical nucleation work. Therefore, this area will not become a new nucleation point, the nucleation rate rapidly decreases to zero. Under the condition of the nucleation rate rapidly decreasing to zero, the relative start times of carbonitride precipitation under three different nucleation mechanisms of grain boundaries (ga), dislocations (da), and homogeneity (a) can be represented by formulas (2.45)–(2.47) [13]:

Homogeneous nucleation:

$$\lg \frac{t_{0.05a}}{t_{0a}} = \frac{2}{3}\left(-1.28994 - 2\lg d^* + \frac{1}{\ln 10} \times \frac{\Delta G^* + 2.5Q}{kT}\right) \tag{2.45}$$

Grain boundary nucleation:

$$\lg \frac{t_{0.05ga}}{t_{0ga}} = 2\left(-1.28994 - 2\lg d^* + \frac{1}{\ln 10} \times \frac{\Delta G_g^* + Q}{kT}\right) \tag{2.46}$$

Dislocation nucleation:

$$\lg \frac{t_{0.05da}}{t_{0da}} = -1.28994 - 2\lg d_d^* + \frac{1}{\ln 10} \times \frac{\Delta G_d^* + \frac{5}{3}Q}{kT}, \tag{2.47}$$

where

Q diffusion activation energy of micro-alloy elements;
k constant.

The initial precipitation curves of different microalloy carbonitrides under different nucleation mechanisms can be calculated from formulas (2.45)–(2.47). Usually, our research takes the PTT curve when the precipitation amount reaches 5% as the initial precipitation curve, and when the precipitation amount reaches 95%, it is considered as the end of precipitation. It can be obtained by formula (2.48) [13]:

$$\lg \frac{t_{0.95}}{t_0} = \frac{1.76644}{n}, \tag{2.48}$$

where n mainly depends on the nucleation and growth mechanism of the microalloy carbonitrides. When the nucleation rate rapidly decreases to zero, the values of n for homogeneous nucleation, grain boundary nucleation and dislocation nucleation are 3/2, 2, and 1, respectively.

2.2.2.2 Ti (C, N) Precipitation

Figure 2.18 is the solidification process of the low carbon micro-alloyed steel containing Ti and Nb (the contents of C, Ti, Nb and N are 0.075, 0.01, 0.025 and 0.003%, respectively). The critical nucleation work of Ti (C, N) under three nucleation mechanisms of grain boundary nucleation, dislocation nucleation and homogeneous nucleation varies with temperature. It can be seen that as the temperature decreases, the critical nucleation work under each nucleation mechanism gradually decreases. Among them, the critical nucleation work is the least during grain boundary nucleation, followed by dislocation nucleation, and the critical nucleation work required for homogeneous nucleation is the most. The main reason is that the atomic arrangement at the grain boundary is irregular, and the crystal structure has a large distortion. Ti (C, N) preferentially precipitates at the grain boundary during the cooling process, thereby reducing the concentration of alloy solution on the grain boundary and forming a concentration difference in the crystal, causing the alloy solute in the crystal to diffuse to the grain boundary. It causes the grain boundary to form a chain microalloy carbonitride to precipitate concentratedly.

Figure 2.19 shows the precipitation PTT curves of Ti (C, N) under the three nucleation mechanisms of grain boundary, dislocation and homogeneous in the composition system of a low carbon micro-alloyed steel. It can be seen that the curves under the three nucleation mechanisms all present C-curve characteristics. Generally, in the research process, we often define the temperature at which the nucleation rate of the carbonitride precipitation process of this type of PTT curve is the highest and the precipitate is easiest to grow as the "nose point" temperature, focusing on the analysis of the "nose point" temperature distribution and its changes under different nucleation mechanisms.

It can be seen that under the conditions of this typical low carbon micro-alloyed steel composition, the "nose point" temperature of Ti (C, N) grain boundary nucleation is about 1290 °C, and the "nose point" temperature of dislocation nucleation

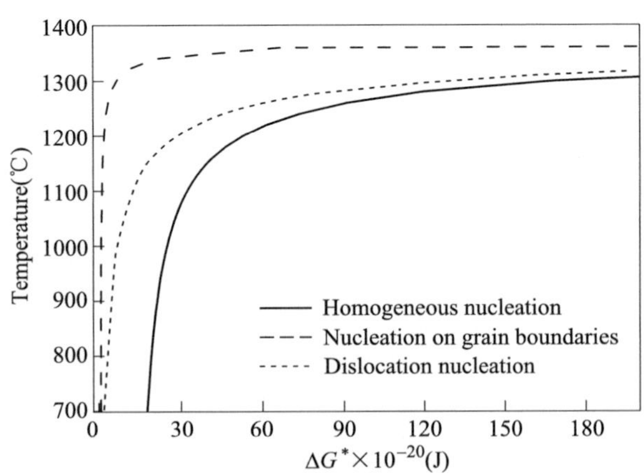

Fig. 2.18 Relative nucleation rate of Ti (C, N) under different nucleation mechanisms

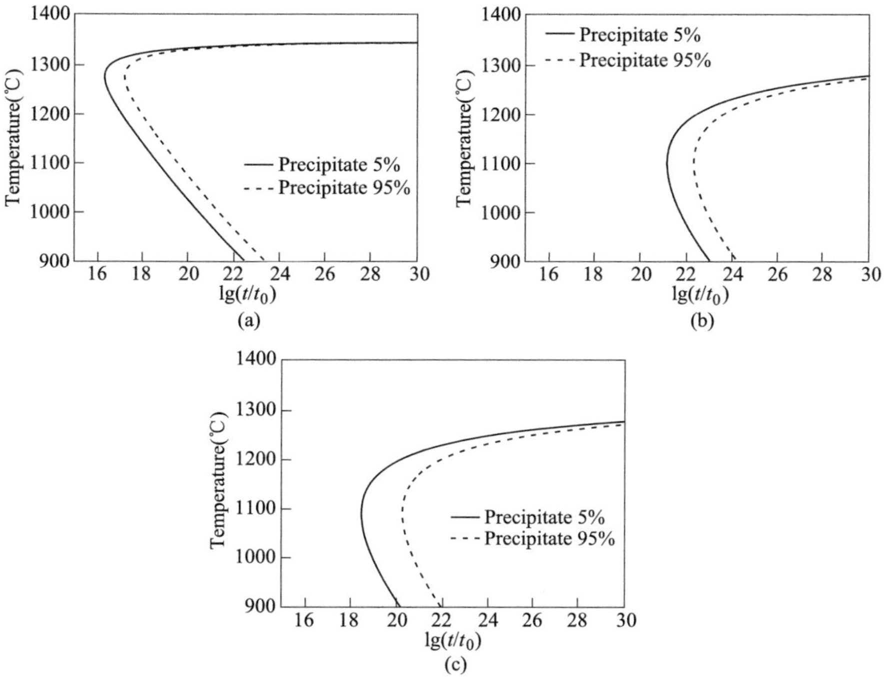

Fig. 2.19 PTT curves of Ti(C, N) precipitation under different nucleation mechanisms. **a** Grain boundary nucleation (GN); **b** Dislocation nucleation (DN); **c** Homogeneous nucleation (HN)

and homogeneous nucleation are slightly lower, both between 1100 and 1120 °C. For the C-curve, near the "nose point" temperature, the precipitate has the longest time from the beginning of precipitation to the completion of precipitation, which is the most favorable temperature range for the growth of the precipitate. Usually, by increasing the cooling rate near the "nose point" temperature, the residence time at this temperature is reduced to inhibit the growth of the precipitation.

From the thermodynamics of carbonitride precipitation, it is known that the precipitation temperature of Ti (C, N) is high, and it significantly increases with the increase of Ti and N content in the steel. Figure 2.20 shows the PTT curves of Ti (C, N) under different nucleation mechanisms at different Ti and N contents. As can be seen from the fig., as the Ti and N contents in the steel increases, the "nose point" temperature of Ti (C, N) precipitation under different nucleation mechanisms continues to increase. When the distribution of Ti and N content in the steel increases to 0.03 and 0.0045%, respectively, the grain boundary nucleation "nose point" temperature has reached 1585 °C. When the N content in the steel increases to 0.006%, the "nose point" temperature of Ti (C, N) precipitation under all nucleation mechanisms has exceeded 1550 °C, under high Ti and N content, Ti (C, N) has gradually precipitated under the condition of molten steel, and even under low Ti and N contents, the Ti (C, N) at the precipitation "nose point" temperature is also in the mold.

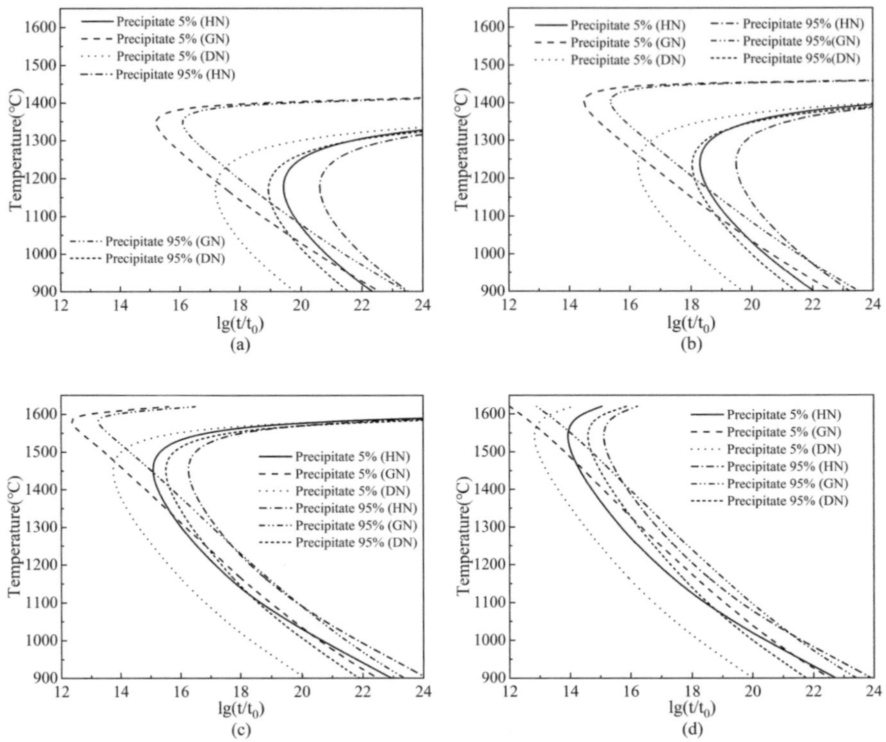

Fig. 2.20 PTT curves of V(C, N) precipitation under different Ti and N content with different nucleation mechanisms. **a** 0.01% Ti-0.0045% N; **b** 0.01% Ti-0.006% N; **c** 0.03% Ti-0.0045% N; **d** 0.03% Ti-0.006% N

2.2.2.3 Nb (C, N) Precipitation

For Nb (C, N) precipitation, the PTT curves of Nb (C, N) under different nucleation mechanisms at typical Nb and N contents in V-bearing steel are shown in Fig. 2.21. As can be seen from the fig., with the increase of Nb and N contents in the steel, the "nose point" temperature of Nb (C, N) precipitation under different nucleation mechanisms changes slightly compared to that of Ti (C, N) precipitation. When the N content in the steel is 0.0045%, with the increase of N content from 0.025 to 0.035% and 0.045%, the grain boundary precipitation "nose point" temperature of Nb (C, N) increases from about 870 to 916 °C and 932 °C, respectively, while the "nose point" temperature of Nb (C, N) precipitation under homogeneous nucleation and dislocation nucleation mechanisms changes very little, which is about 720 °C. Similarly, when the N content in the steel is 0.006%, with the increase of N content from 0.025 to 0.035% and 0.045%, the grain boundary precipitation "nose point" temperature of Nb (C, N) increases from about 890 °C to about 935 °C and 975 °C, respectively, while the "nose point" temperature of Nb (C, N) precipitation under homogeneous nucleation and dislocation nucleation mechanisms is stable at about

760 °C. For the main Nb-bearing steel composition systems at home and abroad, considering the actual production conditions, the key control temperature range for Nb (C, N) dispersed precipitation should be 820–1000 °C.

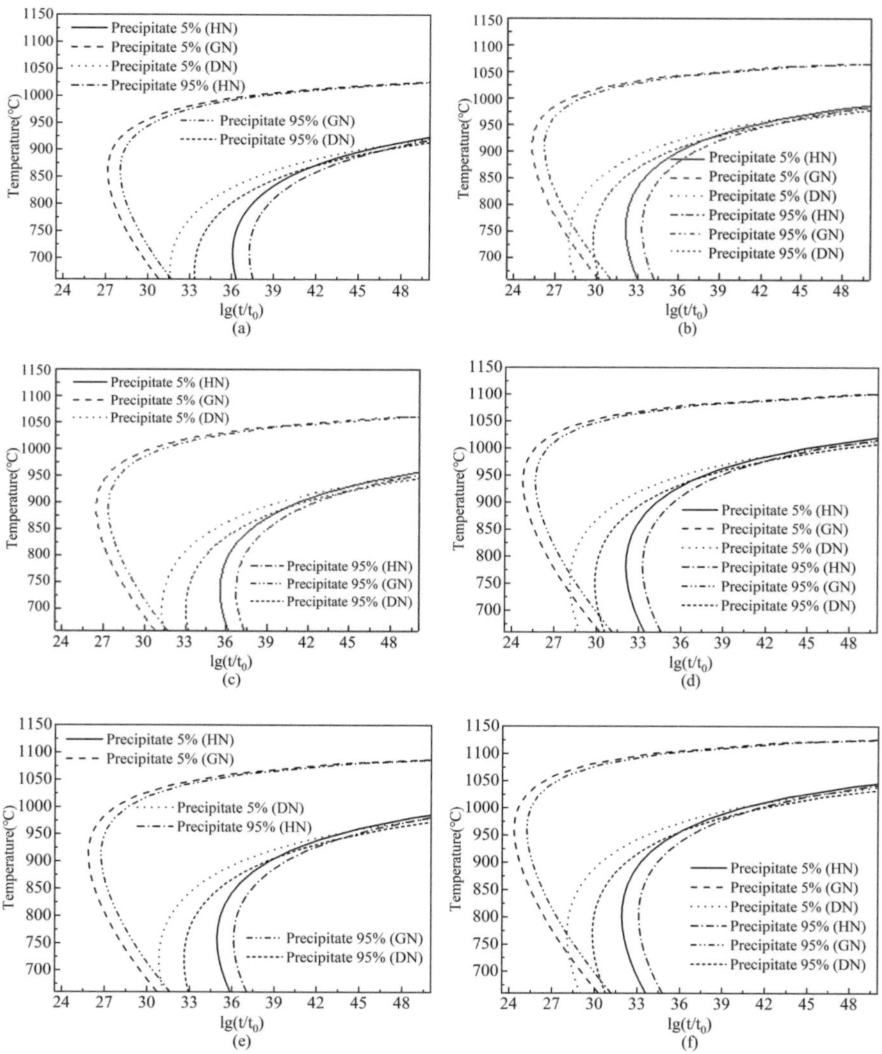

Fig. 2.21 PTT curves of Nb (C, N) precipitation under different Nb and N content with different nucleation mechanisms. **a** 0.025% Nb-0.0045% N; **b** 0.025% Nb-0.006% N; **c** 0.035% Nb-0.0045% N; **d** 0.030% Nb-0.006% N; **e** 0.045% Nb-0.0045% N; **f** 0.045% Nb-0.006% N

2.2.2.4 V (C, N) Precipitation

Figure 2.22 shows the PTT curves of V (C, N) precipitation under different nucleation mechanisms during the solidification process of V-bearing steel at different V and N contents. It can be seen that at lower V and N contents, the PTT curve of V (C, N) precipitation is similar to that of elements such as Ti and Nb, showing a obvious C-curve characteristic. However, when the contents of V and N in the steel increases to a certain amount, its PPT curve shows the characteristic of an S-curve, the PTT curve has two peaks of the fastest precipitation. The main reason for this phenomenon is the significant non-linearity of the phase transition free energy in a certain temperature range. At the same time, it can be seen from the fig. that as the contents of V and N in the steel increases, the ”nose point” temperature of V (C, N) precipitation under different nucleation mechanisms gradually increase. But only the evolution under the grain boundary nucleation mechanism is significant, the increase of the “nose point” temperature under the homogeneous nucleation and dislocation nucleation mechanisms is not obvious.

Specifically for V-bearing micro-alloyed steel with a N content of 0.0045%, when the V content is increased from 0.04 to 0.06% and 0.08%, the “nose point” temperature of V (C, N) grain boundary precipitation increases from about 780 °C to about 810 °C and 835 °C, while the “nose point” temperature of precipitation under homogeneous nucleation and dislocation nucleation only increases from about 580 to 590 °C. When the N content in the steel is 0.006%, and the V content is increased from 0.04 to 0.06% and 0.08%, the “nose point” temperature of V (C, N) grain boundary precipitation increases from about 795 °C to about 825 °C and 850 °C and the “nose point” temperature of precipitation under homogeneous nucleation and dislocation nucleation mechanisms only increases from about 600 to 610 °C. According to this precipitation characteristic temperature, the key control link of carbonitride precipitation in the V-bearing micro-alloyed steel continuous casting process should be in the secondary cooling zone, and the key control temperature range should be 750–850 °C.

2.2.2.5 BN Precipitation

Figure 2.23 is the BN precipitation PTT curve of the B-bearing micro-alloyed steel solidification process at different B and N contents and different nucleation mechanisms. It can be seen that its PTT curve also shows the characteristics of the C- curve. With the increase of B and N content in the steel, the “nose point” temperature of different nucleation mechanisms increases continuously, especially under the mechanism of homogeneous nucleation (HN) and dislocation nucleation, the variation of the “nose point” temperature of BN with B and N content is larger than Ti, Nb and V elements. When the N content in the steel is 0.0045%, the B content increases from 0.0008 to 0.0014% and 0.0020%, The “nose point” temperature of BN grain boundary precipitation increases from about 865 °C to about 900 and 925 °C, the “nose point” temperature under homogeneous nucleation mechanism increases from about 710 to

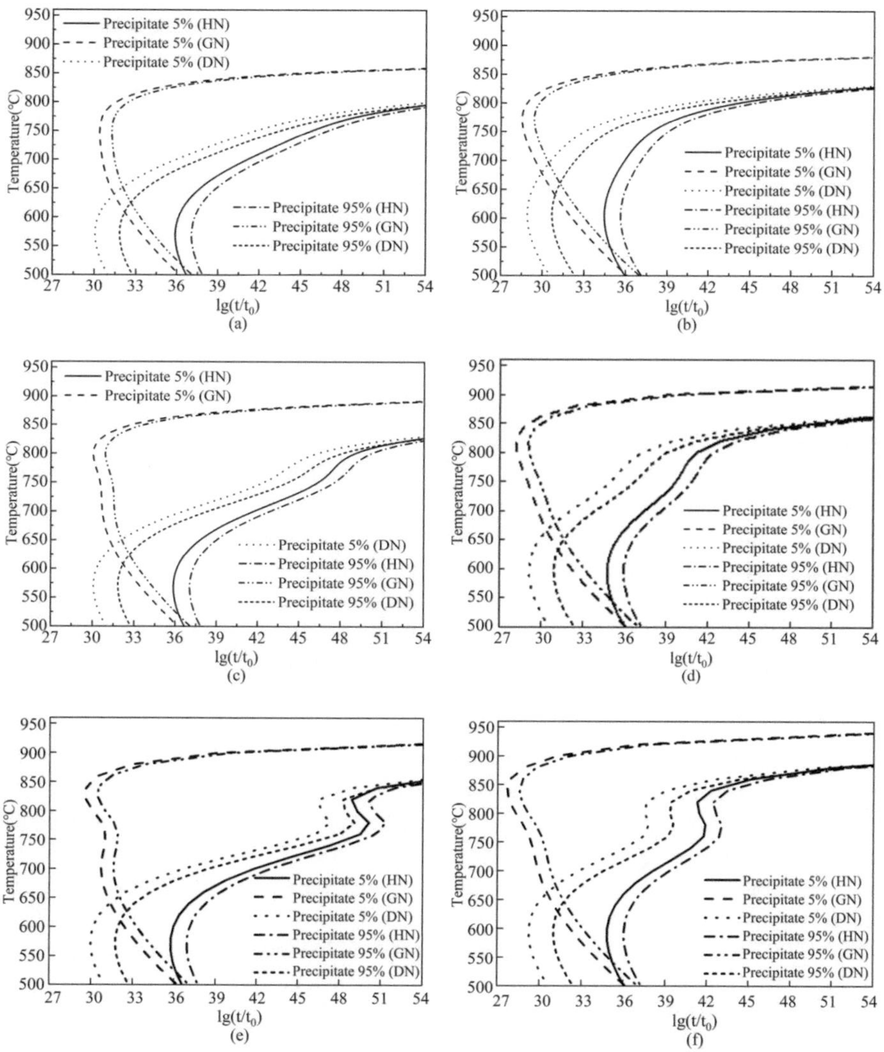

Fig. 2.22 PTT curves of V (C, N) precipitation under different V and N content with different nucleation mechanisms. **a** 0.04% V-0.0045% N; **b** 0.04% V-0.006% N; **c** 0.06% V-0.0045% N; **d** 0.06% V-0.006% N; **e** 0.08% V-0.0045% N; **f** 0.08% V-0.006% N

740 °C and 750 °C, and the "nose point" temperature under dislocation nucleation mechanism increases from 715 to 745 °C and 765 °C. When the N content in the steel is 0.006%, the B content increases from 0.0008 to 0.0014% and 0.0020%, the "nose point" temperature of BN grain boundary precipitation increases from about 875 °C to about 920 and 935 °C, the "nose point" temperature under homogeneous nucleation mechanism increases from about 720 to 750 °C and 770 °C, and the "nose point" temperature under dislocation nucleation mechanism increases from 735 to

765 °C and 775 °C. B has a significant segregation characteristic in the solidification process of steel, the B content in the grain boundary of the steel during solidification is much higher than that under homogeneous nucleation conditions. Therefore, the "nose point" temperature of BN grain boundary precipitation is also higher than that of homogeneous nucleation conditions. According to the solidification and precipitation characteristics of B, its key control cooling link should be in the mold and secondary cooling high temperature zone. The practice shows that the most effective control temperature range of BN dispersion precipitation should be 850–1000 °C.

2.2.2.6 AlN Precipitation

Figure 2.24 shows the AlN precipitation PTT curve under different nucleation mechanisms when the Al and N contents in the steel. It can be seen that the precipitation characteristics of AlN under different nucleation mechanisms are similar to those of B element. The "nose point" temperature of AlN precipitation under grain boundary nucleation, homogeneous nucleation and dislocation nucleation increases with the increase of Al and N contents in the steel. When the N content in the steel is 0.0045%, the Al content increases from 0.02 to 0.03% and 0.04%. The "nose point" temperature of AlN grain boundary precipitation increases from about 825 °C to about 860 °C and 885 °C, respectively. The "nose point" temperature of precipitation under homogeneous nucleation mechanism increases from about 630 °C to about 640 °C and 655 °C respectively, and the "nose point" temperature of precipitation under dislocation nucleation mechanism increases from about 645 to 655 °C and 665 °C, respectively. When the N content in the steel is 0.006%, the Al content increases from 0.02 to 0.03% and 0.04%, the "nose point" temperature of AlN grain boundary precipitation increases from about 855 °C to about 885 °C and 915 °C, respectively, the "nose point" temperature of precipitation under homogeneous nucleation mechanism increases from about 645 °C to about 655 °C and 670 °C, respectively, and the "nose point" temperature of precipitation under dislocation nucleation mechanism increases from about 655 to 680 °C and 705 °C, respectively. According to these precipitation dynamics characteristics and combined with the actual production of slab continuous casting, the key cooling link of AlN precipitation in Al-bearing steel should be concentrated in the high temperature zone of the secondary cooling, and the optimal temperature control range should be 800–950 °C.

2.2.2.7 Determination of Cooling Conditions for Carbonitrides Dispersed Precipitation

Based on the kinetic behavior of the typical microalloy carbonitride precipitation, controlling the cooling rate of the steel structure in the precipitation temperature zone is the main means to refine the size of the precipitates and achieve their dispersed distribution. Ni, as the most widely used alloy in the precipitation strengthening of the new generation of micro-alloyed steel, often strengthens the steel in the form of Nb, Ti

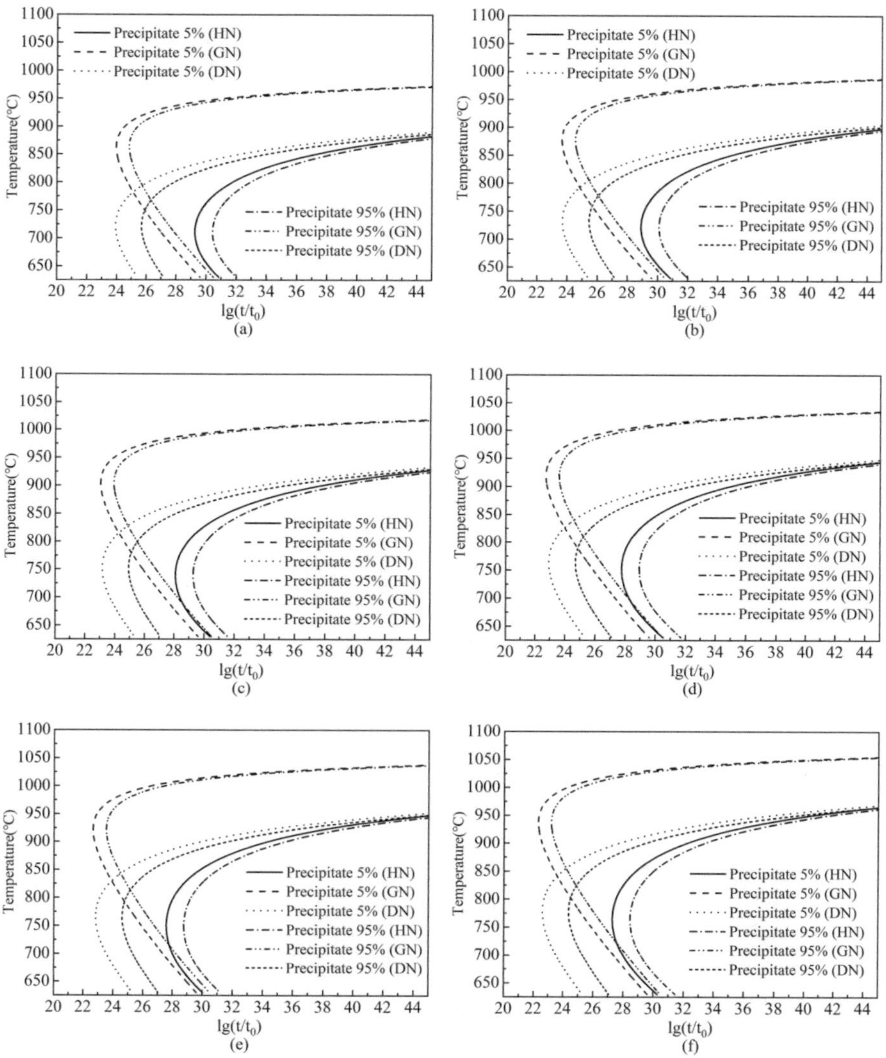

Fig. 2.23 PTT curves of BN precipitation under different B and N content with different nucleation mechanisms. **a** 0.0008% B-0.0045% N; **b** 0.0008% B-0.006% N; **c** 0.0014% B-0.0045% N; **d** 0.0014% B-0.006% N; **e** 0.0020% B-0.0045% N; **f** 0.0020% B-0.006% N

composite precipitation. In this book, a bridge steel containing Nb and Ti as shown in Table 2.2 is used as the representative steel, and the carbonitride precipitation test samples under different cooling rates are prepared by Gleeble-3500 thermal simulator. The size and distribution of carbonitride precipitation under corresponding controlled cooling conditions are detected by transmission electron microscopy, so as to quantitatively explore the cooling rates controlled by carbonitride dispersion

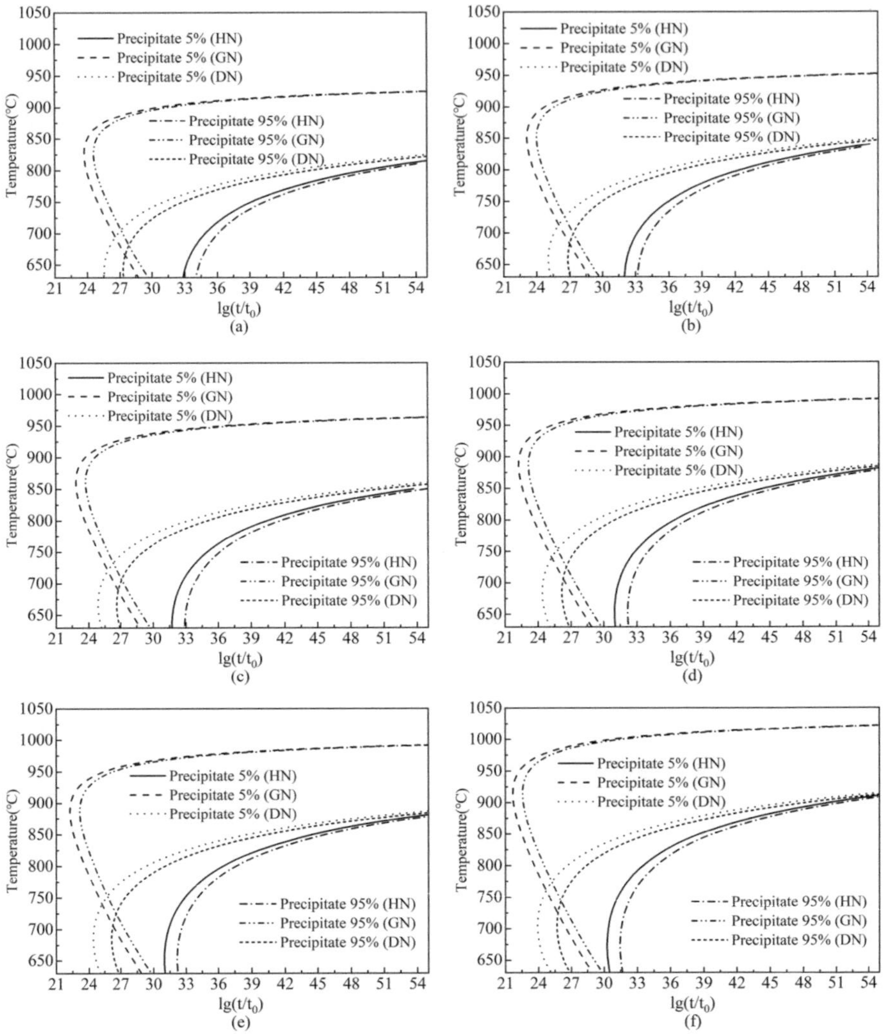

Fig. 2.24 PTT curves of AlN precipitation under different Al and N content with different nucleation mechanisms. **a** 0.02% Al-0.0045% N; **b** 0.02% Al -0.006% N; **c** 0.03% Al-0.0045% N; **d** 0.03% Al -0.006% N; **e** 0.04% Al-0.0045% N; **f** 0.04% Al -0.006% N

precipitation. The controlled cooling curve of the thermal simulation experiment is shown in Fig. 2.25.

Figure 2.26 shows the transmission morphology of the precipitation under cooling rates of 0.5, 1, 3, 4, 5, and 7 °C/s. As can be seen from the fig., when the cooling rate is less than 3 °C/s, the precipitation in the sample are distributed in a chain form, and the size of the precipitation is relatively large, about 150 nm. When the cooling rate is 3 °C/s, although the precipitation still distribute as chain, their size decreases

Table 2.2 Composition of Nb-, Ti- bearing bridge steel (mass fraction, %)

Element	C	Si	Mn	P	S	V	Ti	Nb	N
Content	0.16	0.30	1.05	0.017	0.003	0.006	0.016	0.027	0.003

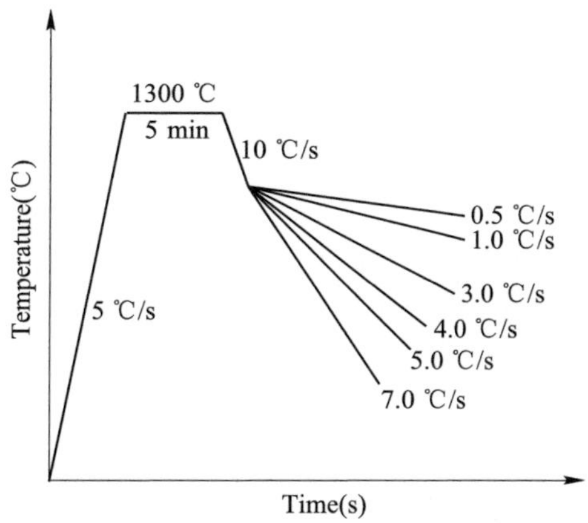

Fig. 2.25 Cooling process of thermal simulation experiment

to less than 100 nm. When the cooling rate reaches 4 and 5 °C/s, the precipitation in the sample gradually become dispersed, and the particles of the precipitation are fine. As the cooling rate further increases, when it reaches 7 °C/s, the size of the precipitation in the test sample is finer, they are mostly within 10 nm and evenly distributed in the matrix. From the above detection results, it can be seen that in the precipitation temperature zone of microalloy carbonitride, when the cooling rate is ≥ 5 °C/s, it can effectively eliminate the chain, large concentrated precipitation at the grain boundaries of the steel structure, and make it dispersed in the matrix of the steel structure, thereby improving hot ductility of the as-cast structure of micro-alloyed steel.

The average size of the micro-alloyed carbonitride of the tested samples at different cooling rates was further counted, and the results were shown in Fig. 2.27. It can be seen from the fig. that as the cooling rate of the steel structure increases, the size of the precipitation decreases significantly. When the cooling rate increases from 0.5 to 4 °C/s, the average size of the carbonitride is refined from nearly 150–8.5 nm. However, when the cooling rate of the steel structure reaches above 5 °C/s, the average size of the precipitation decreases to less than 4 nm, and then the size of the precipitation does not change significantly with the further increase in cooling rate. Therefore, it can be concluded that the cooling conditions for the dispersed precipitation of carbonitride in the corner of the micro-alloyed steel continuous casting slab are: the cooling rate in the precipitation temperature zone is greater than or equal to 5 °C/s.

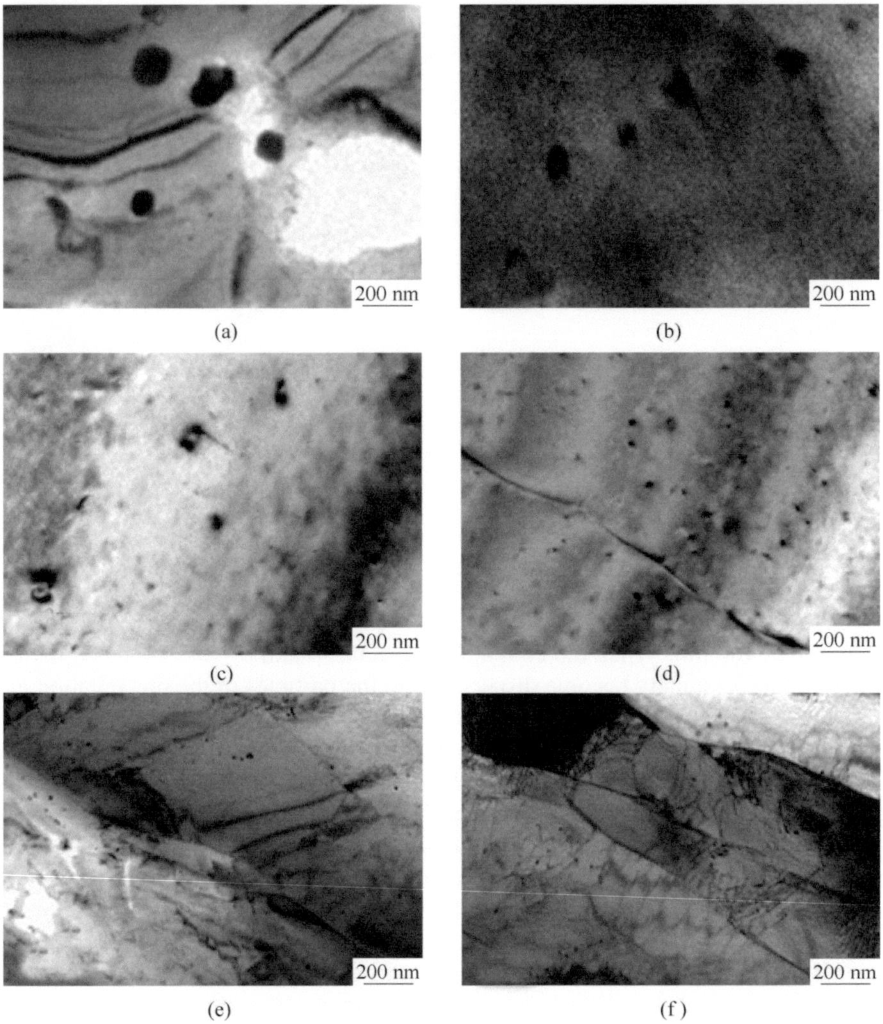

Fig. 2.26 Morphology of the carbonitride precipitation at different cooling rates. **a** 0.5 °C/s; **b** 1 °C/s; **c** 3 °C/s; **d** 4 °C/s; **e** 5 °C/s; **f** 7 °C/s

2.2.3 Critical Control Position for Carbonitrides Precipitation During Continuous Casting

From the thermodynamic and dynamic calculation results of carbonitride precipitation under homogeneous solidification conditions of micro-alloyed steel, it can be seen that although different micro-alloyed steel have different types and contents of alloy and N content in steel, the starting precipitation temperature of microalloy carbonitride is different from the "nose point" temperature. However, in general, the

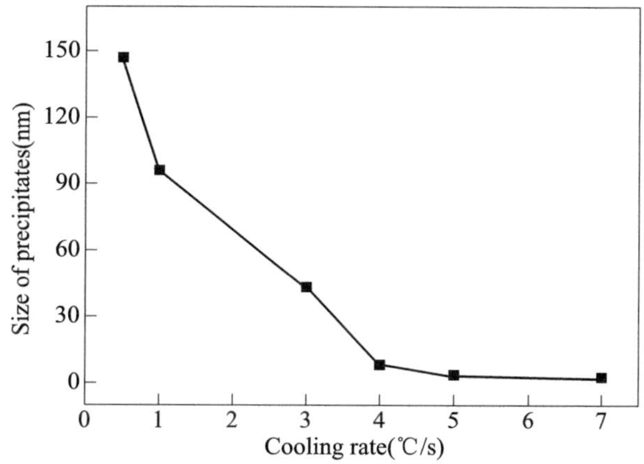

Fig. 2.27 Average size of carbonitride precipitation at different cooling rates

starting precipitation temperature of carbonitrides containing Ti, Nb, B, Al and their composites is close to or more than 1100 °C. And the precipitation end temperature is mostly decrease to about 800 °C. For the solidification of micro-alloyed steel containing V, the starting precipitation temperature of its carbonitrides is generally below 980 °C, and the precipitation end temperature is about 600 °C. In the actual steel solidification process, due to the effect of selective crystallization, the content of microalloy components in the grain boundaries of the solidified structure of the slab will be significantly higher than that under homogeneous solidification conditions, resulting in the starting precipitation temperature of microalloy carbonitrides at the grain boundary and the "nose point" temperature during the solidification process of the actual continuous casting slab. According to the temperature range distribution of precipitates in typical micro-alloyed steel and the evolution of corner temperature field in slab continuous casting production process, the best control link for dispersing the carbonitride precipitation in the corner structure should be the middle and lower part of the mold and the high temperature zone of the secondary cooling casting flow of the continuous caster. Therefore, it is necessary to accelerate the cooling of the corner of the slab in the middle and lower part of the mold and the high temperature zone of secondary cooling to more than 5 °C/s.

2.3 Thermo-mechanical Behavior of Shell Solidification in Mold

From the Sect. 2.2, it can be seen that the microalloy carbonitrides in the corner of the slab, it is first necessary to ensure that the cooling rate of the corner of the slab during solidification in the mold is greater than or equal to 5 °C/s. In addition, to refine the austenite grains in the corner of the slab, a larger cooling rate is also required during the solidification process in the mold. However, in the actual steel continuous casting process, the solidification of the slab in the mold is subject to complex heat transfer and mechanical evolution coupling, with many influencing factors. Quantitatively understanding the thermo-mechanical behavior of typical micro-alloyed steel slab during solidification in the mold is the theoretical premise for solving the efficient heat transfer at the corner of the slab in the mold.

During the solidification process of the slab in the mold, the deformation of the shell is significantly dynamic due to the combined effects of shrinkage stress, ferrostatic pressure of molten steel and taper compensation, which leads to a dynamic uneven distribution of the gap at the interface between shell and copper plate. Affected by this, the heat transfer media such as the air gap and flux film filled in the interface gap also changes dynamically along the height and circumference of the mold, which leads to the dynamic heat transfer in the solidified shell. Therefore, based on the thermal resistance composition and its dynamic change characteristics between the interface of solidified shell and copper plate, developing a heat transfer model at the interface between shell and copper plate that can quantitatively describe the dynamic distribution characteristics of the flux film and air gap in the mold is key to studying the thermo-mechanical behavior of shell solidification in mold.

To accurately describe the heat transfer behavior between the shell and the mold, metallurgical workers at home and abroad have carried out a large amount of research work, among which representative works include:

Savage and Prichard et al. [15] to determine the heat flux distribution during the solidification process of the slab in the mold, first used the heat balance method to measure the surface heat flux of the slab in the static water-cooling mold in the 1950s, and obtained the relationship between the heat flux and the residence time of the shell in the mold:

$$q_{\mathrm{s}} = 2680 - 335\sqrt{\tau}, \left(\mathrm{kW/m^2}\right), \tag{2.49}$$

where τ—The residence time of the shell in the mold, s.

Subsequently, Lait et al. [16] conducted a large amount of actual measurement verification work on the availability of this heat flux model in industrial mold, and obtained its revision:

$$q_s = 2680 - 335\sqrt{60\frac{L}{v}}, \left(\mathrm{kW/m^2}\right), \tag{2.50}$$

where

L The length of the mold, m;
v Casting speed, m/min.

However, studies have shown that the heat flux in the area below the meniscus of the mold obtained by the above two equations is significantly lower than the actual value [17, 18], and therefore cannot truly reflect the solidification heat transfer characteristics of the shell in the mold. For this reason, Samarasekera and Brimacombe et al. [19] measured the temperature of the mold copper plate with thermocouples in several steel companies, calculated the temperature field of the mold copper plate, and obtained the heat flux distribution along the height of the mold using the trial difference method. Because this method is relatively simple to implement, it has become a widely used method for studying the solidification heat transfer behavior of the shell in the mold by metallurgical workers at home and abroad. However, the surface heat flux of the slab in the mold determined by this method can only approximately calculate the heat transfer changes along the height of the mold, and cannot determine the heat flux distributed along the circumference of the slab. Due to the limited number of thermocouple installations, it cannot fully reflect the heat flux evolution of the entire solidification process of the shell. For this reason, Dippenaar and Samarasekera et al. [20] The heat flux distribution along the circumference of the shell was further estimated by the distribution of black and white solidification bands in the macro etching image of the cross section of the continuous casting slab, obtaining an approximate distribution of heat flux along the circumference of the mold. However, the formation of the cross section solidification morphology of the slab is a process of cumulative thermal history. This method cannot determine the circumferential heat flux distribution of the slab at different heights in the mold.

Subsequently, as people's understanding of the existence of air gap in the shell-mold interface and their significant impact on the heat transfer of the solidified shell deepened [21], a large amount of research has been carried out focusing on the dynamic distribution behavior of air gap in the mold and their impact on the heat transfer of the shell. Among them, Dippenaar [20], Hills [22], Wimmer [23] and Perkins [24] divided it into several different heat transfer zones along the circumferential and height directions of the mold, and assigned different heat transfer coefficients, trying to explore the impact of air gap on the heat transfer of solidified shell in the mold. However, in the actual continuous casting process, the air gap distribution at different positions in the mold is not the same, and the fluctuation of steel composition and continuous casting process parameters is large, intensifying its dynamic distribution. It is difficult to accurately determine the suitable heat transfer coefficient of each area under different continuous casting conditions by using the loading heat transfer coefficients method. Because of this, Chen et al. [25] divided the heat transfer between the shell and the mold into contact area and air gap area according to the contact state of the shell-mold interface. The interface heat transfer coefficient of the contact area is determined by the high temperature material contact heat transfer coefficient, and the air gap area is composed of conductive heat transfer and radiation heat transfer, developing an interface dynamic heat transfer model.

The mold flux film in the shell-mold interface not only plays an important role in slab lubrication, but its distribution behavior (especially the dynamic coupling distribution behavior with the air gap) has an important impact on the heat transfer of the mold. For this reason, metallurgical workers further subdivided the heat transfer composition of the shell-mold interface, introducing the mold flux film as the heat transfer medium of the interface into the interface heat flux calculation model [26–35]. Among them, as typical representatives, Moitra and Thomas [34, 35] qualitatively divided the heat transfer process of the shell-mold interface into heat transfer between the shell and the mold flux, air gap conduction, mold flux heat transfer, heat transfer between the mold flux and the copper plate, and radiation heat transfer between the shell and the copper plate, introducing the impact of the mold flux on the heat transfer of the mold into the interfacial heat flux calculation model.

Kim and Han et al. [32, 36] based on the work of Moitra and Thomas et al. it is assumed that the heat transfer between the shell and the mold is composed of heat transfer at the interface between shell and flux film, flux film, air gap, and interface between copper plate and flux film. A heat transfer model describing the mold-shell interface of slab and beam blank was developed. This model innovatively divides the heat transfer within the shell-mold interface, which is closer to the actual continuous casting process of steel. However, in this model, the two researchers only treated the thickness of the mold flux within the interface as a constant distribution, which does not match the dynamic distribution of the mold flux in the mold in the actual continuous casting process along its height and circumferential direction. Therefore, it still cannot accurately describe the dynamic solidification behavior of the shell in the mold.

It can be seen that the currently developed heat flux calculation model of the shell-mold interface lacks effective consideration of the dynamic distribution behavior of the flux film and air gap, making it difficult to accurately explore the temperature evolution of the solidification process of the shell (especially the corner) in the mold. Developing a calculation model that reflects the high temperature phase transition behavior of slab during solidification process, and the dynamic distribution of multiple heat transfer media in the shell-mold interface and thermo-mechanical behavior calculation model of shell solidification under coupling effect with heat transfer. This is a premise for revealing the temperature field evolution of the solidification process of micro-alloyed steel slab in the mold.

In view of this, this chapter will base on the actual cooling structure and cooling process, by developing a dual-phase zone microsegregation model coupled with δ-γ phase transition to determine the physical properties reflecting the high temperature phase transition behavior of steel, the dynamic heat transfer model of the shell-mold interface is developed by using the relationship between the solidification shrinkage deformation of the slab in the mold and the displacement of the inner surface of the mold, the liquid–solid state and thickness distribution of the mold flux, the temperature distribution of the slab surface and copper plate hot face, and the dynamic change of the air gap. Based on this, a coupled calculation model of the solidification thermo-mechanical behavior of the shell in the mold during the actual slab continuous casting process is obtained, to accurately reveal the thermo-mechanical evolution of the solidification process of the shell in the mold.

2.3.1 Mathematical Model

2.3.1.1 Phase Transition Model with Solute Microsegregation

As mentioned above, the phase transition behavior of the slab during the solidification process in the mold directly affects the dynamic shrinkage deformation behavior of the shell, which in turn affects the heat transfer of the slab. The phase transition behavior of steel is closely related to its composition. To determine the high temperature physical parameters of the solidification process of different composition steel, this book uses a solute microsegregation model coupled with phase transition behavior to calculate and determine the solid fraction and the evolution of each phase fraction with temperature during the solidification process of steel. Based on this, the key physical parameters of steel such as thermal conductivity, specific heat and linear thermal expansion coefficient are determined.

A. Model Assumptions

In the actual slab continuous casting process, the heat transfer and mechanical behaviors of the slab in the mold are very complex, it is necessary to propose the following assumptions for the microsegregation model of the slab solidification process:

(1) It is assumed that the shape of dendrite cross section at the solidification front of the slab is a regular hexagon as shown in Fig. 2.28, and it maintains a regular hexagon and continues to grow as the solidification process progresses until it contacts with adjacent dendrite to achieve complete solidification;
(2) Ignoring the diffusion of solute along the axis during the growth of dendrites, the solute elements are evenly distributed in the liquid phase and are in equilibrium distribution at the phase interface;

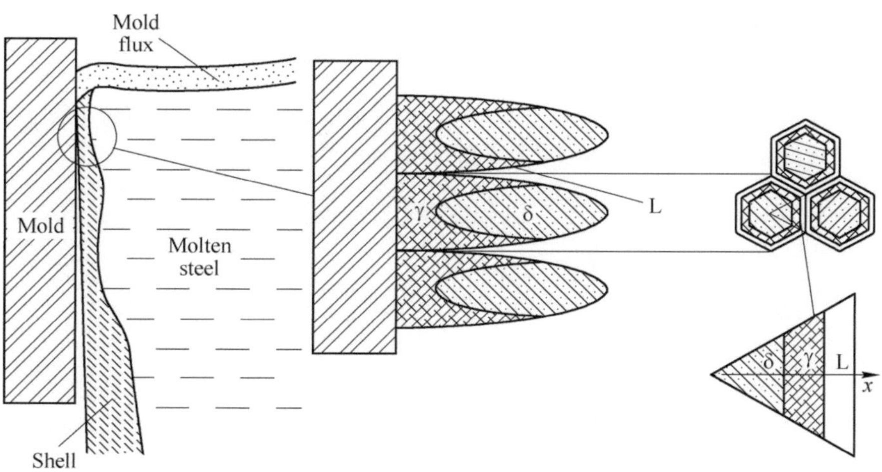

Fig. 2.28 Schematic diagram of the solidification process and model of continuous casting slab

(3) Ignoring the temperature fluctuations caused by the oscillation of the mold and change of the interface heat transfer conditions, the temperature distribution of the dendrite cross section is uniform and only decreases uniformly with time;
(4) During the solidification process of molten steel, the liquid/solid dual-phase zone is the coexistence of three phases or two phases in δ, γ and L, and the movement of the solidification front and the δ/γ phase interface is determined by the solute concentration at the corresponding interface and the current dendrite temperature.

B. Model calculation

Based on the model assumption, according to the symmetry of the regular hexagon, one of the equilateral triangles is taken as the research object, and the diffusion of solute elements is simplified into a one-dimensional diffusion problem along the x-axis. The model also considers the diffusion and distribution of 5 solute elements C, Si, Mn, P, and S during the solidification process accompanied by the δ/γ phase transition. The diffusion of solute elements in the solid phase is controlled by Fick's first law, and the diffusion at the solid/liquid interface is determined by the law of mass conservation and the equilibrium distribution coefficient of solute elements at the interface. The model is divided into 100 nodes with equal spacing, and the basic control equations are as follows [38]:

(1) Diffusion of solute in δ or γ phase:

$$\begin{cases} C_i = C_i + D\frac{\text{flag}_1 L_i (C_{i+1}^0 - C_i^0) - \text{flag}_2 L_{i-1}(C_i - C_{i-1}^0)}{A_i \Delta x}\Delta t \\ D = D_0 \exp(-Q/RT) \end{cases} \tag{2.51}$$

(2) Diffusion of solute at the δ/γ interface:

$$\begin{cases} C_{i+1} = C_{i+1} + \frac{\text{flag}_1 L_{i+1} D^{\gamma}(C_{i+2}^0 - C_{i+1}^0) - \text{flag}_2 L_{i-1} D^{\delta}(C_i^0 - C_{i-1}^0)}{(A_i k_{\delta/\gamma} + A_{i+1})\Delta x}\Delta t \\ C_i = k_{\delta/\gamma} C_{i+1} \end{cases} \tag{2.52}$$

(3) Diffusion of solute at the δ/L or γ/L interface:

$$\begin{cases} C_L = \frac{C_{\text{initial}} \sum_{j=1}^{100} A_j - \sum_{j=1}^{i-1} C_j A_j}{k A_i + \sum_{j=i+1}^{100} A_j}, \\ C_i = k C_L \end{cases} \tag{2.53}$$

where

C_i	the solute concentration node i;
A_i	the area of node i;
L_i	the side length of the triangle corresponding to node i;
Δx	node spacing;
Δt	time step;

k	interphase equilibrium partition coefficient;
$flag_1$, $flag_2$	take 0 only when $i = 1$ and $i = 100$, otherwise take 1;
D_0	diffusion constant;
Q	diffusion activation energy;
R	gas constant;
T	thermodynamic temperature.

The equilibrium partition coefficients and diffusion coefficients of each element in the formulas in each phase are shown in Table 2.3.

In the model, the dendrite spacing is one of the key parameters affecting the microsegregation behavior of the solute. In view of the fact that the size of dendrites in the solidification front of the slab is related to the cooling rate and C content in the actual continuous casting process, the dendrite spacing in the model of this book is selected from the expression of the secondary dendrite spacing under different cooling rates and C contents obtained by regression of the experimental results of Won et al. [39]:

$$\lambda_D = \begin{cases} (169.1 - 720.9w_C)C_R^{-0.4935} & 0 < w_C \leq 0.15 \\ 143.9C_R^{-0.3616}w_C^{(0.5501-1.996w_C)} & w_C > 0.15 \end{cases}, \tag{2.54}$$

where

λ_D	Secondary dendrite spacing, μm;
C_R	Cooling rate, °C/s;
w_C	Content of C element, %.

The cooling rate C_R affects the diffusion time and amount of solute elements in steel.

According to the traditional slab mold control cooling process, the cooling rate in the model is selected as the comprehensive cooling rate of the shell in the mold of 10 °C/s. The principle of phase interface movement is as follows: when the local liquidus temperature T_L obtained from the concentration of solute element in the liquid phase at the solidification front, or the local $\delta \rightarrow \gamma$ phase transition temperature T_{Ar4} obtained from the concentration of solute element at the δ/γ interface is equal to

Table 2.3 Equilibrium partition coefficients and diffusion coefficients of solute elements

Solute element	$k_{\delta/L}$	$k_{\gamma/L}$	$k_{\delta/\gamma}$	δ phase		γ phase	
				D_0(μm²/s)	Q(J/mol)	D_0(μm²/s)	Q(J/mol)
C	0.19	0.34	1.79	0.0127×10^4	81,379	0.0761×10^4	143,511
Si	0.77	0.52	0.68	8.0×10^4	248,948	0.30×10^4	251,458
Mn	0.76	0.78	1.03	0.76×10^4	224,430	0.055×10^4	249,366
P	0.23	0.13	0.57	2.90×10^4	230,120	0.01×10^4	182,841
S	0.05	0.035	0.70	4.56×10^4	214,639	2.40×10^4	223,425

Table 2.4 Values of coefficients m_x and n_x

Item	C	Si	Mn	P	S
m_x	78	7.6	4.9	34.4	38
n_x	– 1122	60	– 12	140	160

the current temperature of the dendrite, the corresponding phase interface moves. The values of T_L and T_{Ar4} are shown in formulas (2.51) and (2.52). The corresponding coefficients m_x and n_x are derived from the binary equilibrium phase diagram of Fe–X (X = C, Si, Mn, P, S), the values are shown in Table 2.4. The finite volume method is used to divide equilateral triangles into 100 nodes, and the C++ language is used to write the program. The relationship between the phase fraction and temperature in the solidification process of steel is obtained, and the relationship between the high temperature physical properties and temperature in the solidification process of typical microalloy is calculated [39].

$$T_L = T_p - \sum m_x C_{L,x} \tag{2.55}$$

$$T_{Ar4} = T_p^{\delta/\gamma} - \sum n_x C_x \tag{2.56}$$

2.3.1.2 Shell-Mold Interface Heat Transfer Model

As mentioned above, the quantitative determination and description of the dynamic heat transfer behavior of the shell-mold interface is the key to study the thermo-mechanical behavior of the shell solidification in the mold. Figure 2.29a provides a schematic of the solidified shell in the mold, where the gap between the copper plate and the solidified shell is filled with mold flux and air gap. In the actual continuous casting production process, the liquid flux flows into the shell-mold interface from the meniscus under the oscillation of the mold. Cooled by the copper plate of the mold, the mold flux solidifies into a glassy state near the copper plate side and then transforms into a crystalline state. Due to the shrinkage of the mold flux during solidification, a large variable interface thermal resistance is formed at the mold copper plate-mold flux interface. In the upper part of mold, because the surface temperature of the initial solidified shell is high, the mold flux near the shell side is still liquid, and the mold flux on the cross section of the flux channel is in a liquid–solid coexistence state. However, with the slab moves downward, the surface temperature of the solidified shell continues to decrease. When the temperature is reduced to or below the solidification temperature of the mold flux, the mold flux is completely solidified. The shell-mold flux interface begins to form the air gap under the continuous thermal shrinkage of the shell, and the structure of air gap and solid flux is formed. In this process, with the advancement of the solidification process of

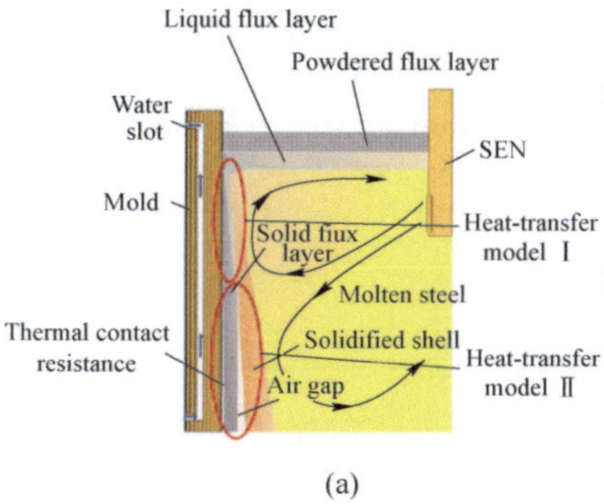

Fig. 2.29 Schematic of the heat transfer in shell/copper interface. **a** Schematic shell solidification in mold; **b** Heat transfer circuit diagram in the interface

the slab, the thickness of the air gap and the state of the mold flux in the interface change along the circumferential and height directions of the mold.

In order to effectively describe the interfacial heat transfer, the following assumptions are proposed:

(1) Ignoring the composition of the glass phase and crystalline phase in the solid flux layer, and the mold flux flowing into the flux channel is only distinguished between the solid flux layer and the liquid flux layer, and the liquid flux has good fluidity.
(2) The state of the mold flux is determined by the temperature of the shell surface and the hot face of the mold and its own melting characteristics, and the thickness distribution of the flux layer is jointly determined by the interface gap of the shell-mold and the state of the mold flux.
(3) Ignore the solidification deformation of the mold flux, and the solidification shrinkage of the mold flux is handled by introducing an interfacial thermal resistance related to the thickness of the mold flux.

From assumption (1), it can be seen that the liquid flux has good fluidity in the mold, so when the shell shrinkages, the liquid flux can fill the shrinkage gap in the shell-mold flux interface in real time under the oscillation of mold. Assumption (2) not only ensures that the thickness of the liquid flux layer increases with the shrinkage of the solidified shell, but also simulates that when the mold flux is solidified at the upper part of the mold, the thickness of the mold flux at the lower part of the interface is not filled by the liquid flux and maintains a constant value, thus forming the air gap. By introducing the thermal resistance of the interface between mold flux and copper plate related to the thickness of the mold flux, assumption (3) can better describe the effect of the increase of the interfacial thermal resistance with the increase of the thickness of the mold flux film.

Therefore, based on the above assumptions, according to the relationship between the surface temperature of the shell and the solidification temperature of the mold flux, the composition of the total thermal resistance between the shell and the mold in the direction of the mold height can be developed as heat transfer mode I (liquid flux layer, solid flux layer, mold-solid flux interfacial thermal resistance composition) and heat transfer mode II (air gap, solid flux layer thermal resistance, mold-solid flux interfacial thermal resistance composition), as shown in Fig. 2.29b (the symbols in the fig. are explained in the formulas below in this book). Therefore, the calculation formula of interfacial heat flux can be expressed as:

$$q = \frac{T_{\mathrm{s}} - T_{\mathrm{m}}}{R_{\mathrm{liq}} + R_{\mathrm{sol}} + R_{\mathrm{air}} + R_{\mathrm{int}}} \tag{2.57}$$

The meanings of the symbols in the formula are as follows:

q Heat flux at the shell-mold interface, W/m^2;
T_{s} Surface temperature of the shell, K;
T_{m} Hot face temperature of the mold copper plate, K;
R_{liq} Thermal resistance of the liquid flux layer, m^2 K/W;
R_{sol} Thermal resistance of the solid flux layer, m^2 K/W;
R_{air} Thermal resistance of air gap, m^2 K/W;
R_{int} Thermal resistance of the interface between mold and solid flux, m^2 K/W.

Among them, because the heat transfer among the liquid flux layer, solid flux layer, and air gap not only involves conduction but also a large proportion of radiation [40, 41], the heat transfer within the shell-mold interface requires to consider both conduction (represented by the superscript c) and radiation (represented by the superscript rad). The corresponding thermal resistance calculation formulas are shown in formulas (2.58)–(2.61) [42]:

(1) Thermal resistance of the liquid flux layer:

$$\begin{cases} R_{\mathrm{liq}}^{\mathrm{c}} = \frac{d_{\mathrm{liq}}}{k_{\mathrm{liq}}} \\ R_{\mathrm{liq}}^{\mathrm{rad}} = \frac{0.75E_{\mathrm{liq}}d_{\mathrm{liq}} + \left(\frac{1}{\varepsilon_{\mathrm{s}}} + \frac{1}{\varepsilon_{\mathrm{f}}}\right) - 1}{\sigma n_{\mathrm{liq}}^2 (T_{\mathrm{sol}}^2 + T_{\mathrm{s}}^2)(T_{\mathrm{sol}} + T_{\mathrm{s}})} \\ \frac{1}{R_{\mathrm{liq}}} = \frac{1}{R_{\mathrm{liq}}^{\mathrm{c}}} + \frac{1}{R_{\mathrm{liq}}^{\mathrm{rad}}} \end{cases} \tag{2.58}$$

(2) Thermal resistance of the solid flux layer:

$$\begin{cases} R_{\mathrm{sol}}^{\mathrm{c}} = \frac{d_{\mathrm{sol}}}{k_{\mathrm{sol}}} \\ R_{\mathrm{sol}}^{\mathrm{rad}} = \frac{0.75E_{\mathrm{sol}}d_{\mathrm{sol}} + \left(\frac{1}{\varepsilon_{\mathrm{f}}} + \frac{1}{\varepsilon_{\mathrm{m}}}\right) - 1}{\sigma n_{\mathrm{sol}}^2 (T_{\mathrm{sol}}^2 + T_{\mathrm{a}}^2)(T_{\mathrm{sol}} + T_{\mathrm{a}})} (T_{\mathrm{s}} > T_{\mathrm{sol}}) \\ \frac{1}{R_{\mathrm{sol}}} = \frac{1}{R_{\mathrm{sol}}^{\mathrm{c}}} + \frac{1}{R_{\mathrm{sol}}^{\mathrm{rad}}} \end{cases} \tag{2.59}$$

or

$$\begin{cases} R_{\text{sol}}^{\text{c}} &= \frac{d_{\text{sol}}}{k_{\text{sol}}} \\ R_{\text{sol}}^{\text{rad}} &= \frac{0.75E_{\text{sol}}\, d_{\text{sol}} + \left(\frac{1}{\varepsilon_{\text{f}}} + \frac{1}{\varepsilon_{\text{m}}}\right) - 1}{\sigma n_{\text{sol}}^{2}\,(T_{\text{x}}^{2} + T_{\text{a}}^{2})(T_{\text{x}} + T_{\text{a}})} (T_{\text{s}} \le T_{\text{sol}}) \\ \frac{1}{R_{\text{sol}}} &= \frac{1}{R_{\text{sol}}^{\text{c}}} + \frac{1}{R_{\text{sol}}^{\text{rad}}} \end{cases} \tag{2.60}$$

(3) Heat transfer in the air gap:

$$\begin{cases} R_{\text{air}}^{\text{c}} &= \frac{d_{\text{air}}}{k_{\text{air}}} \\ R_{\text{air}}^{\text{rad}} &= 0.5\sigma(\varepsilon_{\text{s}} + \varepsilon_{\text{f}})(T_{\text{x}}^{2} + T_{\text{s}}^{2})(T_{\text{x}} + T_{\text{s}}) \\ R_{\text{air}} &= \frac{1}{R_{\text{air}}^{\text{c}}} + \frac{1}{R_{\text{air}}^{\text{rad}}} \end{cases} \tag{2.61}$$

In the formula:

k_{liq} Thermal conductivity of liquid flux, W/(m K);
k_{sol} Thermal conductivity of solid flux, W/(m K);
k_{air} Thermal conductivity of air gap, W/(m K);
T_{sol} Flux solidification temperature, K;
T_{a} Temperature at the mold-flux interface, K;
T_{x} Temperature at the flux-air gap interface, K;
ε_{s} Emissivity of the shell;
ε_{f} Emissivity of the mold flux;
ε_{m} Emissivity of the mold copper plate;
n_{liq} Refractive index of liquid flux;
n_{sol} Refractive index of solid flux;
E_{liq} Extinction coefficient of liquid flux, m^{-1};
E_{so} Extinction coefficient of solid flux, m^{-1};
d_{liq} Layer of liquid flux, m;
d_{sol} Layer of solid flux, m;
d_{air} Thickness of the air gap, m;
σ Boltzmann constant, value is 5.67×10^{-8}, ($\text{W/m}^2\ \text{K}^4$).

Some relevant thermal properties of mold flux, copper, and steel are shown in Table 2.5 [41–43]. Among them, the sum of d_{liq}, d_{sol}, and d_{air} is the gap d_{gap} between the shell and the mold interface,

$$d_{\text{sol}} + d_{\text{li}q} + d_{\text{air}} = d_{\text{gap}} \tag{2.62}$$

Based on the principle that the heat flux through each heat transfer medium layer is equal, the heat transfer mode I and heat transfer mode II calculation formulas can be obtained by combining the equations, as shown in formulas (2.63) and (2.64):

$$\begin{cases} \frac{T_{\text{s}} - T_{\text{sol}}}{R_{\text{liq}}} = \frac{T_{\text{sol}} - T_{\text{a}}}{R_{\text{sol}}} \\ \frac{T_{s} - T_{\text{sol}}}{R_{\text{liq}}} = \frac{T_{\text{a}} - T_{\text{m}}}{R_{\text{int}}} \\ d_{\text{sol}} + d_{\text{d}} + d_{\text{air}} = d_{\text{gap}} \end{cases} \tag{2.63}$$

Table 2.5 Thermal physical parameters of mold flux, copper plate, and shell

Object	Item	Range of values	Default value
Shell	Thermal conductivity k_{liq} (W/(m K))	0.25–0.26	0.26
Liquid flux layer	Absorption coefficient α (m^{-1})	51–350	51
	Refractive index n_{liq}	1.55–1.58	1.58
Solid flux layer	Thermal conductivity k_{sol} (W/(m K))	1.62–1.63	1.625
	Extinction coefficient E_{sol} (m^{-1})	1356–2625	1356
	Emissivity ε_{sol}	0.9	0.9
	Refractive Index n_{sol}	1.59	1.59
Air gap	Thermal Conductivity k_{air} (W/(m K))	0.0472, 0.06, 0.1	0.1
Mold	Emissivity ε_{mold}	0.4	0.4

$$\begin{cases} \frac{T_s - T_X}{R_{air}} = \frac{T_X - T_a}{R_{sol}} \\ \frac{T_s - T_X}{R_{air}} = \frac{T_a - T_m}{R_{int}} \\ d_{sol} + d_{air} = d_{gap} \end{cases} \tag{2.64}$$

As for the thermal resistance R_{int} at the mold-flux interface, there are many influencing factors, including the types and characteristics of the mold flux [44], the cooling rate at the interface between mold flux and copper plate [44], the thickness of the mold flux [31], etc. Different scholars obtain slightly different results due to different research conditions and methods [27–31]. Cho et al. [31, 45] adopt industrial mold flux as the material to study the fitting calculation formula by combining experimental detection and simulation calculation. During the actual slab continuous casting process, considering the inconsistent distribution of the thickness of the mold flux in the mold at different heights and circumferences, the value of the thermal resistance R_{int} at the interface between mold flux and copper plate in the model is calculated by Cho et al. [45].

In addition, the shell surface temperature T_s, the temperature of copper plate hot face T_m, and the interfacial gap d_{gap} between the shell and mold are presented in the coupling calculation process of the thermo-mechanical coupled finite element of the slab-mold system in Sect. 2.3.1.3. Based on the parameters, the heat flux of each node on the shell surface and the mold hot face is dynamically solved by the Monte Carlo method and applied to the finite element model, thus realizing the calculation of the finite element model.

2.3.1.3 Thermo-Mechanical Finite Element Model of Slab-Mold System

In the actual slab continuous casting production, the initial solidified shell first forms near the meniscus and continuously solidifies in the mold, as well as moves downward at a certain casting speed. In this process, the shell not only continuously transfers heat outwards, but also undertakes complex mechanical loads (as shown in Fig. 2.30). So,

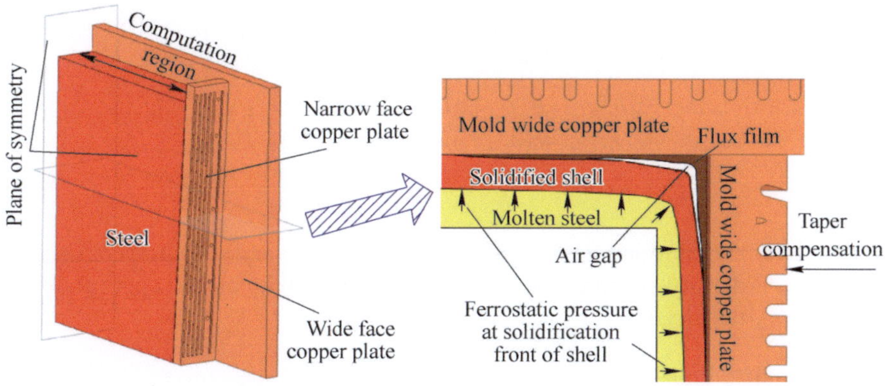

Fig. 2.30 Schematic diagram of mechanical boundary conditions in the mold

the thermal behavior and mechanical behavior interact and affect each other, which is a completely coupled complex process. In order to develop the thermo-mechanical finite element model of the slab under this complex solidification condition, some assumptions have been proposed:

(1) Ignore the heat transfer of the slab and the mold copper plate along the casting direction, and simplify the model to a 2D heat transfer model;
(2) Ignore the influence of molten steel flow on the convective heat transfer of the shell-molten steel interface, and use the effective thermal conductivity of the molten steel;
(3) The thermal conductivity of the mold copper plate is isotropic. The density and specific heat are considered constants. The physical parameters of the steel are only related to the composition and the temperature where it is located;
(4) The mold flux film at the meniscus of the mold is uniformly distributed;
(5) The mechanical behavior of the solidified shell is plane stress and strain, ignoring the deformation of the mold. The contact between the mold copper plate and the shell is rigid-flexible contact;
(6) Ignore the friction between the solidified shell and the mold, and assume the sliding contact between the solidified shell and the mold as smooth contact;
(7) The heat transfer and mechanical behavior of the shell and the mold are symmetrical.

A. Heat Transfer Equation

According to the above symmetry assumption, the 1/4 cross section of the slab and mold are taken as the calculation domain. Since the solidification of the slab and the heat transfer of the mold is a nonsteady state heat transfer process with an internal heat source. Therefore, the heat transfer differential equations of the above process are selected, respectively, the formulas (2.65) and (2.66):

$$\frac{\partial H(T)}{\partial t} = \frac{\partial}{\partial x}\left(\lambda_s(T)\frac{\partial T}{\partial x}\right) + \frac{\partial}{\partial y}\left(\lambda_s(T)\frac{\partial T}{\partial y}\right) \tag{2.65}$$

$$\rho c\frac{\partial T}{\partial t} = \frac{\partial}{\partial x}\left(\lambda_m\frac{\partial T}{\partial x}\right) + \frac{\partial}{\partial y}\left(\lambda_m\frac{\partial T}{\partial y}\right), \tag{2.66}$$

where

$H(T)$, $\lambda_s(T)$ enthalpy and thermal conductivity of steel vary with temperature;
ρ, c and λ_m density, specific heat, and thermal conductivity of the copper plate and the nickel layer.

B. Mechanical Equation

In the actual slab continuous casting process, in addition to heat, elasticity and ductile deformation, the solidification process of steel in the mold is also accompanied by creep. The traditional generalized elastoplastic constitutive equation makes it difficult to accurately describe the above deformation behavior of steel during the solidification process in the mold. In order to fully consider the influence of creep, this model uses the Anand rate-related constitutive equation proposed by Anand and Brown [47, 48]:

$$\dot{\overline{\varepsilon}}_{ie} = A\exp\left(-\frac{Q_A}{T}\right)\left[\sinh\left(\xi\frac{\overline{\sigma}}{s}\right)\right]^{\frac{1}{m}} \tag{2.67}$$

where the evolution equation for s is:

$$\dot{s} = \left(h_0\left|1-\frac{s}{\tilde{s}\left[\frac{\dot{\varepsilon}_{ie}}{A}\exp\left(\frac{Q_A}{T}\right)\right]^n}\right|^{\alpha}\operatorname{sign}\left(1-\frac{s}{\tilde{s}\left[\frac{\dot{\varepsilon}_{ie}}{A}\exp\left(\frac{Q_A}{T}\right)\right]^n}\right)\right)\dot{\overline{\varepsilon}}_{ie} \tag{2.68}$$

The meanings and corresponding values of each parameter in the formula are detailed in Table 2.6.

C. Geometric Model

Based on the symmetry assumption of the above model, the physical structure of a mold from a steel plant is used as a prototype, and the molten steel fills the inner cavity of the mold. A finite element model is developed by a 1/4 cross section of the slab and mold system. The physical model is shown in Fig. 2.31.

The height of the mold is 900 mm. In order to ensure the uniform heat transfer of the copper plate, two deep water slots are arranged on both sides of the fixing bolt. The cooling water inlet is set at the lower part of the copper plate, and the outlet is set at the upper part of the copper plate. At the same time, there is a nickel layer whose thickness gradually changes from top to bottom on the hot face of the copper plate to reduce the wear of the copper plate.

Table 2.6 Explanation and values of parameters in Anand's constitutive equation [49]

Parameter	Meaning	Unit	Value
$\dot{\varepsilon}_{ie}$	Equivalent ductile strain rate		
s_0	Initial value of deformation resistance	MPa	43
Q_A	Activation energy for viscoplastic deformation/Boltzmann constant	K	32,514
A	Pre-exponential factor	s^{-1}	1.0×10^{11}
ξ	Stress multiplier		1.15
m	Strain sensitivity index		0.147
h_0	Hardening/softening constant	MPa	1329
$\tilde{s}$	Saturation value of s at a given temperature and strain rate	MPa	147.6
n	Strain rate sensitivity of strain impedance saturation value		0.06869
α	Strain rate sensitivity index related to hardening/softening		1
$\overline{\sigma}$	Stress	MPa	
s	Deformation impedance	MPa	

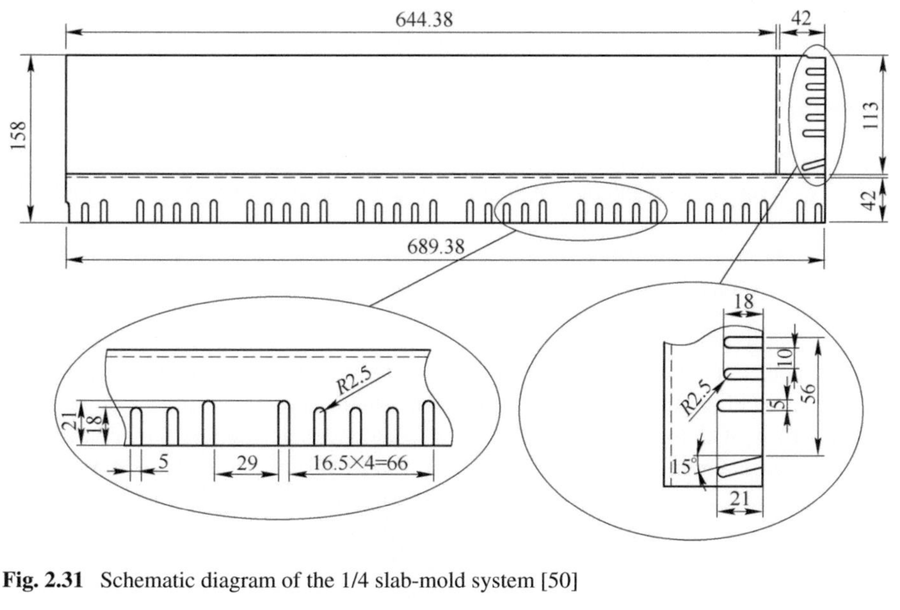

Fig. 2.31 Schematic diagram of the 1/4 slab-mold system [50]

D. High Temperature Physical Properties of Steel

During the solidification and phase transition process of steel, due to the different proportions of different phase structures, the high temperature physical properties are also different. It is a function with temperature and the proportion of each phase structure. The high temperature physical properties of the micro-alloyed steel used in the model developed process are based on the calculation results of the solute microsegregation model of the coupled phase transition behavior in Sect. 2.3.1.1, and are calculated and determined with different temperatures and their corresponding phase fractions.

(a) Thermal Conductivity

Traditional research on the thermo-mechanical behavior of the slab during solidification in the mold generally assumes the thermal conductivity of solid steel to be a constant value of 33.47 W/(m °C) [51]. In order to accurately consider the effects of high temperature phase transition and temperature evolution of specific composition micro-alloyed steel, the high temperature thermal conductivity calculation formula proposed by Harste [52–54] and Jimbo [55] was used to determine the thermal conductivity at different temperatures, as shown in formula (2.69):

$$\begin{cases} K = K_\Gamma F_\Gamma + K_\Delta F_\Delta + K_L F_L \\ K_\Gamma = 21.6 + 9.35 \times 10^{-3}\mathrm{T} \\ K_\Delta = \left(20.14 - 9.313 \times 10^{-3}T\right)\left(1 - A_1 W_\mathrm{c}^{A_2}\right) \\ K_L = 39.0 \\ A_1 = 0.425 - 4.385 \times 10^{-5}T \\ A_2 = 0.209 + 1.09 \times 10^{-3}T \end{cases}, \tag{2.69}$$

where

k	Thermal conductivity, W/(m);
$f_\gamma, f_\delta, f_\mathrm{L}$	The proportions of γ-austenite phase, δ-ferrite phase and liquid phase in the steel solidification process;
t	Current temperature, °C;
w_C	C content, %.

It should be pointed out that under the action of the kinetic energy of the molten steel flowing out of the submerged nozzle or the electromagnetic stirring energy, the molten steel undergoes forced convection, which accelerates the elimination of the superheat of the molten steel. In order to compensate for this part of the heat transfer effect, Thomas et al. [56] and Han et al. [32], respectively, expanded the thermal conductivity of the flowing molten steel in the molten pool to 6.5 times and 6 times the thermal conductivity of the static molten steel. This book also draws on this method, combined with the flow characteristics in the mold, and has expanded the thermal conductivity of the flowing molten steel by 6 times.

(b) Latent Heat of Solidification

It is one of the difficulties in the numerical calculation of steel solidification heat transfer to dealing with the steel solidification front liquid–solid interface which is continuously moving and releasing phase transition latent heat. At present, the commonly used methods for numerical simulation of solidification phase transition latent heat mainly include the front tracking method [57–60] and the fixed grid method [61–65], with the latter being more commonly used. In engineering calculations, the release of phase transition latent heat is usually implemented by fixed grid methods such as the enthalpy method, equivalent specific heat method and post iteration method. In order to accurately consider the latent heat release of the solidification phase transition process of specific composition micro-alloyed steel, this book also uses the relationship between the fraction of each phase in the steel structure and the temperature as the link, and uses the relationship shown in formula (2.70) [52–54] to determine its high temperature latent heat.

$$\begin{cases} H = H_\gamma f_\gamma + H_\delta f_\delta + H_L f_L \\ H_\gamma = 0.43T + 7.5 \times 10^{-5} T^2 + 93 + \alpha_\gamma \\ H_\delta = 0.441T + 8.87 \times 10^{-5} T^2 + 51 + \alpha_\delta \\ H_L = 0.825T - 105 \\ \alpha_\gamma = \frac{37w_C + 1.9 \times 10^3 w_C^2}{44w_C + 1200} \\ \alpha_\delta = \frac{18w_C + 2.0 \times 10^3 w_C^2}{44w_C + 1200} \end{cases}, \tag{2.70}$$

where

H	Heat enthalpy of steel, kJ/kg;
f_γ, f_δ, f_L	The proportion of γ-austenite phase, δ-ferrite phase and liquid phase in the steel solidification process;
T	Current temperature, K;
w_C	Content of C, %.

(c) Density

The volume of the steel solidification process changes with the evolution in phase composition, temperature and other conditions. This book also uses the function calculation formula shown in formula (2.71) with phase fraction and temperature as variables when considering the density evolution [52–54, 66].

$$\begin{cases} \rho = \rho_\gamma f_\gamma + \rho_\delta f_\delta + \rho_L f_L \\ \rho_\gamma = \frac{100(8106 - 0.51t)}{(100 - w_C)(1 + 0.008w_c^3)} \\ \rho_\delta = \frac{100(8111 - 0.47t)}{(100 - w_C)(1 + 0.013w_c^3)} \\ \rho_L = 7100 - 73w_C - (0.8 - 0.09w_C)(t - 1550) \end{cases}, \tag{2.71}$$

where

ρ	Density of steel, kg/m^3;
f_γ, f_δ, f_L	The proportions of the γ-austenite phase, δ-ferrite phase and liquid phase in the steel solidification process;
t	Current temperature, °C;
w_C	C content, %.

(d) Thermal Linear Expansion Coefficient

When the base temperature of the material changes from the reference temperature T_{ref} to the current temperature T, the relative evolution in length can be expressed as:

$$\mathrm{TLE} = \frac{\Delta L(T)}{L(T_{\mathrm{ref}})} = \frac{L(T) - L(T_{\mathrm{ref}})}{L(T_{\mathrm{ref}})} = \frac{L(T)}{L(T_{\mathrm{ref}})} - 1 \tag{2.72}$$

The following relationship exists between density and the cube of length:

$$\frac{\rho(T)}{\rho(T_{\mathrm{ref}})} = \left[\frac{L(T_{\mathrm{ref}})}{L(T)}\right]^3 \tag{2.73}$$

Therefore, the definition of the thermal linear expansion coefficient of material can be obtained as follows:

$$\mathrm{TLE} = \sqrt[3]{\frac{\rho(T_{\mathrm{ref}})}{\rho(T)}} - 1 \tag{2.74}$$

However, in the actual finite element calculation process, it is usually necessary to deal with the thermal linear expansion coefficient as an instantaneous thermal linear expansion coefficient, so it is necessary to further derive the formula (2.70) for temperature, that is:

$$\alpha = \frac{\mathrm{d}_{\mathrm{TLE}}(T)}{\mathrm{d}T} \tag{2.75}$$

It should be pointed out that choosing different reference temperatures T_{ref} will result in different instantaneous thermal linear expansion coefficients, and it will affect the solidification shrinkage behavior of steel in the mold. This book considers that the steel with the solidification front temperature higher than the viscosity temperature LIT (Liquid Impenetrable Temperature) has a certain fluidity and is in a free state. When it is lower than this temperature, the compact dendrites at the solidification front of the slab hinder the filling of the liquid phase, and the further cooling and solidification of the slab will cause significant volume shrinkage, so T_{ref} is chosen as the LIT temperature.

Regarding the definition of the LIT temperature, Clyne et al. [67], Davies and Shin [68] and Kim et al. [69] divided the dual-phase zone into a filling zone and a crack zone, and defined the temperature limit as the temperature corresponding to the solid phase fraction $f_s = 0.9$. Matsumiya et al. [70] define this temperature as the temperature corresponding to the solid phase fraction $f_s = 0.85$. This book comprehensively considers the influence of solute microsegregation on the solidification under the condition of the high cooling rate in the mold and the mesh of the microsegregation model, and selects the temperature corresponding to the solid phase fraction $f_s = 0.884$ as the LIT temperature.

(e) Elastic Modulus and Poisson Ratio

The elastic modulus and Poisson ratio are two important parameters describing the mechanical behavior of material, both of them change with temperature. Among them, the elastic modulus decreases with the increase of steel temperature and is very sensitive to strain rate. There are significant differences between the measured values of different researchers due to different measurement methods and conditions. Mizukami et al. [71] used the "relaxation method" to consider the high temperature creep effect of steel in the measurement of elastic modulus, and the measured data matched well with the plastoelasticity, creep and ductile constitutive equations. In order to better simulate the high temperature mechanical behavior of steel under continuous casting conditions accompanied by creep and other phenomena, this book selected the regression formula of elastic modulus experimental data given by Mizukami et al. and set the Poisson ratio as the regression formula of experimental data given by Uehara et al. [72], as shown in formula (2.76) and (2.77).

$$E = 968 - 2.33t + 1.9 \times 10^{-3}t^2 - 5.18 \times 10^{-7}t^3 \tag{2.76}$$

$$v = 0.278 + 8.23 \times 10^{-5}t \tag{2.77}$$

In the formula:

E Elastic modulus, GPa;
v Poisson ratio;
t Current temperature of the steel, °C.

According to the above formulas, the thermal conductivity, enthalpy, density, and instantaneous thermal linear expansion coefficient of a certain micro-alloyed steel with the composition shown in Table 2.2 during the solidification process with temperature evolution are shown in Fig. 2.32.

E. Thermal properties of copper plate and cooling water

According to the finite element model assumption of this book, the deformation behavior of the mold copper plate is ignored, so only the heat transfer behavior of the copper plate is considered in the calculation model. The thermal properties of the mold and cooling water are shown in Table 2.7.

F. Model Initial and Boundary Conditions

(a) Heat Transfer Initial Conditions

The heat transfer initial conditions of the calculation model mainly include the initial temperature conditions of the slab, the initial conditions of the mold flux thickness, and the initial temperature conditions of the mold copper plate. For the slab, its initial temperature is set to the casting temperature, that is:

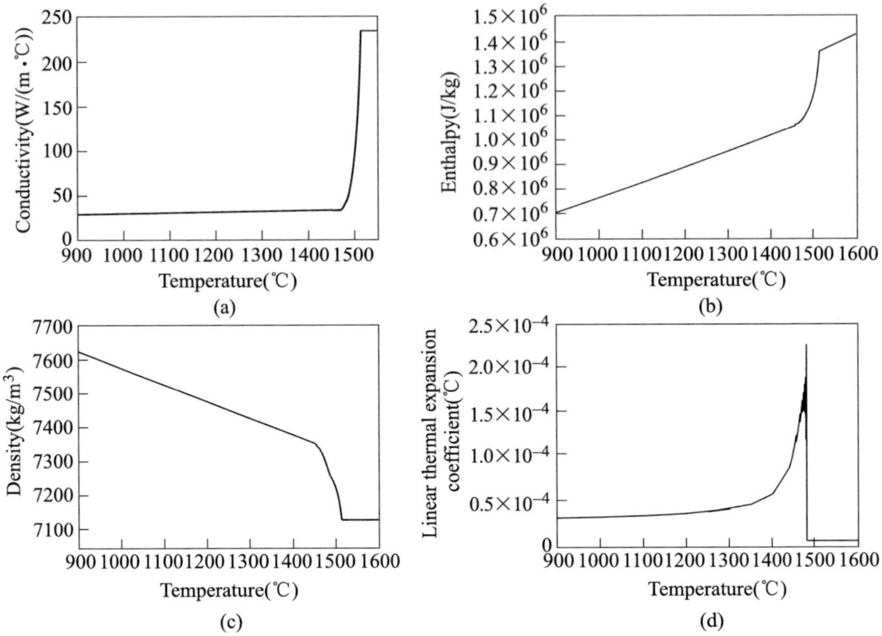

Fig. 2.32 Variation of physical properties of micro-alloyed steel with high temperature [50]. **a** Thermal conductivity; **b** Enthalpy; **c** Density; **d** Linear thermal expansion coefficient [50]

Table 2.7 Thermal properties of copper plate and cooling water

Material	Thermal conductivity (W/(m K))	Specific heat capacity (J/(kg K))	Density (kg/m^3)
Copper	335 (298 K)	410	8940
	315 (393 K)		
Nickel	82.9	460.6	8910
Water	0.597	4187	998

$$T = T_{\text{ic}}(\tau = 0), \tag{2.78}$$

where

T_{ic} Casting temperature, K;
τ Time.

For the initial conditions of the mold flux film, based on the assumption that the mold flux film is evenly distributed at the meniscus, the initial thickness of the mold flux film at the meniscus of the mold is calculated based on the average flux consumption and the slab cross section size.

The initial temperature of the mold copper plate is controlled by the heat transfer conditions at the shell-mold interface, and in turn, the evolution in the temperature of the copper plate hot face has a counter effect on the interface heat transfer, which is an iterative process that tends toward stability. Therefore, this book uses the preprocessing program shown in Fig. 2.33 when initializing the temperature of the mold copper plate. The convergence limit for the calculation of the initialization of the mold copper plate temperature is that the temperature difference between two iterations of the copper plate hot face node is less than 1 K.

In the initial mechanical and boundary conditions, since the steel is in a free state at the meniscus, the initial stress of the shell is set to 0.

(b) Heat transfer boundary conditions

The heat transfer boundary conditions of the slab and the mold mainly include convection heat transfer, conduction heat transfer, radiation heat transfer, and adiabatic boundary conditions (symmetrical surface), as shown in Fig. 2.34.

The heat transfer boundary of the symmetrical surface of the slab and the mold copper plate is:

$$\text{Wide face}: \mathrm{q} = -\lambda \frac{\partial \mathrm{T}}{\partial \mathrm{x}} = 0 \tag{2.79}$$

$$\text{Narrow face}: \mathrm{q} = -\lambda \frac{\partial \mathrm{T}}{\partial \mathrm{y}} = 0 \tag{2.80}$$

The heat transfer between the surface of the slab and the hot face of the mold copper plate is dynamically determined node by node by the shell-mold interface heat flow model developed in Sect. 2.3.1.2.

The convective heat transfer boundary between the mold copper plate and the cooling water is determined by the following formula:

$$q = h_{\mathrm{w}}(T - T_{\mathrm{w}}), \tag{2.81}$$

where T and T_{w} are the surface temperature of the mold water slot and the cooling water temperature, respectively, in K; h_{w} is the convective heat transfer coefficient between the water slot and the cooling water, in W/(m^2 K), determined by the following formula [73]:

$$\frac{h_{\mathrm{w}} d_{\mathrm{w}}}{\lambda_{\mathrm{w}}} = 0.023(\frac{\rho_{\mathrm{w}} u_{\mathrm{w}} d_{\mathrm{w}}}{\mu_{\mathrm{w}}})^{0.8}(\frac{c_{\mathrm{w}} \mu_{\mathrm{w}}}{\lambda_{\mathrm{w}}})^{0.4} \tag{2.82}$$

The physical meanings of the parameters in the formula are as follows:

λ_{w} Thermal conductivity of cooling water, W/(m K);
d_{w} Equivalent diameter of the water slot, m;
ρ_{w} Density of cooling water, kg/m^3;

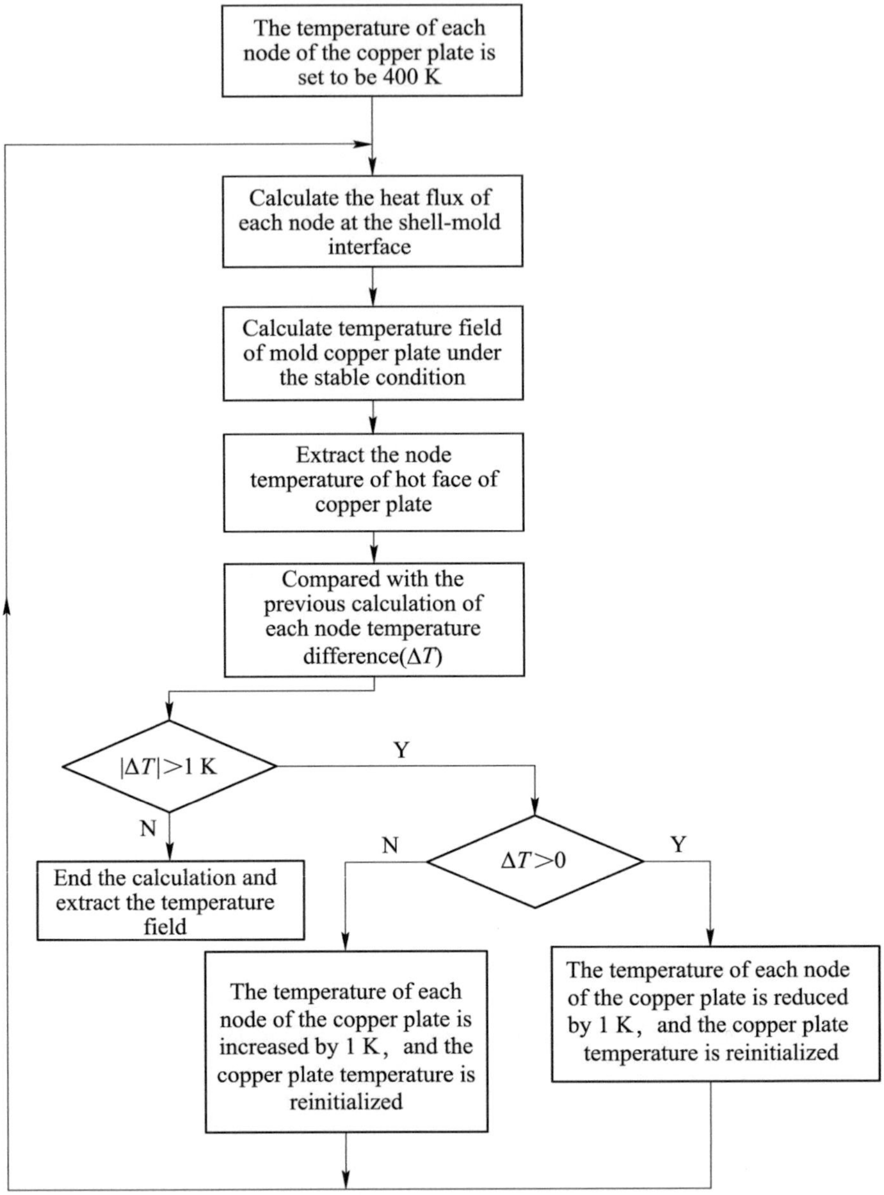

Fig. 2.33 Flowchart of the temperature initialization program for the mold copper plate

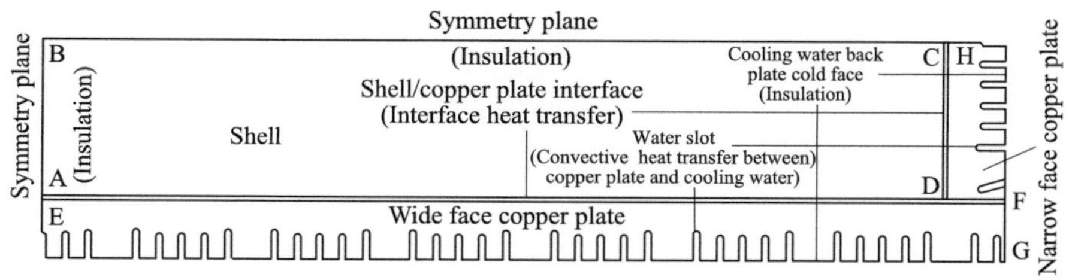

Fig. 2.34 Schematic diagram of heat transfer boundary conditions between shell and copper plate

u_w Flow speed of cooling water, m/s;
μ_w Viscosity of cooling water, Pa s;
c_w Specific heat of cooling water, J/(kg K).

In the actual continuous casting process, the temperature difference between the inlet and outlet of the mold cooling water generally does not exceed 10 °C, and it gradually increases from the exit to the top of the mold. To simplify the calculation model, this book treats the temperature of the cooling water in the mold cooling water slot as linearly changing along its height direction, that is:

$$T_w = T_{in} + \frac{T_{out} - T_{in}}{h_{op}} h_{curr} \tag{2.83}$$

In the formula:

T_{in}, T_{out} the water temperature at the inlet and outlet of the mold, K;
h_{op}, h_{curr} the actual working height of the mold and the current height of the solidified shell in the mold, m.

For the heat transfer boundary of the cold face of the mold copper plate, on the one hand, due to the contact gap between it and the stainless steel back plate, the thermal resistance of the copper plate to the stainless steel back plate is large. On the other hand, the temperature of the cold face of the copper plate is relatively low, and the intensity of radiant heat transfer is small. Therefore, this boundary is treated as adiabatic, and the heat flow expression is as follows:

$$\text{Wide face}: q = -\lambda \frac{\partial T}{\partial y} = 0 \tag{2.84}$$

$$\text{Narrow face}: q = -\lambda \frac{\partial T}{\partial x} = 0 \tag{2.85}$$

(c) Mechanical boundary conditions

In the actual slab continuous casting process, the shell and the mold are subjected to complex mechanical behavior. According to the assumptions in this book, the mechanical boundary conditions of the solidified shell and the mold mainly include displacement boundary conditions and force boundary conditions. Among them, according to the assumption of the symmetry of the slab solidification in the mold, the displacement constraint of the symmetry surface of slab is set to 0, that is:

$$\text{Wide face}: \delta_{\text{y}} = 0 \tag{2.86}$$

$$\text{Narrow face}: \delta_{\text{x}} = 0 \tag{2.87}$$

As for the displacement constraint of the mold, since the deformation of the mold copper plate caused by the ferrostatic pressure and the thermal strain is ignored, the displacement constraint of the wide side copper plate of the mold is set to 0. For the narrow face copper plate of the mold, in order to simulate its taper effect, the total offset of its taper is evenly divided by the height of the mold, and the narrow face copper plate gradually moves toward the center of the wide face as the solidified shell descends. The displacement constraint in the direction perpendicular to it is set to 0. The mathematical expressions are shown in formulas (2.88) and (2.89):

$$\text{Wide face}: \delta_{\text{x}} = \delta_{\text{y}} = 0 \tag{2.88}$$

$$\text{Narrow face}: \delta_{\text{x}} = \frac{\text{l}_{\text{tap}}}{\text{N}}, \quad \delta_{\text{y}} = 0, \tag{2.89}$$

where

l_{tap} is the total offset of the narrow face taper, m;

N is the number of slices.

As for the setting of the ferrostatic pressure boundary condition at the solidified shell front, due to the effect of the ferrostatic pressure on the solidified shell changes with the advancement of the solidification. It's one of the difficulties to accurately define the solidification front of the slab and apply the ferrostatic pressure. Dynamically eliminating the micro-solidified liquid core at the solidification front of the slab and applying the load of ferrostatic pressure to the already solidified slab unit is the most direct and effective method with the clearest physical meaning. Therefore, in the model establishment process of this book, the dynamic elimination of the liquid phase method is used to dynamically apply the ferrostatic pressure to the solidification front of the shell, the magnitude of which is:

$$P = \rho_{\text{molten}} g v_{\text{c}} \tau, \tag{2.90}$$

where

ρ_{molten} Density of molten steel, kg/m^3;
v_c Casting speed, m/s;
τ Time, s.

It should be noted that the definition of the position of the solidification front of the shell in the mold will directly affect the application effect of the ferrostatic pressure, thereby affecting the accuracy of the description of the mechanical behavior of the initial solidified shell. According to the definition of the viscosity temperature LIT of the two-phase zone of steel solidification in this book, LIT is the boundary distinguishing whether the molten steel in the two-phase zone has the ability to flow freely. When the temperature of the slab unit is lower than LIT, the compact dendrite hinders the filling of the molten steel to the grain boundary, and the ferrostatic pressure cannot directly act on the grain boundary. When the temperature of the slab unit is higher than LIT, the slab unit is liquid and has fluidity. Only when the temperature of the slab unit is at LIT, the ferrostatic pressure is just at the junction of the liquid phase and the solidified slab, that is, the ferrostatic pressure directly acts on the solidification front of the slab. Therefore, this book draws on the LIT definition method, defines the solidification front of the continuous casting slab as the temperature corresponding to LIT, that is, the solid phase fraction $f_s = 0.884$ corresponding temperature, and stipulates that when the model performs mechanical analysis, as the continuous casting slab solidification process progresses, units above this temperature are liquid slab units, the model automatically performs dynamic elimination processing on them, and applies the load of the ferrostatic pressure to the solidification front unit.

As for the mechanical constraints between the surface of the slab and the hot face of the mold copper plate, due to the relatively low strength of the high-temperature solidified shell in the mold, under the action of ferrostatic pressure, the shell will tend to bulge toward the side of the mold copper plate. If the contact surface between the shell surface and the hot face of the mold copper plate is not constrained with the corresponding treatment, the phenomenon of "shell surface nodes penetrating the mold copper plate" will occur, and contact mechanics analysis treatment needs to be done on the slab surface and the hot face of the mold copper plate. In the model establishment assumption, the mold copper plate is assumed to be non-deformable, and the shell has hot ductility and creep behaviors in the mold, it is "softer" than the mold copper plate, so the hot face of the mold copper plate and the shell surface are subjected to rigid-flexible contact analysis.

G. Calculation Process

Based on the above-mentioned heat transfer and mechanical control equations of the slab and copper plate and their boundary conditions, specific component micro-alloyed steel high-temperature thermal properties, etc., a thermo-mechanical coupling finite element calculation model of 1/4 slab and mold system is established, as shown in Fig. 2.35. Among them, the heat flow required for model calculation on the shell surface and copper plate hot face is obtained by iterating the heat flow model at the shell/mold interface, and the parameters required for this interface heat flow model T_{s}, T_{m} and d_{gap} are obtained from the previous calculation results of the finite element model, and the detailed thermo-mechanical behavior rules of the shell in the mold during the continuous casting process of typical component micro-alloyed slab are obtained by iterative solution.

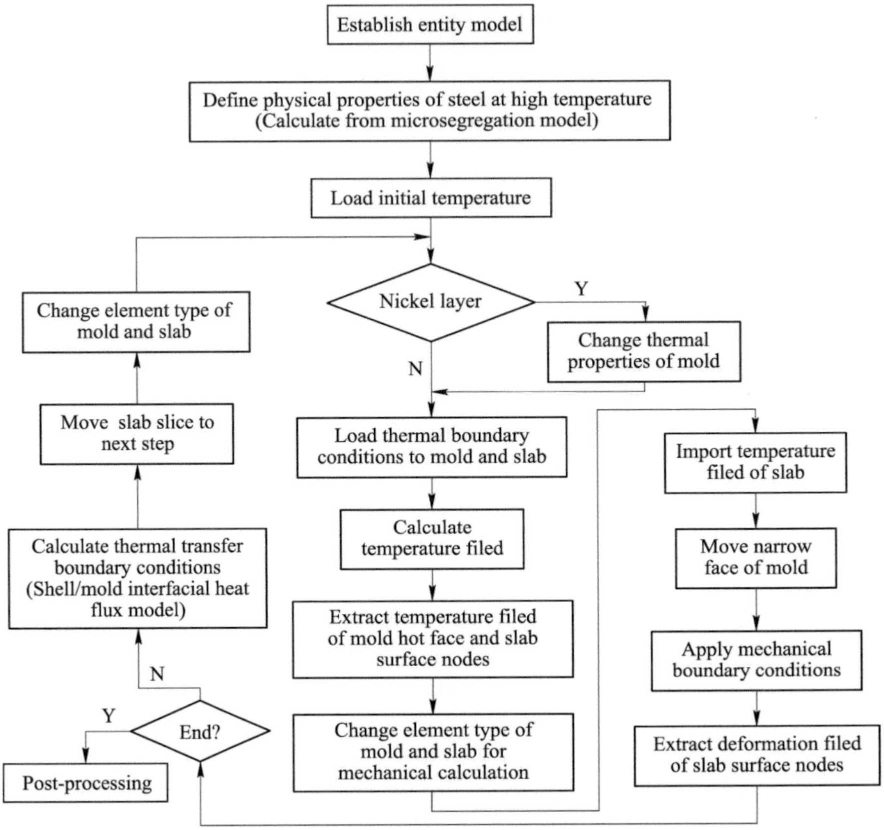

Fig. 2.35 Flowchart of thermos-mechanical coupling process of steel solidifying in continuous casting mold

2.3.1.4 Model Verification

The mold in the steel continuous casting process is a high-temperature and relatively closed "black box". The thermal behavior of the shell solidification process in the mold largely depends on the thermocouples installed in the wide and narrow copper plates of the mold, which monitor the copper plate temperature in real-time during the casting process to indirectly reflect the solidification heat transfer behavior of the shell inside. Therefore, the temperature detected by the mold thermocouple has also become the main means for scholars at home and abroad to verify the effectiveness of the mold and its internal shell solidification heat transfer model. In addition, the thickness change of the shell of the mold breakout, the thickness of the mold flux film of the mold, etc., are also important ways to verify the effectiveness of the model.

Figure 2.36 shows the actual measured values of the thermocouple temperature of the mold copper plate in the process of continuous casting production of a certain carbon steel wide-thick slab in a steel plant, and the copper plate temperature distribution at the thermocouple under the continuous casting production conditions calculated by the above model. As can be seen from Fig. 2.36a, affected by the distribution of the water slot, the copper plate temperature at the installation position of the wide-face thermocouple of the mold changes periodically along its circumference, and the copper plate temperature at the bolt installation position is about 10 °C higher than the temperature at the water slot distribution. As shown in Fig. 2.36b, near the center of the narrow face of the mold, due to the uniform distribution of the water slot and its consistent depth, the temperature distribution of the mold copper plate is also relatively uniform, and the temperature distribution at the installation depth of the internal thermocouple of the copper plate has no obvious fluctuation. However, in the off-corner area 43 mm away from the corner of the narrow face of the mold, due to the presence of bolts, the copper plate temperature in this area increases to a certain extent due to the increase in the spacing of the water slot and then decreases again due to the appearance of the water slot. From the actual measured temperature of the thermocouple, the actual measured temperatures of the upper and lower rows of thermocouples on the wide and narrow faces of the mold in the continuous casting process of certain carbon steel fluctuate around 118 and 107 °C and 116 and 102 °C, respectively, and the calculation results of the model are in good agreement with the actual measured values.

Figure 2.37a is a comparison of the change in the thickness of the shell at the center of the wide and narrow faces of a certain carbon steel wide-thick slab along the height direction of the mold calculated by the above thermo-mechanical coupling finite element model with the actual measured thickness of the shell. The actual measured values of the shell are taken from the breakout shell under the corresponding simulation conditions, as shown in Fig. 2.37b. As can be seen from the fig., at the height of 0–400 mm below the meniscus of the mold, the calculated value of the shell thickness of the finite element model is smaller than the actual measured value, and as the shell moves down, its thickness gradually increases, and the calculated value gradually matches the actual measured value. The main reason for the phenomenon of smaller calculated values of the shell in the height range of 0–400 mm below the meniscus is that during the actual solidification process of a certain carbon steel in the mold, the molten steel in the area near the meniscus continuously releases superheat and solidification latent heat through the mold copper plate until it initially solidifies into a slab, and then the solidified shell descends at a certain speed. However, in the calculation model in this book, the initial condition of the calculation is the molten steel with a uniform temperature of 25 °C superheat, and the molten steel near the side of the mold lacks the process of fully releasing superheat and solidification latent heat in the area near the meniscus. However, as can be seen from the fig., the simulation results are in good agreement with the results of Hanao et al. [75]. In addition, overall, the thickness of shell calculated by the model is in quite good agreement with the measured value.

Figure 2.38 shows the comparison between the calculated and measured thickness of the mold flux film on the interface of the shell/mold during the continuous casting production process of carbon steel. In particular, the mold flux film in Fig. 2.38b is

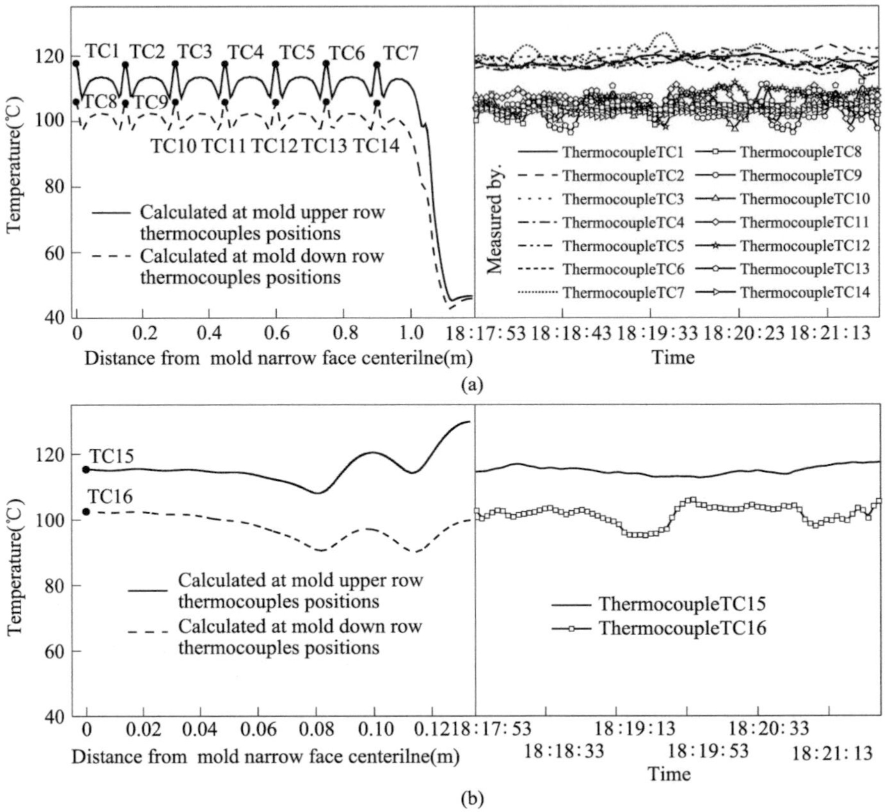

Fig. 2.36 Comparison of the calculated and measured temperature of the slab copper plate [50]. **a** Wide face; **b** Narrow face

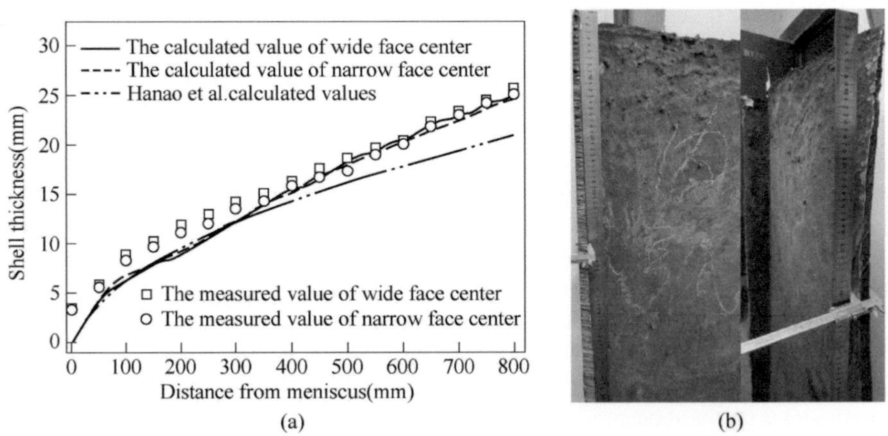

Fig. 2.37 Comparison of the calculated and measured shell thickness. **a** Data comparison; **b** Mold breakout shell [74]

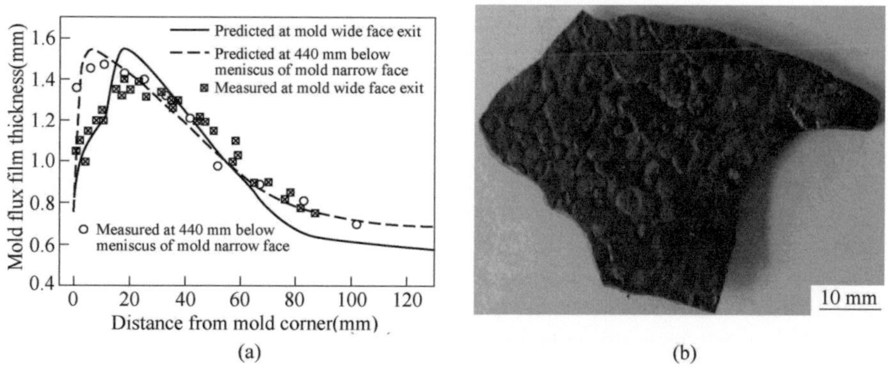

Fig. 2.38 Comparison of predicted and measured thickness of mold flux [74]. **a** Mold flux thickness; **b** Morphology of mold flux film

taken from the area near the wide face angle of the copper plate under the corresponding mold steel breakout condition. As can be seen from the fig., both the distribution trend and thickness of the mold flux film are in good agreement with the calculated and measured values of this book's model. Therefore, the shell solidification thermo-mechanical coupling finite element model developed in this book can accurately quantify and predict the heat transfer and mechanical behavior of specific composition slab in the mold during the continuous casting process, providing a theoretical basis for understanding the high-temperature heat transfer behavior of the slab corner during the continuous casting process of micro-alloyed steel and developing efficient heat transfer molds for the corner.

2.3.2 Shell Dynamic Shrinkage and Deformation Behavior

The above finite element model is used to calculate the process of continuous casting of micro-alloyed steel conventional slab in a steel plant (the structure diagram of the mold is shown in Fig. 2.34). The morphology of the shell shrinkage and deformation at different heights in the mold after 5 times magnification is shown in Fig. 2.39. As can be seen from the fig., in the initial stage of shell solidification, the deformation of thermal shrinkage is not significant due to the overall high temperature of the initial solidified shell. The wide and narrow face of the shell are in good contact with the corresponding faces of the copper plate as shown in Fig. 2.39a. However, as the solidified shell moves downward, the temperature of the solidified shell drops rapidly, the effect of thermal shrinkage gradually becomes significant. At this time, due to insufficient taper compensation of the narrow face copper plate of the mold, the narrow face shell does not get the support of the copper plate, and the wide-narrow face off-corner of shell and corner area begins to deform and deviate from the mold copper plate, forming a noticeable shell-mold interface gap, as shown in

Fig. 2.39b, c. When the shell moves downward to about 110 mm from the mold exit, due to the solidification shrinkage of the slab along the center direction of the wide face gradually slows down and the continuous compensation effect of the taper of the narrow face copper plate, the off-corner and corner area of the narrow face of the slab gradually fit with the copper plate to eliminate the gap, as shown in Figs. 2.39d and 2.40a. For the wide face of the slab, due to the small compensation along the thickness direction and the continuous thermal shrinkage effect of the shell along the center direction of the narrow face, the shrinkage and deformation of the slab wide face off-corner and corner area continue to increase, resulting in a larger interface gap in the slab wide face off-corner and corner area.

In addition, as can be seen from Fig. 2.40, in the actual continuous casting production process, in order to compensate the shrinkage of the narrow face of the slab in the center direction of the wide face, a larger taper is often required when adopting the traditional flat-plate mold with a linear single taper of the narrow face. On the one hand, it will cause serious wear of the narrow face of the mold copper plate, as shown in Fig. 2.40b. On the other hand, it will also increase the friction between the solidified shell and the mold copper plate, thereby increasing the risk of tearing of the narrow face and corner of the slab.

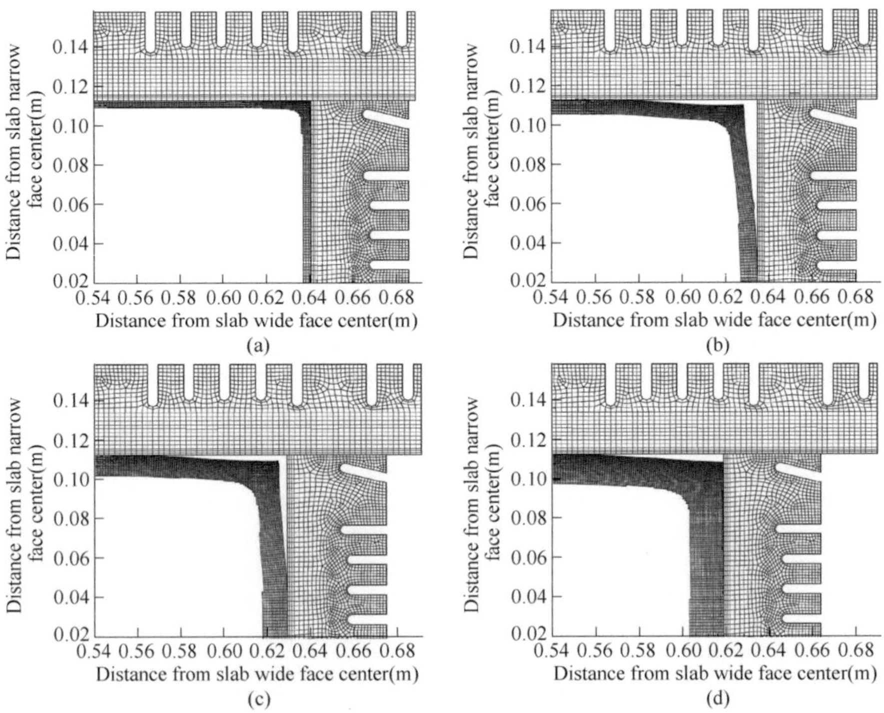

Fig. 2.39 Deformation of the solidified shell at different positions below the meniscus [74]. **a** 100 mm; **b** 300 mm; **c** 500 mm; **d** The mold exit

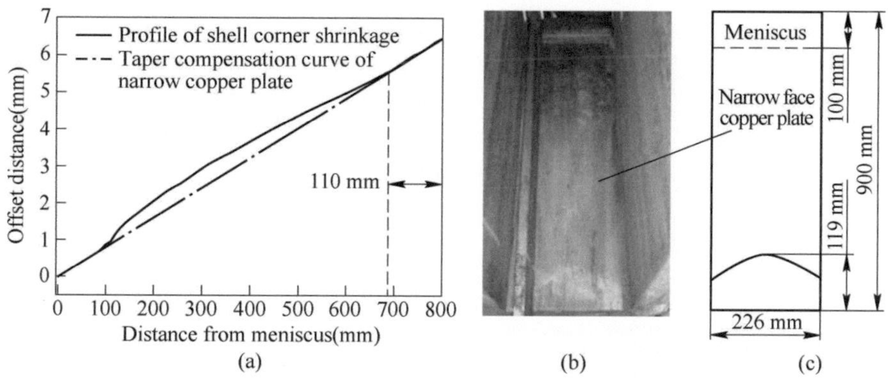

Fig. 2.40 Shrinkage curve of shell corner and wear of narrow copper plate. **a** Shell corner shrinkage profile; **b** The wear topography of mold narrow face copper plate; **c** The wear profile of mold narrow face copper plate

Influenced by the dynamic shrinkage and deformation of the shell in the mold, the large interface gap at the corner of the mold and its vicinity will cause the thick mold flux film and air gap to be concentrated in the off-corner and corner area of the wide and narrow faces of the shell, thus significantly reducing the heat transfer rate of the shell off-corner and corner area. This will lead to the formation of coarse austenite structure on the surface of the solidified shell and the concentrated precipitation of microalloy carbonitrides along the grain boundary, thereby greatly reducing the ductility of the slab off-corner and corner area, which is not conducive to crack controlling. Due to the lack of effective support of wide and narrow copper plates, the initial solidified shell has excessive deformation, which causes obvious tensile stress at the solidification front of the shell and significantly increases the sensitivity of subsurface crack at the off-corner and corner area of the slab. The distribution of crack sensitivity at the solidification front of the shell near the corner of the slab and the subsurface crack near the corner of the slab frequently occurred in the actual continuous casting production process are shown in Fig. 2.41 [76].

2.3.3 *Shell-Mold Interface Heat Transfer Behavior*

The thermal behavior of interface between the shell and copper plate in the mold is a key link affecting the heat transfer efficiency and uniformity of the shell to the mold. From the dynamic shrinkage and deformation behavior of the shell in the mold, it can be seen that the zone near the corner of the solidified shell is compensated by unreasonable taper, and the obvious interface gap is formed during the solidification process of the shell. This will cause the dynamic concentration distribution of the air gap and the mold flux film in the interface, thereby affecting the heat transfer in the shell off-corner and corner. Therefore, this section will conduct a deep analysis

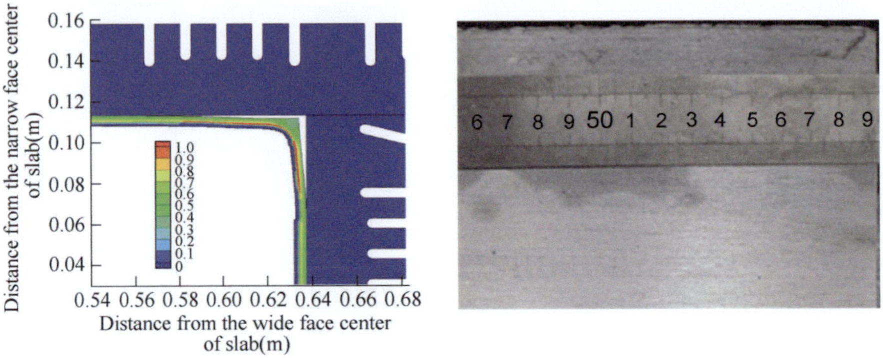

(a) (b)

Fig. 2.41 Distribution of crack sensitivity coefficient at the solidification front of the shell 300 mm below the meniscus and the morphology of subsurface cracks in the corner of the wide face in slab. **a** Distribution of shell crack sensitivity coefficient; **b** Morphology of subsurface cracks in the corner of the wide face in slab

of the distribution of key factors affecting the interfacial heat flux, such as the air gap, status and thickness of mold flux, and composition of thermal resistance at the corner and off-corner of the shell.

2.3.3.1 Mold Flux Film Distribution

Figure 2.42 is the distribution of mold flux film along the height direction of the mold in the range of 0–120 mm and 0–100 mm from the wide corner and the narrow corner of the shell under the condition of dynamic shrinkage and deformation of the solidified shell in the mold. It can be seen that in the traditional slab continuous casting process, the mold flux film in the mold is mainly concentrated in the zone of 0–60 mm from the wide and narrow face corners of the slab due to the separation of the slab corner and off-corner from the copper plate.

Specifically, in the height direction of the mold, due to the high temperature of the solidified shell in the upper part of the mold, the thermal shrinkage is not significant, and the thickness distribution of the mold flux film along the wide and narrow directions of the mold remains basically stable. However, as the shell continues to move downward, due to the significant increase in the shrinkage of the solidified shell, the liquid mold flux with good fluidity continuously fills the interfacial gap under the oscillation of the mold, and the corner point of the slab at 180 mm below the meniscus first reaches the maximum value. At this time, because the slab temperature in the off-corner of the wide and narrow face is still higher than the solidification temperature of the mold flux, the liquid mold flux continues to fill the interfacial gap in the corresponding zone, so that the thickness of the mold flux film in the off-corner of the wide and narrow faces is significantly higher than other zones. This causes the distribution of mold flux film near the mold corner is thickened first and then thinned

Fig. 2.42 Distribution of mold flux thickness of slab corner [50]. **a** Wide face; **b** Narrow face

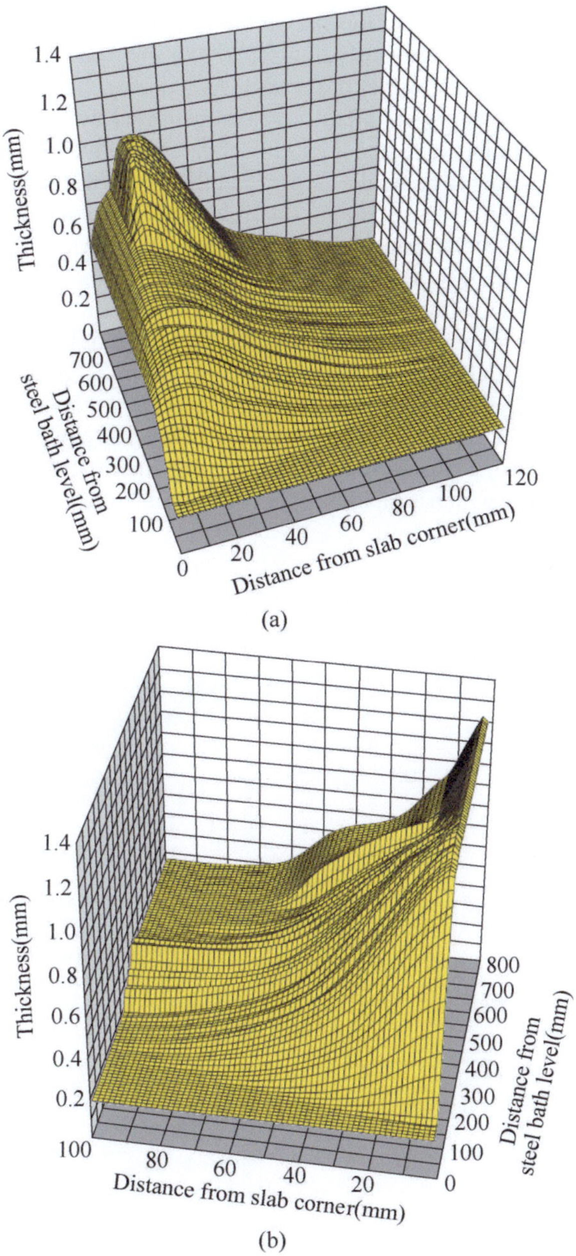

from the corner of the slab to the center of the wide and narrow faces. Meanwhile, in the dynamic distribution process of the mold flux film, because the shrinkage of the slab to the center of the wide face during the solidification process in the middle of the mold is not effectively compensated by the taper of the narrow face copper plate of the mold, the thickness of the mold flux film in the corner of narrow face and off-corner is generally larger than that in the vicinity of the corner of the wide face.

Figure 2.43 shows the distribution of the thickness of the liquid and solid mold flux film along the mold height from the corner of the slab toward the center of the wide face and narrow face at 0, 10, 30 and 50 mm from the corner of the slab (corresponding to the wide face and narrow face of the slab, respectively, referred to as positions of a, b, c, d and a′, b′, c′, d′). It can be seen from the fig., due to the rapid cooling of the mold copper plate, the filling form of the mold flux in the shell-mold interface is mainly solid flux film, and the thickness of the solid flux and liquid flux near the corner of the narrow face is generally thicker than that the wide face.

For the corner of the wide face and the off-corner of slab, except that the corner point a, which is rapidly cooled by 2D heat transfer, so that the mold flux is completely

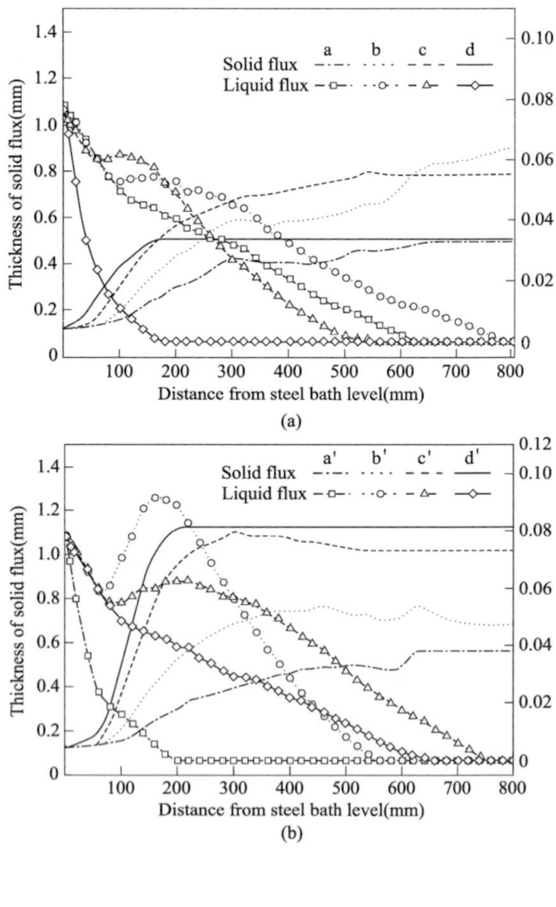

Fig. 2.43 Liquid flux and solid flux distribution of slab corner [50]. **a** Wide face; **b** Narrow face

solidified at the earliest and the thickness does not change, the thickness of the solid flux layer at b, c and d increases at a certain speed with the descending of the slab, then tends to be gentle, and finally reaches a stable state. For the corner of the narrow face and the off-corner of slab, when the solidified shell moves down to 60–200 mm below the meniscus, the thickness of the solid flux film in this height range also rapidly increases due to the accelerated solidification shrinkage of the shell along the center of the wide face, and then the change is slowed down due to the combined effect of the solidification shrinkage of the shell and the continuous compensation of the taper of the narrow face. Finally, it is completely solidified near the exit of the mold and stabilized. It can also be seen from the fig. that under the influence of the above factors, the thickness of the liquid flux at c′ in the off-corner 30 mm away from the corner also increased significantly at 60–200 mm below the meniscus.

Figure 2.44 shows the distribution of the proportion of the thermal resistance of the liquid flux film and solid flux film at a, b, c and d on the wide face and a′, b′, c′ and d′ on the narrow face to the total thermal resistance at the shell-mold interface. It can be seen that the proportion of the thermal resistance of the liquid flux to the total thermal resistance is higher than that of the solid flux near the meniscus. When liquid flux exists, the total thermal resistance of the shell-mold interface does not change significantly due to the absence of obvious air gap on the surface of the solidified shell, and the proportion of the thermal resistance of the liquid flux layer to the total thermal resistance decreases more smoothly. However, when the liquid flux completely solidifies, air gap begins to form from the corner of the slab and then extends to the center of its wide face and narrow face, so that the total thermal resistance of the interface in the corresponding zone to increase or decrease sharply with the change of air gap thickness, and the proportion of the thermal resistance of the solid flux layer to the total thermal resistance also shows a large fluctuation. The concentrated distribution of the thick mold flux film at the corner of the solidified shell and its vicinity significantly hinders the heat transfer at the corner of the solidified shell, which is not conducive to the efficient heat transfer at the corner of the solidified shell.

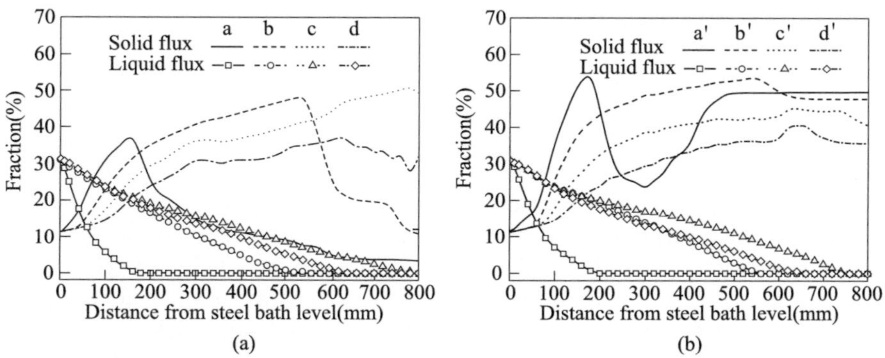

Fig. 2.44 Proportion of liquid and solid flux thermal resistance to total thermal resistance of slab corner [50]. **a** wide face; **b** narrow face

2.3.3.2 Air Gap Distribution

Figure 2.45 shows the 2D distribution of the air gap near the corner of the slab under the deformation conditions in the mold. It can be seen that the air gap is initially formed at the corner of the solidified shell at 180 mm below the meniscus, and mainly concentrated in the wide and narrow face areas within 0–20 mm from the corner of the shell. For the wide face corner of the slab, due to the insufficient compensation for the shrinkage of the solidified shell by the taper of the copper plate and the premature solidification of the mold flux, the gap of the continuous shrinkage of the wide face corner of the shell lacks further filling of the mold flux, making its air gap distribution compared to that of slab narrow face show a continuous and rapid growth characteristic. When the slab moves to the exit of the mold, the maximum thickness of the air gap at the wide face corner of the slab reaches about 0.8 mm.

At the same time, as can be seen from Fig. 2.45, as the solidified shell moves downward, the air gap at the wide face corner of the slab shows a trend of continuously expanding toward the center of the wide face from the corner. During the expansion process, because the solidification of the mold flux in the off-corner of the slab lags behind that of the corner, the flux film continuously fills the shrinkage gap, and the air gap rapidly decreases along the expansion direction.

It is worth noting that, as can be seen from Fig. 2.45, at 600 mm below the meniscus of the mold, the air gap thickness at the wide face corner of the solidified shell grows rapidly. The main cause for this phenomenon is: at this height, due to the slow shrinkage of the narrow face shell in the direction of the wide face center, under the continuous compensation of the taper of copper plate, the squeezing effect between the narrow face shell and the copper plate becomes gradually obvious, forcing the narrow face corner of the shell to turn toward its center, thereby causing the gap between the shell and the mold interface at the wide face corner to increase. Also, at this height, the temperature of the wide face corner of the slab has decreased below the solidification temperature of the mold flux, the rapidly growing interfacial gap cannot be filled by the liquid flux, making the total thickness of the slab flux film relatively stable, the increased interfacial gap thickness all turns into the air gap, thereby accelerating the growth of the air gap, which is not conducive to the efficient heat transfer of the slab corner at the lower part of the mold.

For the narrow face corner of the solidified shell, although its air gap also forms mainly within 0–20 mm from the slab corner, its growth characteristics are quite different from those of the wide face corner. It is mainly concentrated in the range of 180–670 mm below the meniscus on the height of the mold. At the same time, in the initial stage of air gap formation, although the shrinkage intensity of the solidified shell toward the center of the wide face is greater than the taper compensation of the narrow face copper plate, due to the effective filling of the gap by a large amount of mold flux, the growth rate of the air gap at the slab corner has increased rapidly and reached its maximum at 370 mm below the meniscus, but its thickness is significantly smaller compared to the wide face corner. Then, as the slab moves downward, due to the decrease in the shrinkage amplitude of the solidified shell and the taper

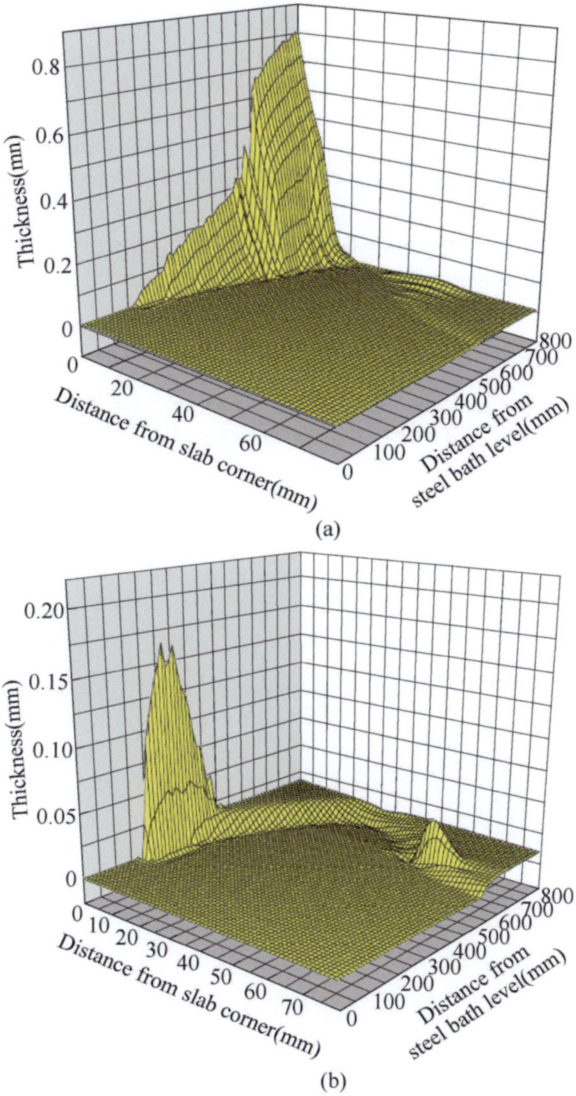

Fig. 2.45 Distribution of air gap thickness of slab corner [50]. **a** Wide face; **b** Narrow face

compensation of the copper plate, the air gap gradually decreases and tends to stable distribution.

2.3.3.3 Heat Flux Distribution

Influenced by the concentration and dynamic distribution of the mold flux film and air gap at the slab corner, the heat flux at the interface between shell and copper plate also changes significantly. Figure 2.46 shows the distribution of the interface heat flux along the height of the mold under the dynamic distribution conditions of the mold flux film and air gap within the interface between shell and copper plate, respectively, at 0–120 mm and 0–100 mm away from the corner of the slab. As can be seen from the fig., the interfacial heat flux significantly decreases at 0–20 mm away from the corner of the solidified shell. The main reason is that the temperature of the solidified shell in this area is relatively low, and the thickness of the mold flux film and air gap in the slab shrinkage gap is large, significantly increasing the interfacial thermal resistance and hindering the efficient heat transfer at the slab corner.

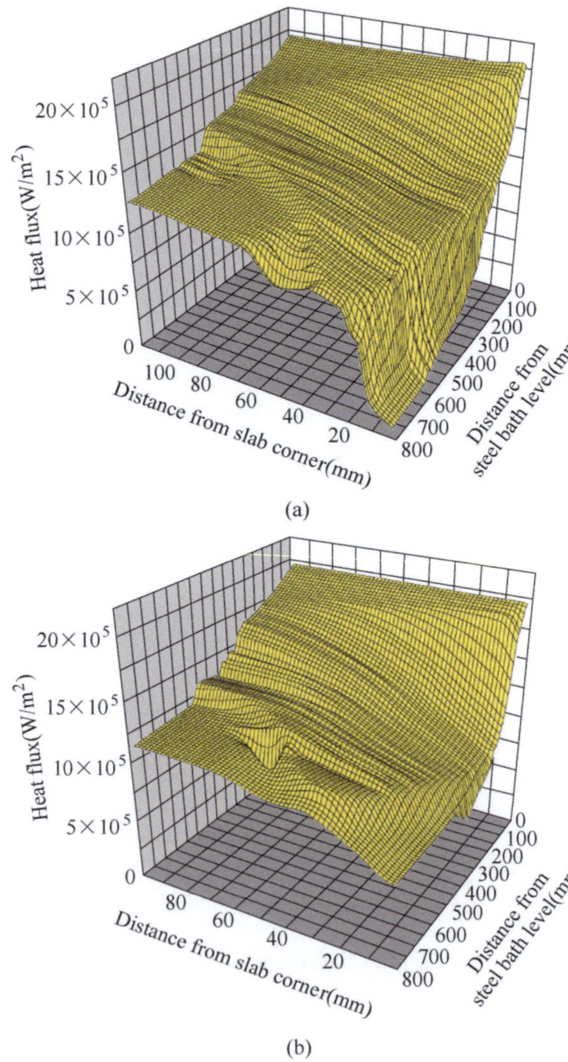

Fig. 2.46 Heat flux distribution between mold and slab interface [50]. **a** Wide face; **b** Narrow face

It is worth noting that since the thickness of the air gap formed near the corner of the wide face of the solidified shell is larger than that near the narrow face corner (see Fig. 2.45), the decrease of heat flux at the wide face corner is more significant than that at the narrow face corner. The interfacial heat flux at the narrow face corner is significantly affected by the dynamic distribution of the air gap in the range of 180–550 mm below the meniscus, and then tends to stable due to the significant reduction in the thickness of the air gap. The occurrence of this phenomenon reflects that the taper design system of the traditional flat-plate mold is not the most suitable for micro-alloyed steel continuous casting, and it also indicates that the wide face corner in the mold also needs effective taper compensation or prevention of the slab corner from twisting during the solidification process in the lower part of the mold, which exacerbates the interfacial gap at the wide face corner.

2.3.4 Solidified Shell Temperature Field

Figure 2.47 shows the temperature distribution of the shell of 100, 300 and 500 mm below the meniscus and at the exit and the copper plate of the mold at the cross section under the heat transfer conditions of the shell-mold interface. As can be seen from the figure, during the initial solidification stage of the molten steel (0–100 mm below the meniscus), the temperature of the solidified shell is relatively high, and due to the relatively uniform distribution of the mold flux film within the shell-mold interface, the temperature distribution of the solidified shell along its circumference is also relatively uniform, only the corner temperature rapidly decreasing due to the effect of 2D heat transfer, as shown in Fig. 2.47a. As the shell continues to move downward, the air gap begins to appear at the corner of the slab, and the amount of mold flux filling also gradually increases, gradually reducing the heat transfer rate in the corner area of the slab, and at 300 mm below the meniscus, "hot spot" gradually appear in the off-corner of the wide and narrow faces, and the distribution tends to expand as the shell moves downward.

Figure 2.48 shows the distribution of surface temperature along the height direction in the area near the corner of the wide and narrow faces of the solidified shell in the mold. At the top of the mold, the solidified shell near the corner (0–10 mm area from the corner of the shell) is affected by 2D heat transfer, and the surface temperature drops quickly until the slab moves downward to 200 mm below the meniscus. This is due to the generation of a certain thickness of air gap and the concentration of thick mold flux at the corner of the mold, which gradually slows down. For the off-corner within 10–20 mm from the corner of the slab, due to its uniform interface heat flux distribution in the upper part of the mold, the surface temperature evolution is relatively smooth. As the solidified shell continues to move downward, its surface temperature decreasing rate slows down, and the slowing trend gradually extends toward the center of the wide and narrow face of the slab. When the slab moves downward to the mold exit, the surface temperature of the shell off-corner is 120 and 61 °C higher than the corresponding wide and narrow face centers, respectively. The

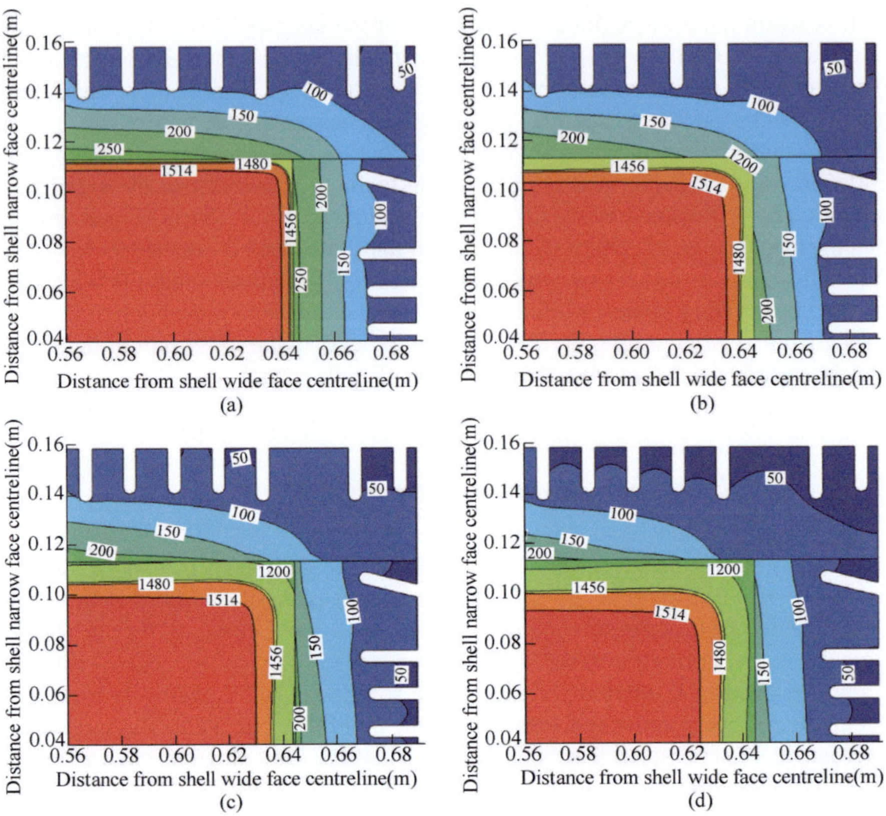

Fig. 2.47 Temperature fields of shell and copper plate at the distances below the meniscus. **a** 100 mm; **b** 300 mm; **c** 500 mm; **d** Mold exit

range has also expanded from the initial 10–20 mm width to a range of 10–75 mm and 10–60 mm from the corner, becoming the "hot spot" of the solidified shell in the mold, which is very easy to cause the slab to form the subsurface crack defects in the slab off-corner as shown in Fig. 2.49.

Affected by the above-mentioned non-uniform heat transfer in the mold, the growth of the shell along the circumference shows obvious non-uniformity. According to the definition of the viscous temperature LIT, when the shell temperature drops to this temperature, there will be no flowing liquid phase between the dendrites of the corresponding solid fraction, so the solid fraction corresponding to this temperature can be determined as the solidification front of the slab. Therefore, calculates the distribution of the shell thickness at the mold exit along the mold circumference with a solid fraction $f_s = 0.884$ as the solidification front of the shell, as shown in Fig. 2.50. As can be seen from the fig., due to the slow heat transfer in the off-corner of the wide and narrow faces of the slab, the thickness of the shell in the off-corner of the wide and narrow faces of the slab at the mold exit is 2.17 and

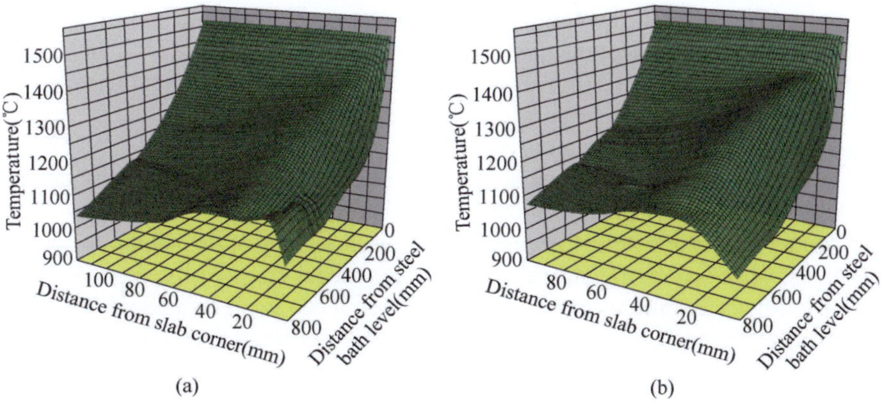

Fig. 2.48 Distributions of slab surface temperature of slab corner [50]. **a** Wide face; **b** Narrow face

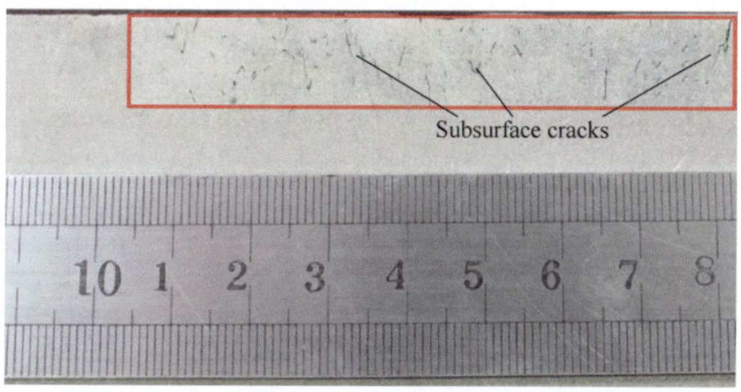

Fig. 2.49 Subsurface cracks below slab off-corner

1.13 mm smaller than the corresponding face center, respectively, greatly reducing the uniformity of the slab solidification in the mold.

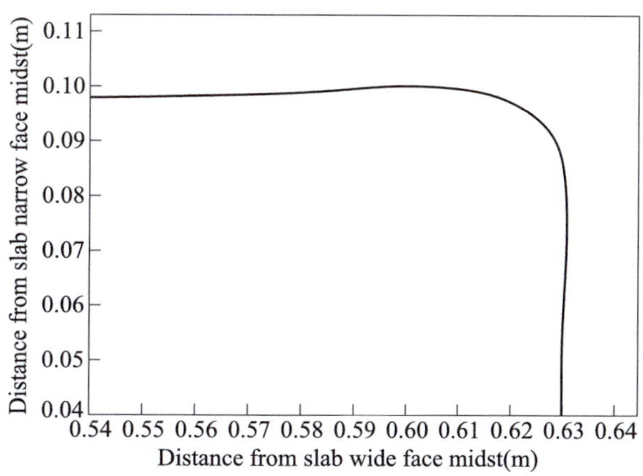

Fig. 2.50 Shell profile at mold exit under conventional mold taper

Figure 2.51 shows the evolution curve of the corner temperature and cooling rate of the solidified shell under the above mold heat transfer conditions. As can be seen from the fig., when the slab moves downward to 150 mm below the meniscus, the corner temperature of the shell drops to about 1150 °C. According to the thermodynamic behavior of the precipitation of microalloy carbonitrides containing Nb, B, Al, etc., in Sect. 2.2.1. the grain boundary structure of the slab corner at this temperature has begun to gradually precipitate microalloy carbonitrides. However, in the lower part of the mold and below, the cooling rate of the corner of the solidified shell is less than 5 °C/s (averaging only about 3.5 °C/s). According to the detection results of microalloy carbonitrides precipitation at different cooling rates in Sect. 2.2.2.3, it cannot meet the cooling rate condition of ≥ 5 °C/s for the dispersion precipitation of microalloy carbonitrides. This has led to the current situation where carbonitrides precipitate along the grains and embrittle the grain boundaries during the solidification process of the corner of the micro-alloyed steel slab. Therefore, according to the heat transfer and shrinkage deformation rule of the shell in the mold, a curved inner cavity mold structure that can effectively compensate for the dynamic shrinkage characteristics of the solidification process of the shell in the mold should be developed. This will promote the mold flux film and air gap in the corner of the solidified shell and its nearby area to be thin and uniform distributed, thereby increasing the cooling rate of the corner of the solidified shell in the lower part of the mold to ≥ 5 °C/s. This will fundamentally "resolve" the causes of frequent transverse cracks in the corner of the micro-alloyed steel slab due to the chain precipitation of microalloy carbonitrides along the austenite grain boundaries embrittling the grain boundaries and the formation of coarse austenite grains during solidification. It will also promote the uniform growth of the solidified shell in the mold.

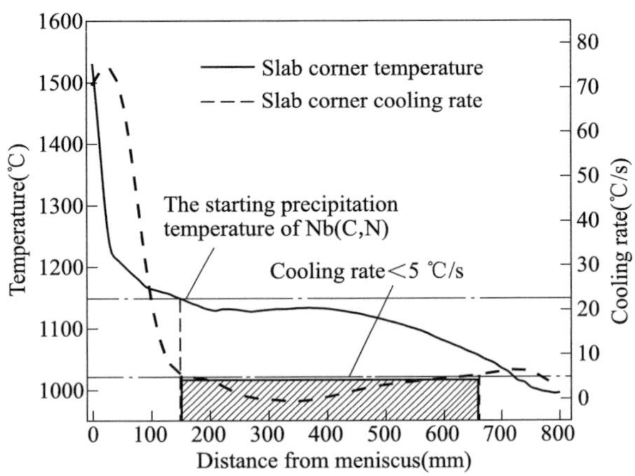

Fig. 2.51 Evolution of corner temperature and cooling rate of the slab in mold

2.4 Thermo-mechanical Behavior of Slab in Secondary Cooling

Secondary cooling is another key cooling link for slab continuous casting. About 95% of the heat will be transferred out in the secondary cooling zone during continuous casting. The heat transfer behavior may directly affect the surface and internal quality of the slab. Specifically for the corner of the slab, under the secondary cooling control structure and continuous casting process conditions of the slab continuous caster, its temperature slowly drops from about 900–1000 °C to 700–820 °C. In this process, on the one hand, the micro-structure of the slab corner has changed significantly, gradually transitioning from a single austenite structure to a low ductility dual-phase structure of austenite + grain boundary ferrite; on the other hand, for micro-alloyed slab containing such as Nb, B, Al, as known from Sect. 2.2, the precipitation of these microalloy elements carbonitrides at the low temperature end is mainly completed in the high temperature zone of secondary cooling, and the precipitation of microalloy carbonitrides containing V will be completed in the entire secondary cooling zone. The characteristic of slow heat transfer in the corner of the slab will cause the microalloy carbonitrides to precipitate at the austenite grain boundaries, thereby embrittling the grain boundaries of the structure. Affected by the above two factors, the ductility of the slab corner structure will be significantly reduced. At the same time, during the solidification process of slab in the secondary cooling zone, the temperature fluctuation of the slab corner (especially in the high temperature zone) is severe, and it needs to undergo deformation processes such as bending and straightening to withstand complex mechanical effects. The low ductility slab corner is prone to induce grain boundary cracking in its structure under excessive stress or strain, forming slab transverse corner crack defects. Therefore, this section will focus on the continuous casting process of conventional slabs as the main research object, analyze the thermo-mechanical behavior rules of the slab secondary cooling solidification process, and lay a theoretical foundation for the control of slab corner structure and the development of its process and device technology.

2.4.1 Mathematical Model

In the actual slab continuous casting process, the cooling and solidification of slab in the secondary cooling zone, heat transfer and loss along the casting direction and perpendicular to the casting direction. In order to accurately describe the solidification thermo-mechanical behavior of the slab in various areas of secondary cooling, this book will take a conventional slab of Nb high strength steel actually produced by a domestic steel plant as the research object, and develop a 3D thermo-mechanical coupling finite element calculation model for the solidification of the full-cast slab.

2.4.1.1 Model Assumptions and Entity Model

The secondary cooling zone of the slab continuous caster is a bow segment structure. During the solidification process of the slab, in addition to thermal stress, it also undergoes complex stress effects such as casting roller clamping, bending, and straightening, and there are many influencing factors. In order to effectively develop the complex thermo-mechanical calculation model of the slab solidification process in the secondary cooling casting flow, this book has made the following assumptions for the model development process:

(1) Ignoring the influence of molten steel flow on the solidification and heat transfer of slab, the effective thermal conductivity of molten steel is adopted.
(2) The material of the slab is isotropic during the solidification process, and its high temperature physical properties only change with temperature and its phase composition.
(3) The heat transfer and stress of the slab solidification process are symmetrical, and the cooling water in the same cooling zone of secondary cooling is evenly distributed. The heat transfer between the slab and the casting roller, the radiation heat transfer of the slab is included in the comprehensive heat transfer coefficient.
(4) Ignoring the deformation of the bow segment casting roller, the contact between the casting roller and the shell is treated as rigid-flexible contact.

Based on the above assumptions, a 1/2 3D entity model is developed and discretized for the casting flow roller array and slab in a steel plant with the center line of the wide surface as the symmetrical surface, as shown in Fig. 2.52. In the actual slab continuous casting process, the straightening and reduction of the slab are mostly completed by two bow segments. In order to comprehensively consider the influence of the reduction of the slab before and after the straightening and reduction process on the stress and evolution of the slab, the length of the slab during modeling is 4500 mm along the casting speed direction. In addition, because the deformation of the slab along the direction of the casting flow is relatively small, the mechanical boundary conditions such as the displacement of the cross section at both ends of the slab are treated as symmetrical surfaces, as shown in Fig. 2.52b (the upper and lower cross sections of the slab are symmetrical surfaces). The roller array arrangement of the casting flow is set according to the actual roller array and the roller gap process adopted by the continuous casting steel grade.

2.4.1.2 Slab Thermo-Mechanical Finite Element Model

A. Heat Transfer and Mechanical Control Equations

Based on the assumptions of the previous chapter, the 3D solidification heat transfer control equation of the slab is developed as shown in (2.91):

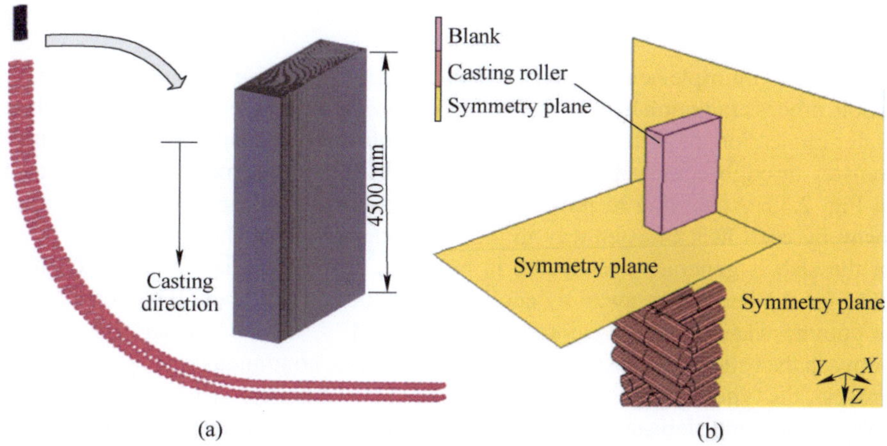

Fig. 2.52 3D finite element model of slab solidification process. **a** Slab and casting flow roller apron; **b** Schematic diagram of model symmetry plane setting

$$\rho C\frac{\partial T}{\partial t} = -\rho v C\frac{\partial T}{\partial z} + \frac{\partial}{\partial x}\left(k_{\text{eff}}\frac{\partial T}{\partial x}\right) + \frac{\partial}{\partial y}\left(k_{\text{eff}}\frac{\partial T}{\partial y}\right) + S_{\text{o}} \tag{2.91}$$

where

- ρ Density, kg/m^3, the determination method is in Sect. 2.3.1;
- C Heat capacity, J/(), in the actual modeling process, it is converted into heat enthalpy, the method of taking heat enthalpy is in Sect. 2.3.1.
- T Temperature, °C;
- k_{eff} Thermal conductivity, W/(m), the determination method is in Sect. 2.3.1;
- S_{o} Internal heat source term, W/m^3.

In aspects of the mechanical control equation, since the slab undergoes not only elastic deformation but also ductile deformation and creep during the solidification and deformation process in the secondary cooling casting flow, in order to fully consider the mechanical characteristics of the slab such as elasticity, ductility and creep, adopts the Anand rate-related constitutive equation proposed by Anand and Brown, as shown in Eqs. (2.67) and (2.68).

B. Initial Conditions

The initial temperature of the slab is the casting temperature of the molten steel, and both stress and strain are 0.

C. Heat Transfer Boundary Conditions

(a) Mold

The heat transfer boundary conditions of the slab in the mold are detailed in Sect. 2.3.1.

(b) Secondary Cooling Zone

The slab has multiple heat transfer ways in the secondary cooling zone, including the whole process radiation heat exchange of the slab, the heat taken away by the cooling water immersion, the heat taken away by the cooling water evaporation, and the heat transfer through contact between the slab and the bow segment roller, as shown in Fig. 2.53. According to previous research results, the proportion of heat taken away by each heat transfer way in the total heat output of the secondary cooling of the slab is generally considered to be about 25% for radiation heat exchange, 25% for the heat taken away by cooling water immersion, about 33% for the heat of cooling water droplet evaporation, and about 17% for the contact heat transfer between the roller and the slab. However, in the actual continuous casting production process, the types of nozzles, nozzle arrangement, water flow density, cooling water temperature, and slab surface conditions of different casters are all different, so the heat transfer proportion of the above heat transfer way is not constant, and it is difficult to quantitatively determine the heat transfer coefficient of each heat transfer way. Therefore, metallurgist at home and abroad often determine the heat transfer coefficient of the slab in the secondary cooling zone through the comprehensive heat transfer coefficient, and through a large number of measurements and research, they have obtained the comprehensive heat transfer calculation formula for the secondary cooling of the slab as shown in formulas (2.92)–(2.98).

Ishiguro et al.

$$h = 0.581W^{0.541}(1 - 0.0075T_{\mathrm{w}}) \tag{2.92}$$

Shimada et al.

$$h = 1.57W^{0.55}(1 - 0.0075T_{\mathrm{w}}) \tag{2.93}$$

Nozaki et al.

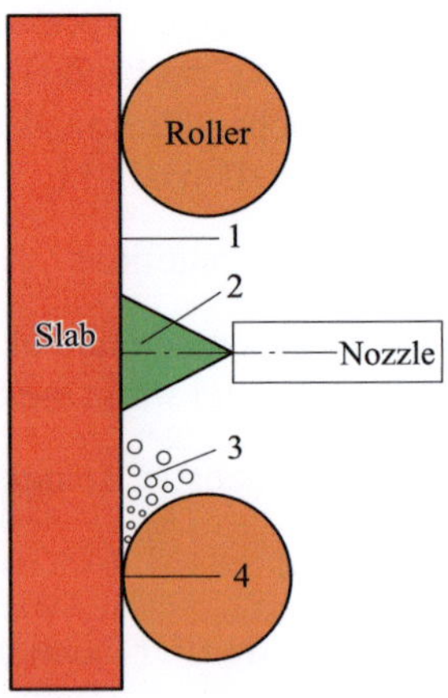

Fig. 2.53 Heat transfer way on the surface of the slab in the secondary cooling zone 1—Radiation heat transfer; 2—Cooling water heat conduction; 3—Cooling water evaporation hear transfer; 4—Rollers contact heat transfer

$$h = 1.57W^{0.55}(1 - 0.0075T_{\mathrm{w}})/\alpha \tag{2.94}$$

Buist et al.

$$h = 0.61W^{0.395} \; 4.5\mathrm{L}/\left(\mathrm{m}^2\,\mathrm{s}\right) \le W < 20\mathrm{L}/\left(\mathrm{m}^2\,\mathrm{s}\right) \tag{2.95}$$

Nozaki et al.

$$h = 0.3925W^{0.55} \times (1 - 0.0075T_{\mathrm{w}}) \tag{2.96}$$

Air water nozzles, etc.

$$h = 0.35W + 0.13 \tag{2.97}$$

Bolle, etc.

$$h = \begin{cases} 0.423W^{0.556} & 1\mathrm{L}/\left(\mathrm{m}^2\,\mathrm{s}\right) < W < 7\mathrm{L}/\left(\mathrm{m}^2\,\mathrm{s}\right),\ 627\,^{\circ}\mathrm{C} < T_{\mathrm{s}} < 927\,^{\circ}\mathrm{C} \\ 0.36W^{0.556} & 0.8\mathrm{L}/\left(\mathrm{m}^2\,\mathrm{s}\right) < W < 2.5\mathrm{L}/\left(\mathrm{m}^2\,\mathrm{s}\right),\ 727\,^{\circ}\mathrm{C} < T_{\mathrm{s}} < 1027\,^{\circ}\mathrm{C} \end{cases} \tag{2.98}$$

where

h Heat transfer coefficient, the unit is kW/(m^2) except where noted;
T_{s} Slab surface temperature, °C;
T_{w} Secondary cooling water temperature, taken as 30 °C;
W Cooling water flow density, L/(m^2 s);
α Cooling correction factor related of roller.

Similarly, the above formulas for calculating the heat transfer coefficient of the slab surface are all based on specific research conditions. The heat transfer coefficients determined by different formulas vary greatly, so they cannot be used directly. Therefore, adopts an inverse calculation model of bisection method based to the measured temperature of the slab surface in the process of determining the comprehensive heat transfer coefficient of the slab surface in each zone of secondary cooling, that is, infrared temperature measurement is performed on the slab surface at the exit of each bow segment, and then the model is used to inverse calculate to obtain the comprehensive heat transfer coefficient of the slab in different cooling zones. The calculation flowchart is shown in Fig. 2.54.

Table 2.8 shows the length of each zone of secondary cooling of the slab continuous caster and the amount of cooling water in each cooling zone under different casting speeds during the production process of Nb-bearing high strength steel. It was measured that at casting speed of 1.5 m/min, the temperatures at the exit of bow segments 3–8 were 999, 982, 975, 962, 948 and 932 °C, respectively. The difference calculation of the cooling zones was carried out, and the results are shown in Table 2.9.

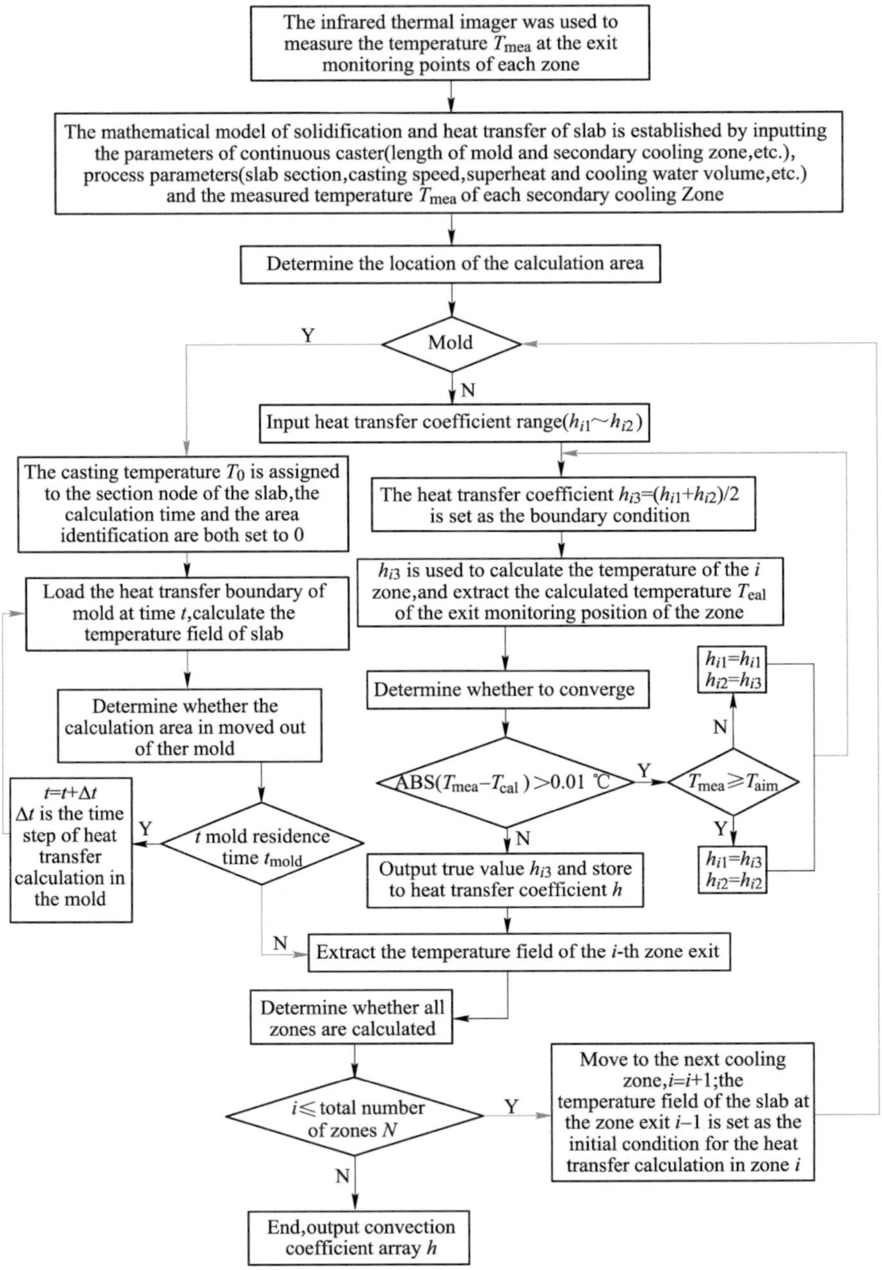

Fig. 2.54 Flowchart of slab heat exchange coefficient in secondary cooling zone based on inverse calculation model of bisection method

Table 2.8 The cooling zones and the corresponding cooling water volume of a slab continuous caster

Cooling zones	Secondary cooling zones length (m)	Water volume corresponding to different casting speed (L/min)		
		1.0 (m/min)	1.2 (m/min)	1.5 (m/min)
1 NL/NR	0.8–1.50	107	128	160
1 I/O	0.8–1.04	169	203	254
2 I/O	1.04–1.60	249	299	374
3 I/O	1.60–2.71	275	330	413
4 I/O	2.71–4.45	244	293	366
5 I/O	4.45–6.37	171	206	257
6 I	6.37–8.29	62	74	93
6 O	6.37–8.29	74	89	112
7 I	8.29–12.13	89	106	133
7 O	8.29–12.13	124	149	186
8 I	12.13–16.02	64	77	96
8 O	12.13–16.02	96	116	145
9 I	16.02–20.78	60	72	90
9 O	16.02–20.78	102	122	152
10 I	20.78–27.92	79	83	101
10 O	20.78–27.92	157	166	203
11 I	27.92–35.06	79	79	81
11 O	27.92–35.06	157	157	161

Table 2.9 Measured and interpolated temperatures of slab surface in each secondary cooling zone

Steel grade	Exit measured and interpolated temperatures (conventional) in each zone (°C)										
	Zone 1	Zone 2	Zone 3	Zone 4	Zone 5	Zone 6	Zone 7	Zone 8	Zone 9	Zone 10	Zone 11
High strength steel	1125	1095	1070	1045	1025	1005	990	975	960	930	900

Based on the slab secondary cooling zone exit measured and interpolated temperatures shown in Table 2.9, using the calculation flowchart shown in Fig. 2.55, the slab surface heat transfer coefficient under different casting speeds is iteratively determined. In determining the initial heat transfer coefficient of the slab surface (h_{i1}–h_{i2}), considering that in the actual slab continuous casting process, the convective heat transfer coefficient of the secondary cooling zone is usually within the range of 50–1000 W/(m^2 °C), determines h_{i1} and h_{i2} initial values are set at 1 W/(m^2 °C) and 1500 W/(m^2 °C). The convergence criterion is determined when the absolute value of the difference between the temperature of the secondary cooling zone and the

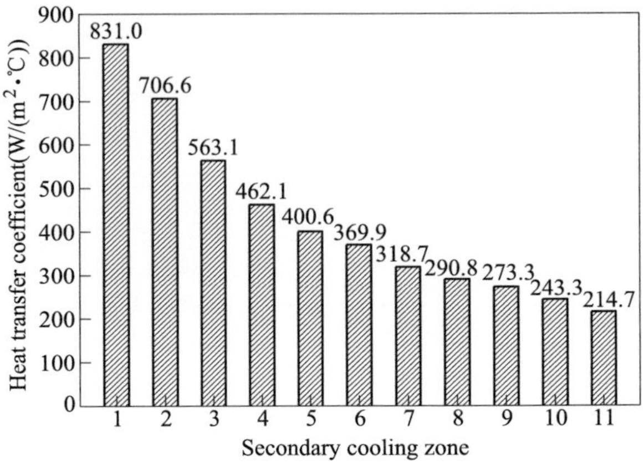

Fig. 2.55 Heat transfer coefficient of each zone of secondary cooling of Nb-bearing high strength steel slab

target temperature obtained from solving the continuous casting solidification heat transfer model is less than or equal to 0.01 °C, that is, $|T_{\text{cal}} - T_{\text{aim}}| \leq 0.01$ °C, the iteration stops.

Figure 2.55 is the heat transfer coefficient of each zone of the wide face of the continuous casting process of Nb-bearing high strength slab in determined according to the above inverse calculation model of bisection method based on the measured temperature of the slab surface. As can be seen from the fig., from the secondary cooling zone 1 to zone 11, the heat transfer coefficient of the slab wide face decreases from 831.0 W/(m^2 °C) to 214.7 W/(m^2 °C), and it decreases rapidly at the beginning, but the rate of decrease slows down in the later stages. Based on the calculation results of the heat transfer coefficient of the slab surface, the model calculation process will dynamically apply the heat transfer coefficient of the slab surface in the secondary cooling zones.

The heat transfer boundary conditions for each symmetry plane of the slab are as follows:

$$q = -\lambda \frac{\partial \tau}{\partial x} = 0 \tag{2.99}$$

The mechanical boundary conditions of the slab in the casting flow are as follows:

(1) Displacement constraint on the wide face symmetry plane:

$$\delta_y = 0 \tag{2.100}$$

(2) Displacement constraint on the cross-sectional symmetry plane at both ends:

$$\delta_z = 0 \tag{2.101}$$

(3) The mechanical boundary between the slab surface and the hot surface of the mold copper plate is described in Sect. 2.3.1.3. The contact surface between

the slab surface and the bow segment roller is calculated using the rigid-flexible contact analysis algorithm. The contact body is defined as illustrated in Fig. 2.52b. The implementation Segments of the soft reduction at the slab solidification end is No.8 segment and No.9 segment, with a total of 6 mm reduction amount distributed uniformly for the two segments.

2.4.2 *Slab Heat Transfer in Secondary Cooling*

Figure 2.56 shows the 3D temperature field distribution of the slab in each zone during secondary cooling, calculated using the slab secondary cooling solidification thermo-mechanical coupling finite element model for continuous casting production of Nb-bearing high strength steel. As can be seen from Fig. 2.56a, due to the high heat transfer rate of the slab in the mold, a large temperature gradient exists in the casting direction. At the same time, because the slab corner undergoes 2D heat transfer, its temperature decreases fastest, and the temperature of the slab corner after exiting the mold is about 970 °C. When the slab moves downward to the second cooling zone, due to the rapid spray cooling on both the wide and narrow faces of the slab in this cooling process, the surface of the slab cools rapidly, as shown in Fig. 2.56b. However, when the slab moves downward to the exit of the second cooling zone, due to the lack of spray water cooling on the narrow face of the slab, its surface quickly reheats. For the wide face, since the amount of secondary cooling water is still large, the temperature of the inner and outer arc surfaces of the slab continues to decrease, but gradually turns into uniform distribution, until the slab exits the No.4 zone of secondary cooling, as shown in Fig. 2.56c, d. After the slab moves downward to the exit of the No.5 zone of secondary cooling, although the amount of secondary cooling water is greatly reduced, because the solidified thickness at the slab corner is already thick, heat transfer from the liquid core to the corner cannot compensate for the heat taken away by the spray, roller contact on the wide face and radiation heat transfer on the narrow surface, the temperature of the slab corner continues to decrease, until the slab exits the continuous caster, as shown in Fig. 2.56e–j. The final solidification position of the slab is about 21.6 m below the meniscus.

Figure 2.57 shows the evolution of the temperature fields at the corner, wide face center, narrow face center, and slab center under the above calculation conditions. As can be seen from the fig., compared to the slab corner, the surface temperature of the wide and narrow face centers is generally higher, only decreasing rapidly in the mold and at the foot roller, then due to the lack of cooling water spray on the narrow face of the slab and the reduction of cooling water in each cooling zone of the wide face, the surface temperature of the wide and narrow face centers of the slab quickly increases. When the slab enters the bender, the temperatures of the wide and narrow face centers of the slab reach 1120 and 1250 °C, respectively, when the surface structure of the slab has not yet precipitated microalloy carbonitrides, and exists in the form of austenite single phase, so the ductility of surface structure is high. During the bending process of the slab, although the outer arc is subject to

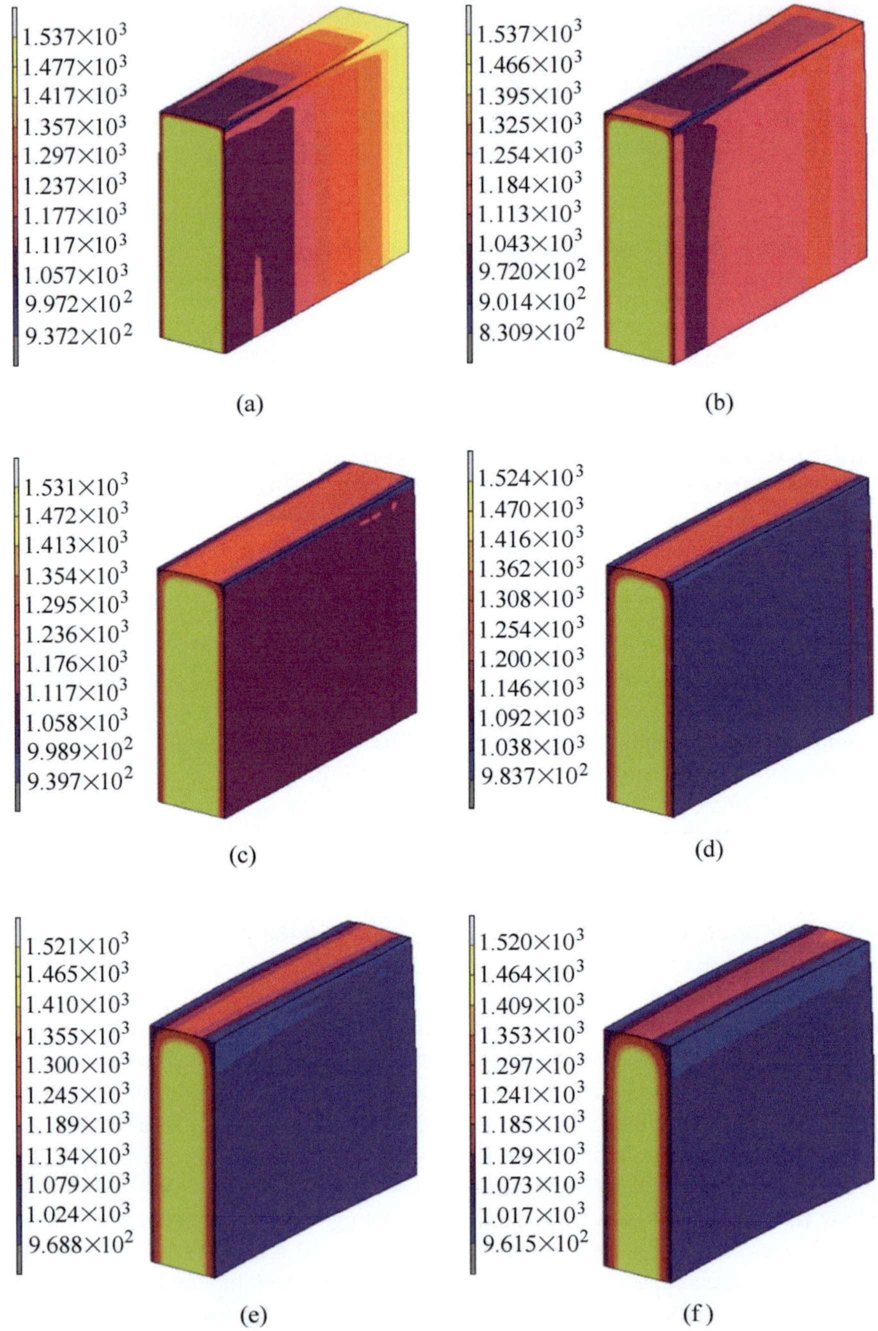

Fig. 2.56 3D temperature field of slab during solidification process at the exit of each cooling zone of secondary cooling. **a** Exit of the mold; **b** zone 2; **c** zone 3; **d** zone 4; **e** zone 5; **f** zone 6; **g** zone 7; **h** zone 8; **i** zone 9; **j** zone 10

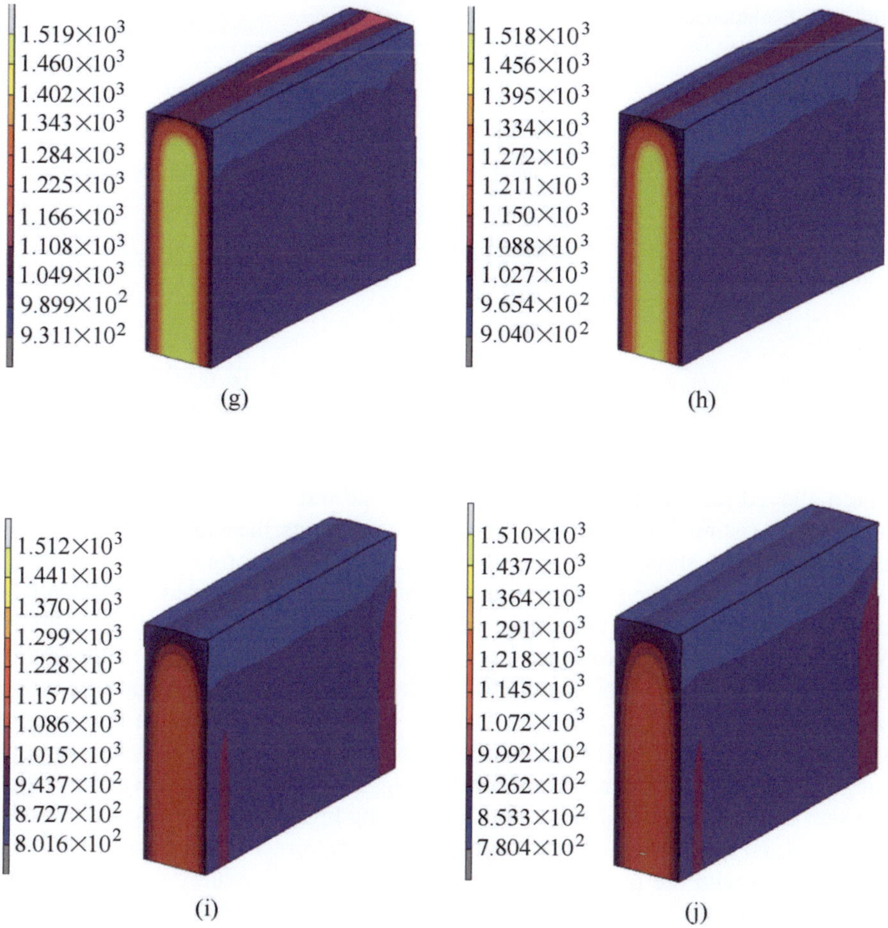

Fig. 2.56 (continued)

tensile force if the accuracy of roller gap meets the requirements, the outer arc wide face of the slab generally does not form surface transverse crack.

When the slab enters the straight segments, the temperatures at the centers of the wide and narrow faces have decreased to 990–960 °C and 1050–995°C, respectively. As can be known from the thermodynamic behavior of microalloy carbonitrides precipitation under typical Nb content in Sect. 2.2.1.2, the Nb (C, N) precipitation at the grain boundaries of the surface structure has exceeded 60% at the temperature and is approaching the "nose point" of carbonitrides precipitation. This means that the Nb (C, N) at the grain boundaries of the slab surface structure has precipitated to a certain extent. However, the slab surface structure at this temperature still exists in the form of a single austenite structure, and ductility is still high, as shown in Fig. 2.58. Therefore, generally speaking, in the continuous casting production of

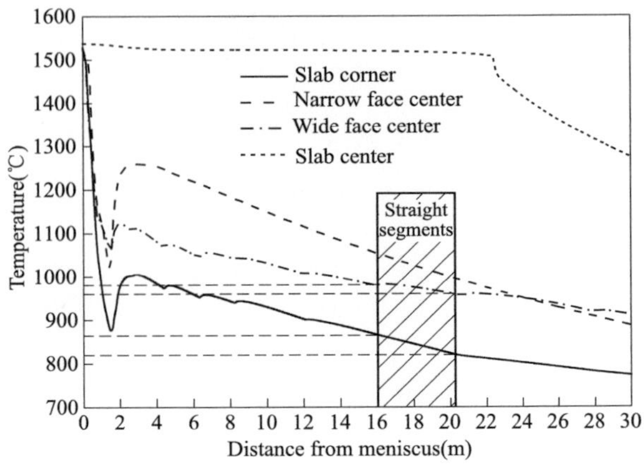

Fig. 2.57 Evolution of the temperature of slab surface

micro-alloyed steel slab containing Nb, Al, B and other alloy elements, when the accuracy of casting flow roller gap meets the requirements, the wide and narrow faces of the slab during straightening process generally does not form transverse cracks.

For the corner of the slab, its temperature decreases in the mold and secondary cooling high temperature zone (foot roller and No.2 zone of secondary cooling). When the slab exits the narrow face foot roller zone of the mold, its corner temperature decreases to a minimum of about 870 °C, then due to reduced amount of cooling water on both the narrow and wide faces, it experiences a noticeable temperature increase. It then oscillates and gradually cools to the straight segments temperature range of 860–820 °C. As can be seen from Fig. 2.59, the cooling rate evolution of the slab corner in the mold and secondary cooling high temperature zone, after the slab exits the mold, the average cooling rate of its corner is only about 1.92 °C/s. Combined with the precipitation kinetics of typical microalloy carbonitrides in Sect. 2.2, under this low cooling rate condition, the Nb (C, N) at the grain boundaries

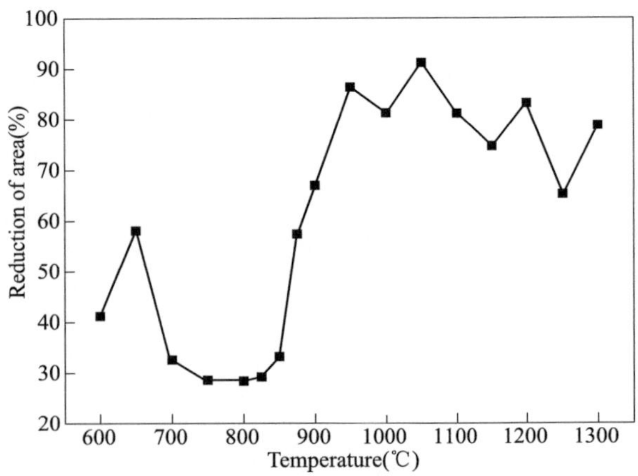

Fig. 2.58 Evolution of RA with temperature of a high strength steel

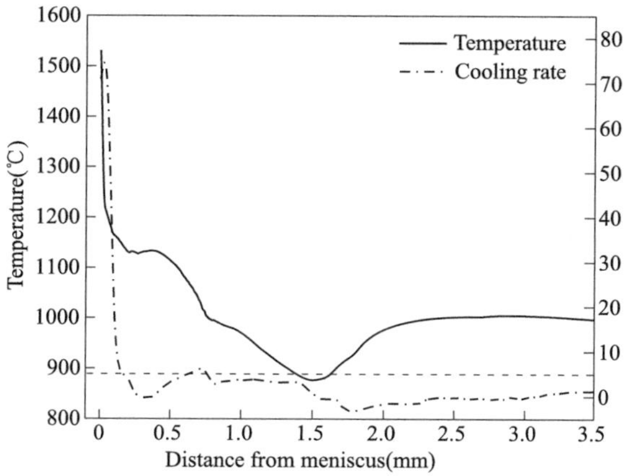

Fig. 2.59 Evolution of temperature and cooling rate of the slab corner in the mold and high temperature zone of secondary cooling

of the slab corner structure will further precipitate large quantities in the secondary cooling high temperature zone, based on precipitation in the lower part of the mold, and grow into large size Nb (C, N) in chain form, causing the ductility of the slab corner structure entering the straight segments significantly decrease.

Meanwhile, as can be seen from Fig. 2.57, the temperature of the slab corner entering the straight segments is only 860–820 °C. The micro-structure of the slab at this temperature has changed from a single austenite phase to a low ductility dual-phase structure of austenite + grain boundary ferrite, further reducing the ductility of the corner structure of the slab. During the straightening process, the slab is prone to cracking along the grain boundary due to low ductility and expands to form transverse cracks at the slab corner.

2.4.3 Mechanical Behavior of Slab in Secondary Cooling Zone

In actual slab continuous casting production, when the accuracy of the device such as arc and roller gap of continuous caster meets the requirements, the maximum force on the slab during solidification occurs in the bender, straight segments, and reduction zone of solidification end. Among them, in the bender of the continuous caster, due to the high cooling intensity of the slab, the bending process of the slab is subject to significant clamping and bending force from bow segment casting rollers but also undergoes a large thermal stress. In the straightening process and reduction process of solidification end, although the cooling intensity of the slab is relatively small and its thermal stress is not significant, with its thick solidification shell and low surface temperature, especially at the corner of the slab, the slab undergoes a strong mechanical stress. In addition, in these three key positions, the slab undergoes significant ductile deformation, so this book mainly investigates the strain and stress evolution of the slab in the bender, straight segments, and reduction zone of solidification end in the secondary cooling casting flow.

2.4.3.1 Mechanical Behavior of Slab During Bending

Figure 2.60 shows the thermal strain, ductile strain, and equivalent stress distribution of the slab in the bender during the continuous casting process of a certain high strength Ni-bearing micro-alloyed steel under the above heat transfer and casting flow roller gap conditions. As shown in Fig. 2.60a, in the bender of the slab, due to the 2D rapid cooling on the wide and narrow faces in these zones, foot roller zone, and secondary cooling zone, the temperature is lower, which leads to the thermal strain of the slab mainly concentrated in the corner. After entering the bender, due to the lack of cooling water spray cooling on the slab narrow face, the shell temperature significantly increases. However, the wide face of the slab is subject to rapid cooling water continuous spray cooling, and the temperature is lower than the narrow face, thus, the thermal strain on the slab wide face is relatively stronger than on the narrow face.

Figure 2.60b is a distribution map of ductile strain during the slab bending process. Due to the lower temperature and higher deformation resistance at the slab corner, ductile deformation is not concentrated at the slab corner but is primarily located at the solidification front near the narrow and wide face of the slab, under the casting roller clamping. This phenomenon indicates that if the design of the bender transition of the continuous caster is unreasonable or the roller gap accuracy cannot meet the requirements, the slab is prone to defects such as inner corner cracks and subsurface cracks. Zones with lower equivalent ductile strain values are primarily on the wide face of the slab, which is not under the clamping force of the bow segment casting roller, the off-corner on the wide face of the slab, and the narrow face of the slab.

Under the combined action of thermal strain and ductile strain, the equivalent stress of the slab during the bending process is shown in Fig. 2.60c. It can be seen from the fig. that the stress during the bending process of the slab is mainly concentrated at the corners, and the stress at the outer arc corner is the highest. However, according to the temperature evolution at the corner of the slab above, the temperature at the corner of the slab in the bender has gradually returned to above 950 °C from 870 °C, and the ductility of the structure at the slab corner is higher (as shown in Fig. 2.58). Therefore, under normal production conditions, the occurrence of transverse corner crack at the outer arc corner of the slab in the bender of the continuous caster is relatively low.

2.4.3.2 Mechanical Behavior of Slab During Straightening

Figure 2.61 shows the distribution of 3D ductile strain and equivalent stress field during the straightening process of a Ni-bearing high strength steel slab under the above heat transfer and continuous straightening conditions. In this straightening

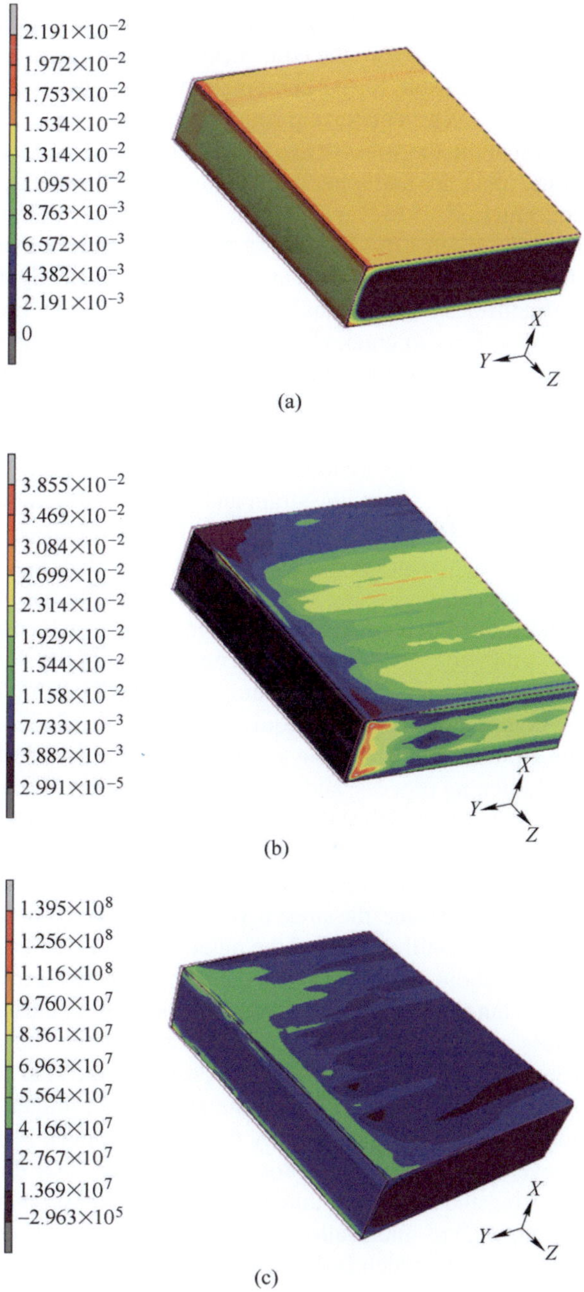

Fig. 2.60 Stress and strain fields of the slab in the bender (unit: Pa). **a** Thermal strain; **b** Ductility strain; **c** Equivalent stress

process of the slab, the ductile strain mainly occurs at the solidification front and corner. Among them, the equivalent ductile strain at the inner arc corner of the slab is the largest due to the combined effect of the clamping and straightening force of the bow segment casting rollers, as shown in Fig. 2.61a. Under the effect of this deformation, the stress of the slab is mainly concentrated at the corner and its vicinity of the inner arc and outer arc, and the stress distribution of the inner arc is the most concentrated. However, the peak stress appears at the outer arc of the slab corner, reaching about 175 MPa. We know that the stress on the inner arc of the slab during the straightening process is tensile stress. Affected by the precipitation of microalloy carbonitrides and the transition of the structure, the low ductility structure at the slab corner is easy to crack along the grain boundary due to the large and concentrated tensile stress during the straightening process, and expands into transverse corner cracks. For the outer arc of the slab, since the stress is mainly compressive stress, even if the stress peak appears in the outer arc, the occurrence of transverse corner crack is still relatively low.

Figure 2.62 shows the strain and stress contour along the casting direction of the centerline of the slab wide face, 1/4 of the width direction and the narrow face of the slab during the straightening process of a Ni-bearing high strength steel slab under the above straightening conditions. The ductile deformation of the slab is mainly concentrated at the solidification front and the slab corner, while the stress is mainly concentrated at the inner and outer arc corner of the slab, and the surface stress of the slab is the largest under the clamping force of casting roller, as shown in Fig. 2.62f.

Figure 2.63 shows the equivalent stress distribution of the slab cross section at the entrance, the clamping and non-clamping states of the casting roller during the straightening process, and at the exit of the straight segments. The stress of the outer arc corner of the slab at the entrance is greater than that of the inner arc corner, and the maximum value is 164 MPa. During the straightening process, due to the combined effect of the contact reaction force under the action of the casting roller clamping and straightening force, the stress at the inner arc corner of the slab significantly increases, but the peak still appears at the outer arc corner, and the maximum equivalent stress value is about 157 MPa. When the slab is separated from the casting roller, the stress of the inner and outer arc corners of the slab is released to a certain extent, but the stress peak at the outer arc corner is still about 100 MPa. After the slab exits the straight segment, the stress at the corner of the slab is further reduced because the inner arc and the outer arc of the slab are no longer affected by the straightening force (but are still clamped by the casting roller). From the stress evolution of the above slab straightening process, it is shown that the occurrence of intergranular cracking and expansion into cracks is the greatest when the corner of the continuous casting slab of micro-alloyed steel contacts with the casting roller at the moment of straightening, which is the main external cause of the corner crack of the slab.

Fig. 2.61 Evolution of stress and strain of the slab during the straightening process (unit: Pa). **a** Plastic strain; **b** Equivalent stress

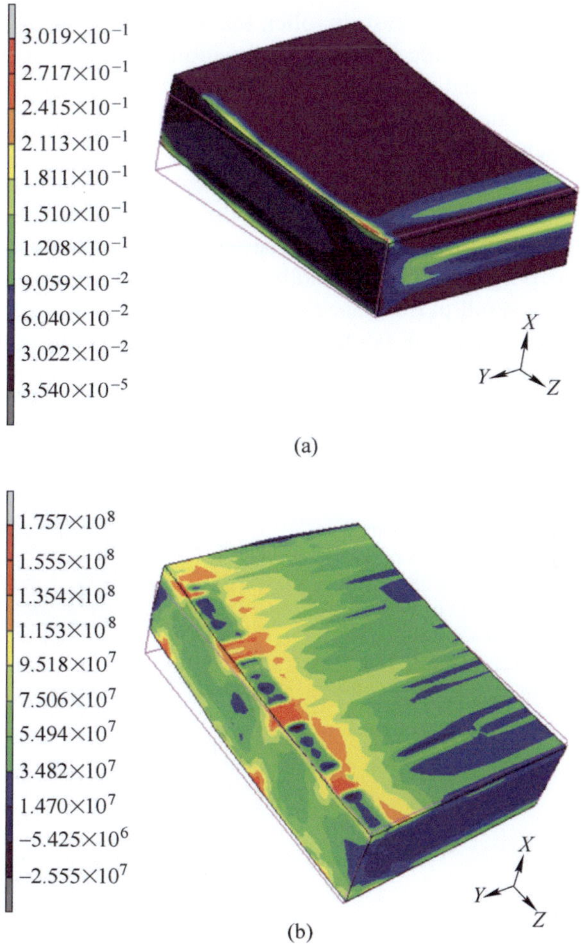

2.4.3.3 Mechanical Behavior of Slab During Reduction

In the process of continuous casting production of Nb-bearing high strength steel slab, the reduction positions are the No.8 and No.9 segments of bow segment, the corresponding casting flow is 18.0–22.0 m from the meniscus of the mold, and the total reduction is 6 mm, which is evenly distributed in two reduction segments. Figure 2.64 shows the displacement changes in the center of the inner arc wide face, the transverse 1/4 and the corner of the slab along the thickness direction during the soft reduction process at the solidification end. As can be seen from the figure, due to the periodic reduction of the bow segment casting roller, the inner arc surface of the slab is gradually reduced along the thickness direction. In this process, due to the low temperature of the slab corner, the shrinkage along the thickness direction in the casting flow is larger than that in the middle of the wide face. Under the same reduction of the casting roller, the reduction of the slab corner relative to the center of the slab wide face and 1/4 position is small. The 1/4 position of the slab width direction and the center of the slab wide face are affected by the reduction of casting roller, and the change of the reduction along the thickness direction is relatively synchronous.

Figure 2.65 shows the stress changes at the typical position of the inner arc surface of the slab under the soft reduction at the end of the solidification. In the process of slab reduction, the change of surface stress at the center of the inner arc wide face and the 1/4 position of the width direction has the same trend, and the stress is similar. The overall performance is periodic fluctuation with the contact and separation of the casting roller. For the slab corner, due to its low temperature during the reduction process, the deformation resistance significantly increases, reaching a maximum of

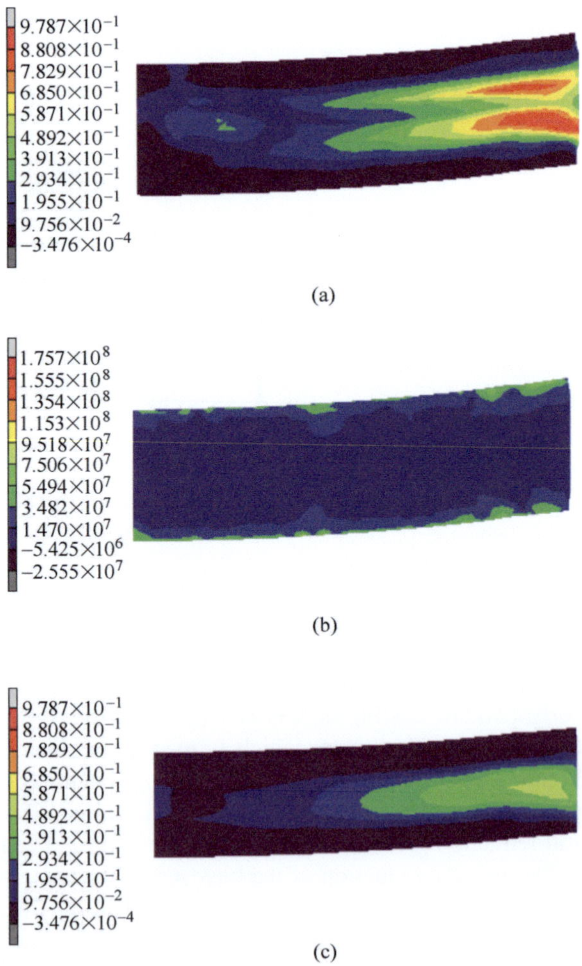

Fig. 2.62 Stress/strain contour of the slab during straightening process (unit: Pa). **a** Strain of the wide face centerline; **b** Stress of the wide face centerline; **c** Strain of 1/4 position of wide face; **d** Stress of 1/4 position of wide face; **e** Strain of the narrow face; **f** Stress of the narrow face

Fig. 2.62 (continued)

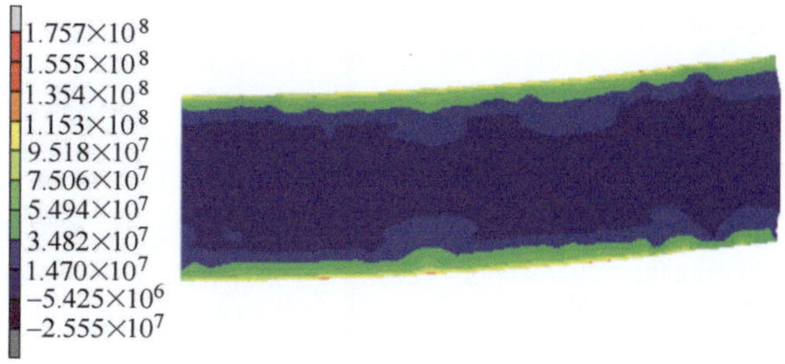

(d)

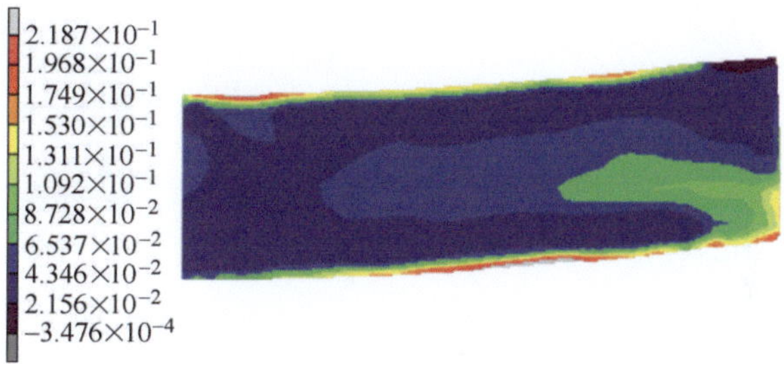

(e)

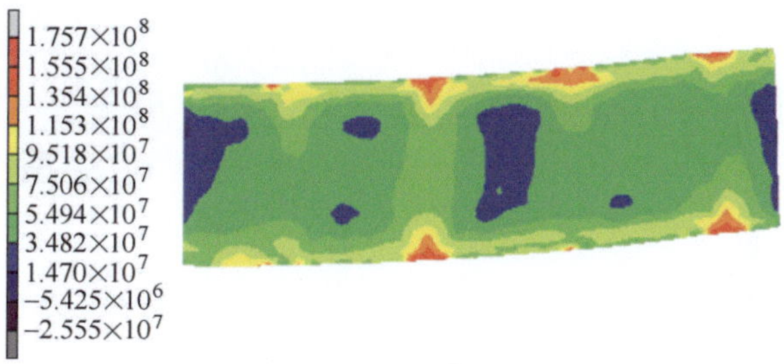

(f)

about 180 MPa. When the slab is separated from the casting roller, the stress releases, and presents a periodic fluctuation of peak and trough.

Figure 2.66 shows the distribution of stress and ductile strain in the cross section of the slab when the No.6 roller of the No.8 bow segment is reduced and when the roller is released. In the process of bow segment casting roller reduction, the ductile strain at the solidification front of the inner arc side of the slab is the largest. At the same time, due to the transverse extension deformation of the slab, the solidification front of the slab is subject to transverse tensile stress, thus it is prone to form inner crack defects. When the slab is separated from the roller reduction, although the surface of the slab has been separated from the roller reduction, it is still subject to the continuous reduction of the former and latter casting rollers. Therefore, the solidification front of the inner arc side of the slab still maintains ductile deformation, but the ductile

Fig. 2.63 Stress contour of the slab cross-section during straightening process (unit: Pa). **a** Entrance of the straight segments; **b** During the straightening process with rollers clamping; **c** During the straightening process when roller separation; **d** Exit of the straight segments

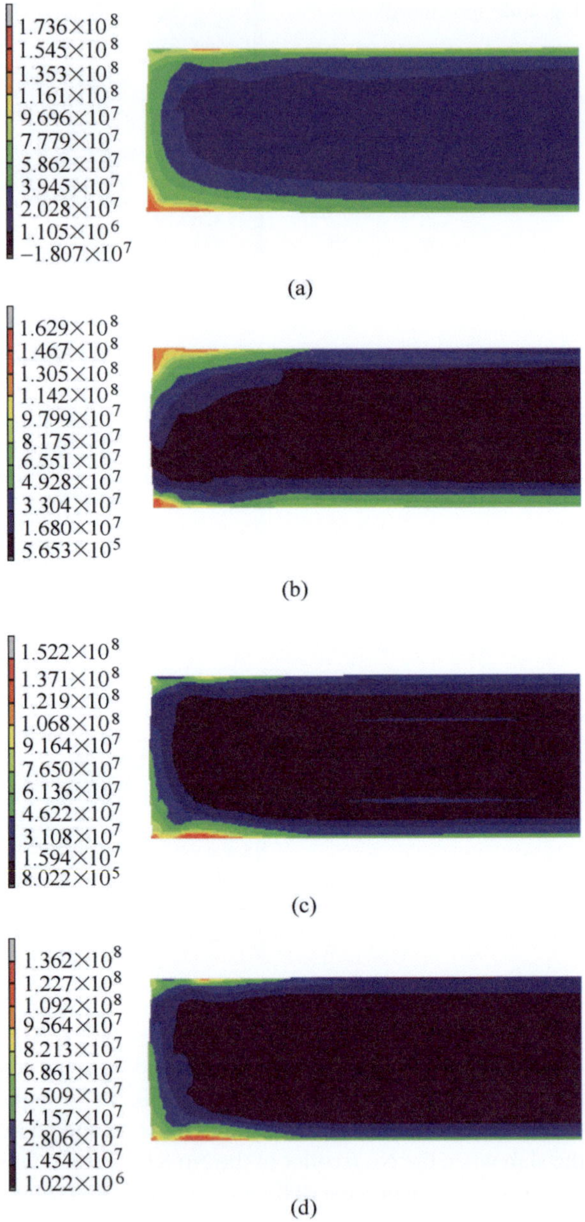

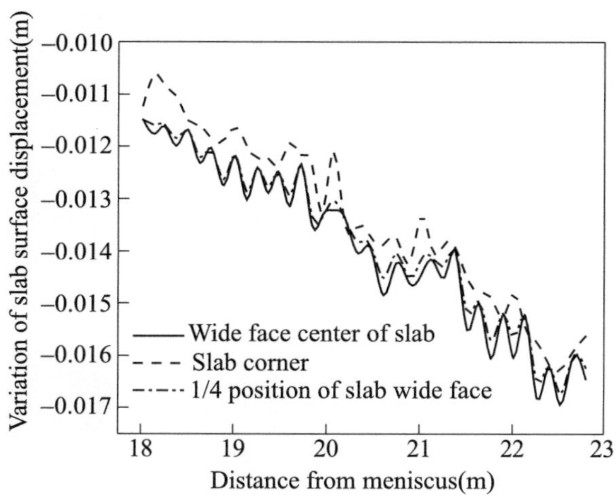

Fig. 2.64 Displacement variation of slab wide face along the thickness direction during reduction process

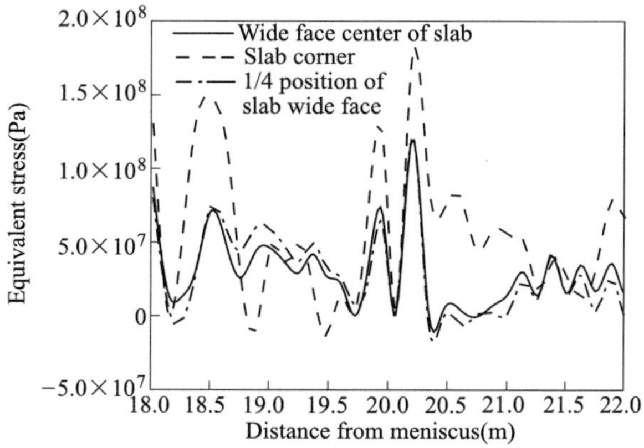

Fig. 2.65 Stress evolution of the slab during reduction process

deformation is significantly lower than the ductile strain by the reduction roller, as shown in Fig. 2.66b.

Due to the low temperature at the corner of the slab and the large deformation resistance during the reduction process, the stress distribution of the slab during the reduction process is significantly different from the ductile deformation. In the process of roller reduction and separation, the stress is mainly concentrated in the slab corner and its vicinity, and the stress in the inner arc corner is generally greater than that in the outer arc corner. At the same time, due to the implementation of the slab reduction process, when the roller and slab contacts and transmits the reduction, for the slab corner not only generates tensile stress along the casting direction, but also has a certain degree of extension deformation along the casting direction, which also generates tensile stress on the slab corner. As can be seen from Fig. 2.57, during this reduction process, the temperature at the corner of the slab is about 810–845 °C, the Nb-bearing carbonitride has basically finished precipitating, and the austenite grain

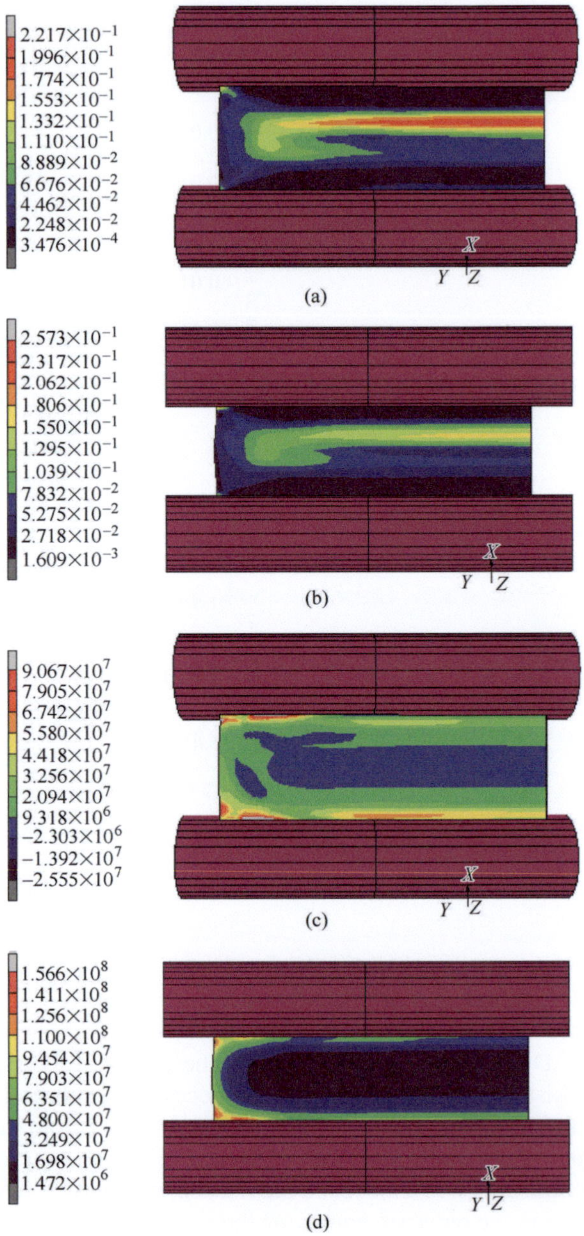

Fig. 2.66 Stress/strain during slab reduction. **a** Equivalent plastic strain during reduction with roller; **b** Equivalent plastic strain during reduction without roller; **c** Equivalent stress during roller reduction; **d** Equivalent stress during reduction without roller

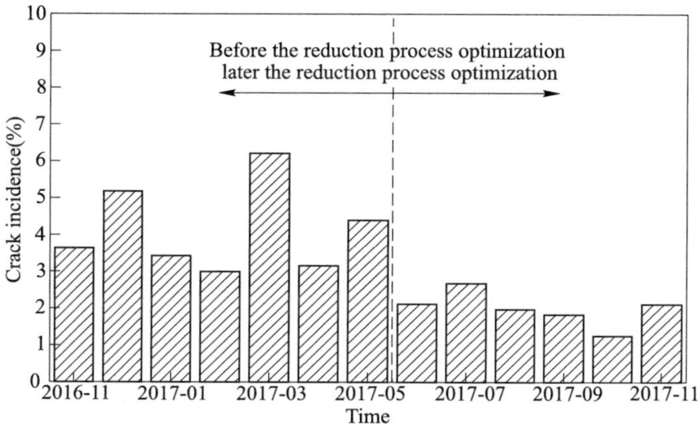

Fig. 2.67 Incidence of slab corner crack of micro-alloyed steel before and after the optimization of the soft reduction process

boundary has fully formed the proeutectoid ferrite film, and the structure of the slab corner is in the low ductility zone. At this time, a large reduction at the solidification end of the slab is carried out, especially when the reduction zone overlaps with the straight segments, it is easy to form transverse crack defects at the inner arc corner of the slab.

Figure 2.67 shows that the occurrence of corner crack in micro-alloyed steel slab is above 3.2% when No.8 segment (straight segment) and No.9 segment of a steel plant implement No.8 segment of centralized reduction and No.9 segment of maintain reduction to reduce the internal segregation of the slab. By evenly distributing the end reduction to 2 bow segments and appropriately increasing the casting speed to avoid straight segments to reduce, the corner crack rate of the micro-alloyed steel continuous casting slab is reduced to below 2.0% on average. Therefore, for the continuous casting production of micro-alloyed steel, the reduction should be evenly distributed as much as possible during the solidification end reduction process of the slab, and the overlap between the reduction zone and the straight segments should be avoided to reduce the incidence of corner cracks in the inner arc of the slab.

2.5 Development of Slab Corner Efficient Heat Transfer Mold and Its Application

According to the formation mechanism of transverse crack at the corner of micro-alloyed steel continuous casting slab, refining the austenite grain size during the solidification process of the slab corner, and dispersing the precipitation of its microalloy carbonitrides, to overall improve the hot ductility of the slab corner structure, is the key to fundamentally eliminating the formation of transverse cracks at the slab corner. Among them, the grain size of the austenite structure at the corner of the slab is limited by the solidification characteristics of the steel composition and its solidification heat transfer in the mold. For carbonitrides with higher precipitation temperatures such as Ti, Nb, B, Al, etc., according to the thermodynamic and kinetic behavior of typical microalloy carbonitride precipitation in Sect. 2.2 and the temperature evolution at the corner of the slab in the traditional conventional slab mold in Sect. 2.3, it is known that the high temperature precipitation zone of the carbonitrides in the slab corner structure is located in the middle and lower part of the mold. Increasing the cooling rate of the slab corner in the middle and lower part of the mold, making its overall cooling rate reach 5 °C/s or above, is the key to dispersing the high temperature carbonitride precipitation and refining the original austenite grains.

However, as can be known from the calculation results of the solidification thermo-mechanical behavior of the shell in the traditional conventional slab mold in Sect. 2.3, the corner and its vicinity in the middle and lower part of the mold cannot be efficiently compensated the solidification shrinkage of the shell due to the copper plate, resulting in the concentrated distribution of thick air gap and mold flux film at the corner and its vicinity of the solidified shell, which is the key to severely hindering the efficient heat transfer at the slab corner. Therefore, developing an internal cavity structure of the mold can efficiently compensate for the solidification of the shell, ensuring that the slab contact with the copper plate throughout the solidification process, uniformly and significantly reducing the distribution of air gap and mold flux film at the corner and its vicinity of the solidified shell, is the fundamental way to increase the heat transfer rate at the corner of the mold. For this reason, based on this slab solidification control idea, this book proposes a new type of corner high efficiency heat transfer novel convex-structure mold.

2.5.1 *Design of Novel Convex-Structure Mold*

As can be known from the calculation results of the solidification thermo-mechanical behavior of the shell in the traditional slab mold in Sect. 2.3, the shrinkage of the slab in the mold during the solidification process is large in the width direction, and it does not vary uniformly from upper to lower part along the mold, overall showing the characteristics of fast solidification shrinkage in the upper part of the mold and slow shrinkage in the middle and lower part. Affected by this, the copper plate of traditional flat-plate mold cannot compensate for the real-time shrinkage and deformation of the solidified shell, resulting in a larger gap in the narrow face corner and its vicinity of the slab. At the same time, the shell lacking copper plate support on the narrow face gradually forms a "bulging" shape under the ferrostatic pressure of the molten steel. As the slab continues to move downward, due to the slowdown of solidification shrinkage, it causes the extrusion effect between narrow face and copper plate, causing the overall harder solidified slab corner to twist toward the center of the narrow face of the slab, exacerbating the gap formation in the wide face corner

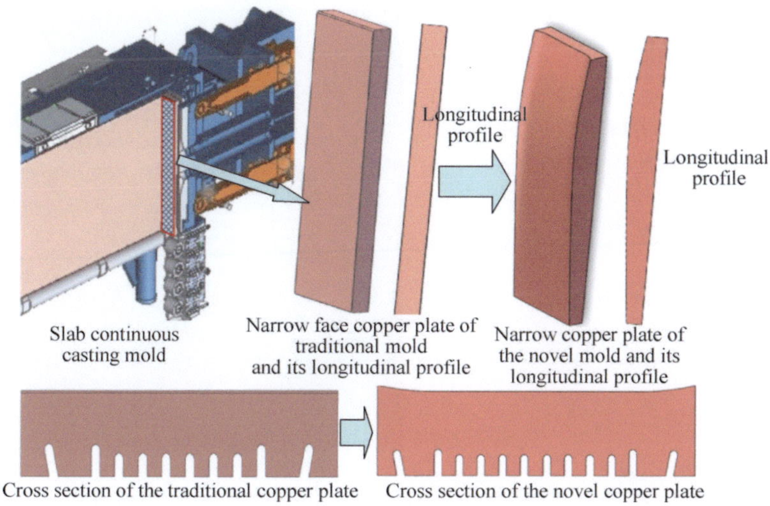

Fig. 2.68 Schematic diagrams of the novel convex-structure mold and the traditional flat-plate mold

and its vicinity of the slab in the middle and lower part of the mold. For this reason, according to the shrinkage and deformation characteristics of the solidified shell in the mold, this book has developed a narrow face copper plate curved inner cavity compensation structure mold that forms a fast compensation in the upper part, slow compensation in the middle and lower part, and efficiently conforms to the shrinkage and deformation of the shell, as shown in Fig. 2.68. Because this mold can effectively improve the heat transfer of the slab corner, we call it novel convex-structure mold.

According to the shrinkage of the shell at different parts of the narrow face of the slab in the mold at different heights along the center direction of the wide face (the shrinkage of the corner and middle area of the narrow face of the shell in the upper part of the mold is large, the shrinkage in the lower part is small, and the shrinkage of the corner is generally larger than that of the middle area) and the wear characteristics of the lower part of the narrow face copper plate of the mold in the practical continuous casting production process of slab (in the practical continuous casting production, the wear of the narrow face copper plate at the lower part of the mold is generally severe, as shown in Fig. 2.69), the narrow face copper plate of the mold is divided into three parts, which consists of the center and two edges, and the central part and edge parts are designed along the height direction of the mold to match the upper fast compensation, middle and lower slow compensation of the shell solidification characteristics, respectively, corresponding to the compensation for the solidification shrinkage of the middle and corner areas of the narrow face of the slab toward the center of wide face.

Among them, the compensation curve of the middle area of the copper plate from upper to lower is given by the dynamic shrinkage value of the solidified slab in Sect. 2.3.2, as shown in Fig. 2.40. The compensation curve of the edge area of the

Fig. 2.69 Wear morphology of the lower part of the narrow face copper plate of mold

copper plate (corresponding to the compensation of the corner area of the narrow face of the shell) is set from upper to low according to the principle of more compensation first and less compensation later. The transition from the edge area to the middle area of the copper plate is achieved by gradually decreasing the maximum compensation amount along the edge to the compensation amount of the middle area, making the inner surface of the copper plate slightly concave in the transverse direction, as shown in the vertical view in Fig. 2.68. Depending on the thickness of the slab produced, the boundary line between the middle and edge areas of the copper plate is usually designed to be 30-45 mm from the edge of the narrow face copper plate.

Taking a certain steel company as an example, the main structure of the narrow face copper plate of the corner high efficiency heat transfer mold is shown in Fig. 2.70. On the transverse surface of this copper plate, it is divided into three parts: the middle area and edge areas, with a boundary 35 mm from the edge. In the direction of the height of the mold, the copper plate is divided into 45 parts from the mold upper to the lower, each part being 20 mm high, and each height of the copper plate edge and middle area is given a curved compensation value. The copper plate production process is based on the compensation values of the edge and middle area copper plates at different heights, and the curved structure is made from upper to lower and from the edge to the middle area using a computer numerical control (CNC) machine tool.

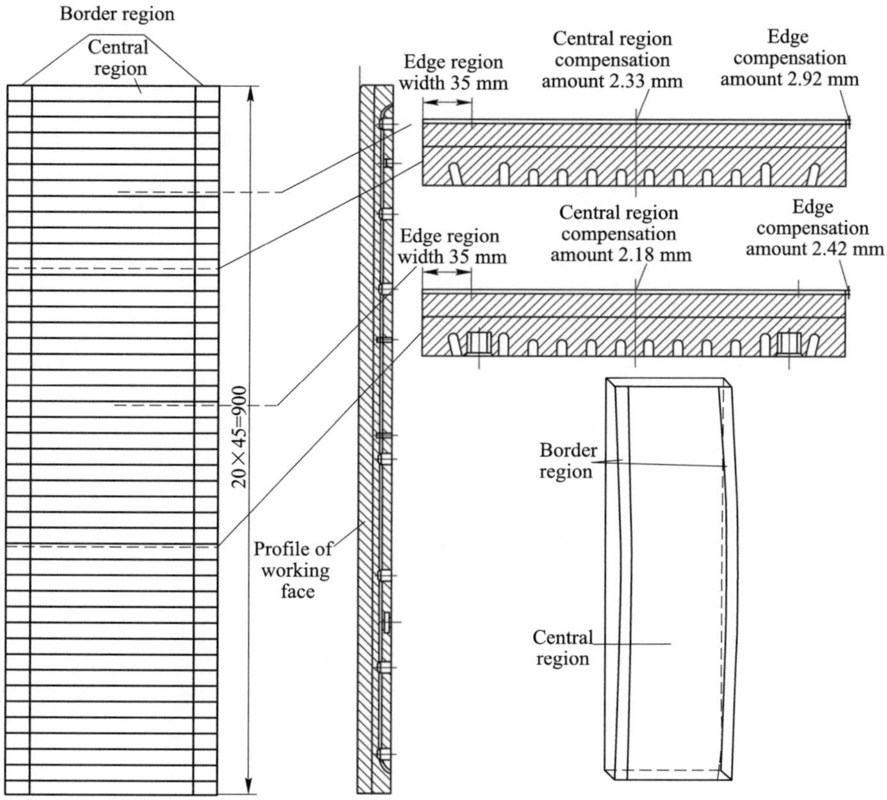

Fig. 2.70 Machining drawing of the novel convex-structure mold copper plate

2.5.2 Thermo-Mechanical Behavior of Shell in Novel Mold

2.5.2.1 Shell Deformation

Figure 2.71 shows the deformation of the solidified shell at different heights under the meniscus during the solidification process of a high strength steel component shown in Sect. 2.3 in a novel convex-structure mold. Compared with the traditional flat-plate mold, the novel convex-structure mold provides more sufficient compensation for the narrow face of the slab throughout the solidification process. Especially in the upper and middle part of the mold, the narrow copper plate is closely attached to the shell. Only when the slab solidifies to the lower part of the mold, a slight gap appears at the corner and off-corner of the narrow face of the slab, thus preventing excessive wear of the lower part of the narrow copper plate of the mold.

At the same time, because the narrow side of the slab during the solidification process has received a good shrinkage compensation effect from the copper plate, it restricts the corner of the slab from twisting toward the center of its narrow side,

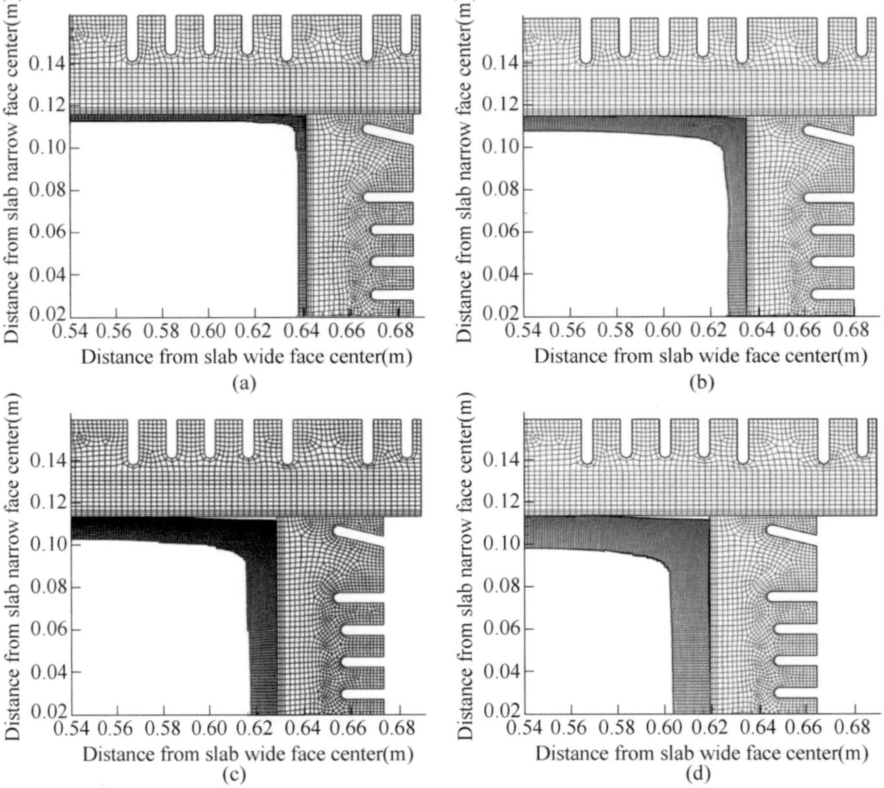

Fig. 2.71 Deformation of the slab at different heights below the meniscus in the novel convex-structure mold. **a** 100 mm; **b** 300 mm; **c** 500 mm; **d** Mold exit

and the solidification shrinkage gap at the corner of the wide side of the slab is significantly reduced compared to the traditional flat-plate mold. The evolution of the shrinkage gap along the height direction of the mold at the corner of the wide and narrow faces of the slab during the solidification process is shown in Fig. 2.72. It can be seen that the gap between the corners of wide and narrow face, and the corresponding copper plate is small, thereby effectively inhibiting the concentration of flux film and air gap in the corner of the mold and its vicinity, achieving efficient heat transfer at the corner of the slab.

2.5.2.2 Air Gap Distribution

Figure 2.73 shows the air gap distribution at the corners of wide and narrow face inside the mold during the production process of micro-alloyed steel slab under the novel convex-structure mold. It can be seen that under the structure of the novel convex-structure mold, the air gap at the corner of the wide face of the shell is

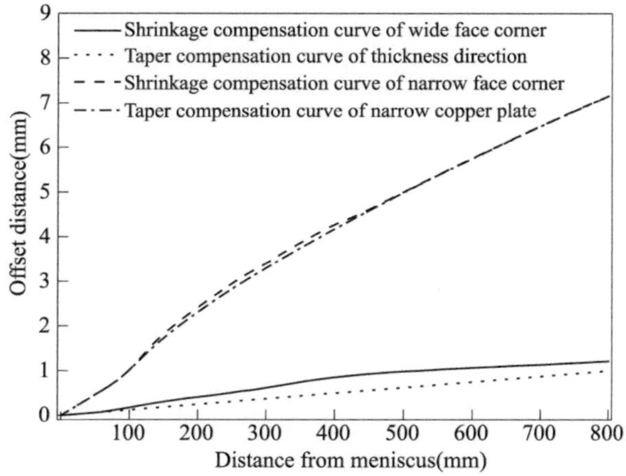

Fig. 2.72 Distribution of solidification shrinkage at the slab corners in the novel convex-structure mold

significantly reduced, with the maximum thickness decreasing by 65% compared to the traditional flat-plate mold, and the distribution trend of the air gap has also changed significantly. It shows a trend of first increasing and then decreasing from the meniscus of the mold to its exit, with the maximum value appearing at 450 mm under the meniscus. At the same time, it can be seen that the air gap at the corner of the narrow face of the shell is basically eliminated, thus the heat transfer efficiency of the corner of the shell will be greatly improved.

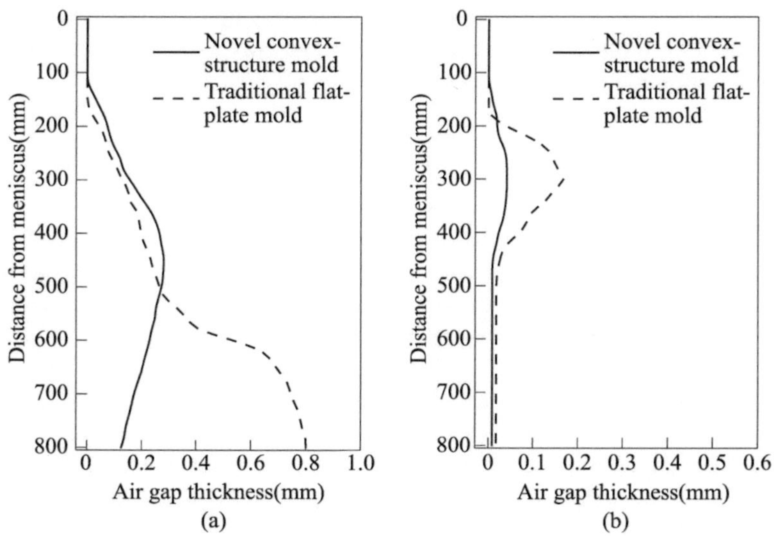

Fig. 2.73 Distribution of air gap thickness in the vicinity of the corner of the solidified shell in different mold. **a** Wide face; **b** Narrow face

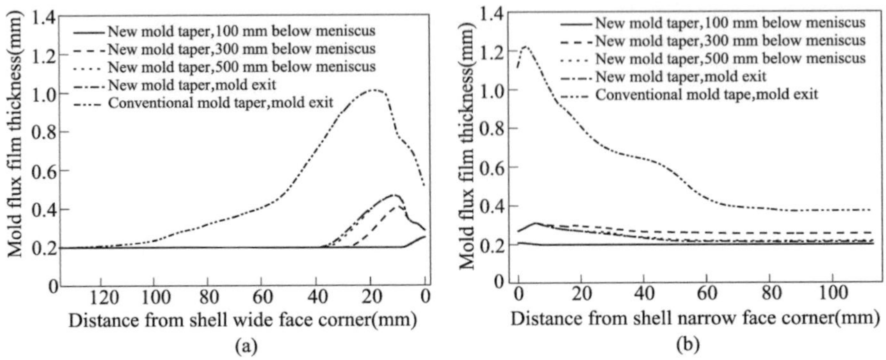

Fig. 2.74 Mold flux film distribution around shell corner in different mold. **a** Wide face; **b** Narrow face

2.5.2.3 Mold Flux Film Distribution

Figure 2.74 shows the thickness distribution of mold flux at different heights under the conditions of the novel convex-structure mold. It can be seen that when producing certain high strength steel with the novel convex-structure mold, the thickness of mold flux at the corners of the shell is significantly reduced, and the corresponding maximum values of thickness of mold flux is also reduced to 46.5 and 24.4% of the maximum thickness of the flux film under the traditional flat-plate mold. In addition, the distribution of the flux film along the circumference of the mold is more uniform, and the concentrated distribution zone near the corner is significantly reduced, which further improves the heat transfer conditions at the corners of the shell, providing conditions for rapid heat transfer at the corner of the slab.

2.5.2.4 Heat Flux of Distribution

Figure 2.75 illustrates the heat flux distribution at the interface between shell and copper plate at the center and off-corner of the wide and narrow faces, as well as at the corner of the slab during the continuous casting process of a specific high strength steel using the above novel convex-structure mold. The distribution trend of heat flux density along the height direction of the interface between shell and copper plate within the novel convex-structure mold is generally similar to that of the traditional mold, decreasing from the meniscus to the mold exit. However, compared with the traditional flat-plate mold, the heat flux distribution on the narrow side of the shell under the novel mold is more uniform. The heat flux at the center and off-corner of the narrow face in the middle and lower part of the mold has significantly increased. Specifically, at the corner of the slab, the heat flux in the middle and lower part of the mold has nearly doubled, which greatly accelerates heat transfer at the corner. This effectively inhibits the issue of corner reheat during the continuous casting process

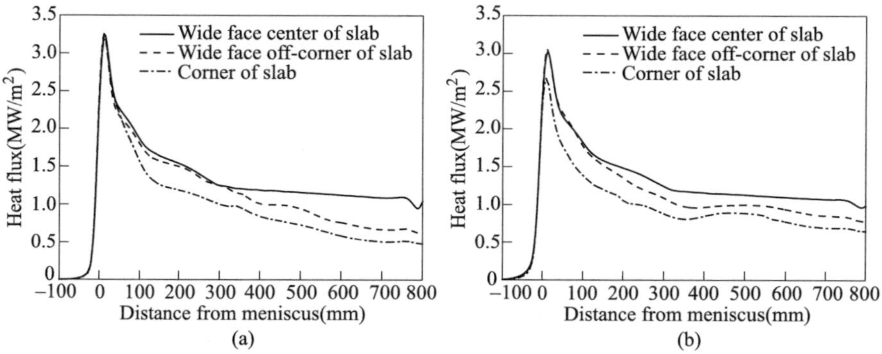

Fig. 2.75 Heat flux distribution at the interface slab between copper plate in the novel convex-structure mold. **a** Wide face; **b** Narrow face

with traditional mold, thereby preventing the coarsening of the austenite structure at slab corner and inhibiting the precipitation of large size carbonitrides along the austenite grain boundaries at high temperatures.

2.5.2.5 Shell Surface Temperature

Figure 2.76 illustrates the lateral distribution of surface temperature at various heights of the solidified shell during the continuous casting production of a specific high strength steel slab under the conditions of the novel convex-structure mold. As observed, under the conditions of the novel convex-structure mold, due to the obvious acceleration of heat transfer rate at the corner of the solidified shell, the surface temperature at the slab corner drops to approximately 800 °C at the mold exit, achieving rapid cooling at slab corner. Additionally, after adopting the novel mold, the elimination of the concentrated air gap and flux film in the off-corner of the wide and narrow faces, the disappearance of "hot spot" in these areas. This improvement promotes solidification of high ductility of the surface structure at the off-corner, and prevents the occurrence of off-corner defects in thick slab.

Figure 2.77 illustrates the evolution of the temperature and cooling rate at the slab corner under the traditional mold and the novel convex-structure mold along the height direction of the mold. As it can be seen that under the cooling conditions of the novel convex-structure mold, the cooling rate of the solidified shell corner in the temperature range between the precipitation temperature of TiN and Nb (C, N) or BN and other carbonitrides can exceed 60 °C/s. The cooling rate at the slab corner from the starting precipitation temperature position of Nb (C, N) or BN and other carbonitrides (150 mm below the meniscus) to the mold exit reach above 10 °C/s throughout, significantly exceed the minimum requirement of 5.0 °C/s for effective dispersion of carbonitrides in micro-alloyed steel slab. Consequently, this enhanced cooling regime theoretically facilitates the dispersion of microalloy carbonitrides

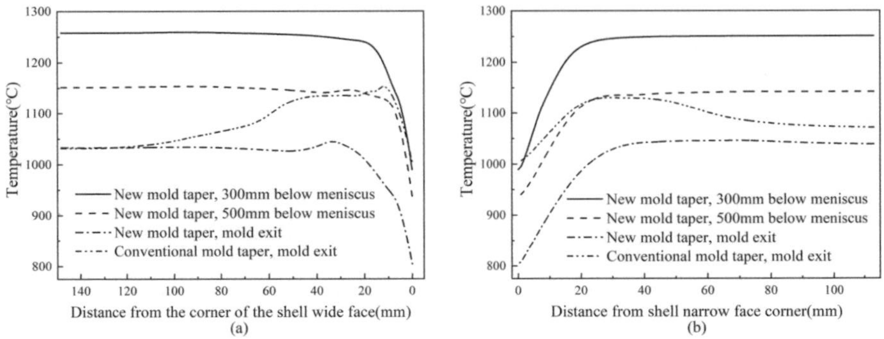

Fig. 2.76 Surface temperature distribution near the solidified shell corner under the novel convex-structure mold. **a** Wide face; **b** Narrow face

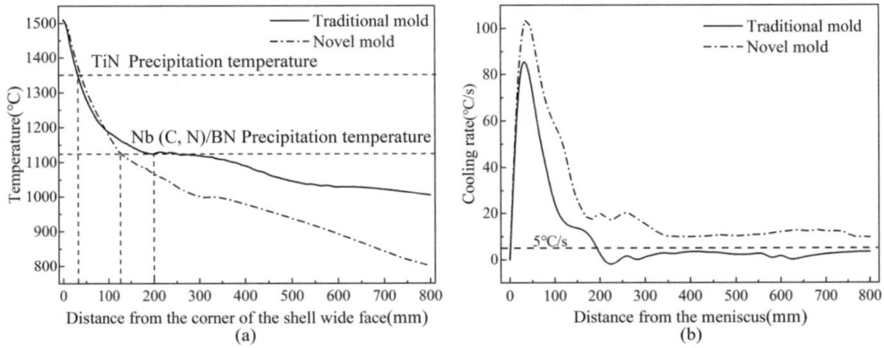

Fig. 2.77 Evolution of temperature and cooling rate of the solidified shell corner in different mold. **a** Temperature; **b** Cooling rate

within the slab corner structure and the refinement of the original austenite grains formed during solidification at slab corner.

2.5.3 Application of the Novel Mold and Its Effect

As the cooling rate at the corner of the mold increases, its intuitive effect is the precipitation distribution of microalloy carbonitrides and the change of grain size. Figure 2.78 shows the morphology of the precipitations at 5 mm below the surface of the slab corner during the continuous casting production of the above high strength steel (Nb, Ti-bearing) by adopting the traditional flat-plate mold and the novel convex-structure mold. In the continuous casting production process of the traditional flat-plate mold, due to the formation of thick air gap and thick flux film in the corner of the mold and its vicinity, the efficient heat transfer in the corner of the slab

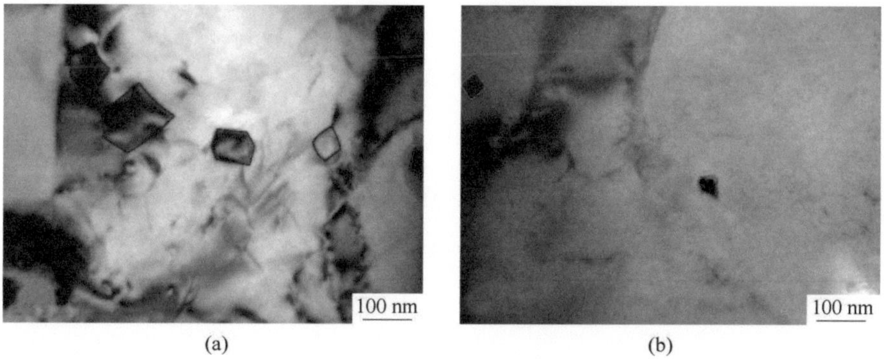

Fig. 2.78 Morphology of micro-alloyed carbonitrides precipitation of the slab corner produced by different mold. **a** Traditional flat-plate mold; **b** Novel convex-structure mold

is seriously hindered, resulting in the chain and large size concentrated precipitation of carbonitride at the austenite grain boundary. The average size of the precipitates reached 87.6 nm, as shown in Fig. 2.78a. However, when adopting the novel convex-structure mold, its narrow face is attached to the shell during the whole solidification process of the slab, and the heat transfer is efficient, which significantly increases the cooling rate in the corner of the slab to more than 5 °C/s, thereby significantly refining the precipitation size of the microalloy carbonitrides. The size of the precipitation at 5 mm below the slab corner is only about 38.7 nm, and the distribution is dispersed, as shown in Fig. 2.78b. This will significantly improve the ductility of the slab corner structure.

From the perspective of grain refinement, Figs. 2.79 and 2.80 are the production process of traditional flat-plate mold and the novel convex-structure mold, respectively, (the production conditions of the contrast sample are only in the different mold and other continuous casting processes are the same), and the room temperature casting micro-structure at 3 and 5 mm under the corner structure of the slab. Among them, the right fig. is the grain boundary outline fig. of the metallographic structure of the left fig. It can be seen from the fig. that the high strength steel is continuously cast by the traditional flat-plate mold. Due to the dynamic reheat or slow heat transfer at the corner of the slab in the middle and lower part of the mold, the ferrite grain transformed from austenite is relatively large, the average ferrite grain size at 3 mm below the corner of slab is 73.7 μm. However, the average size of ferrite grain at the same position of the slab corner is only 35.5 μm, and the average size of grain is reduced by 51.9% under the same conditions of other production processes, and the degree of refinement is more than one time. Similarly, the size of ferrite grains at 5 mm under the corner of the slab is also refined from 75.1 μm in the traditional flat-plate mold to 37.0 μm in the novel convex-structure mold, the degree of refinement is also 50.7%, and the degree of refinement is also more than one time.

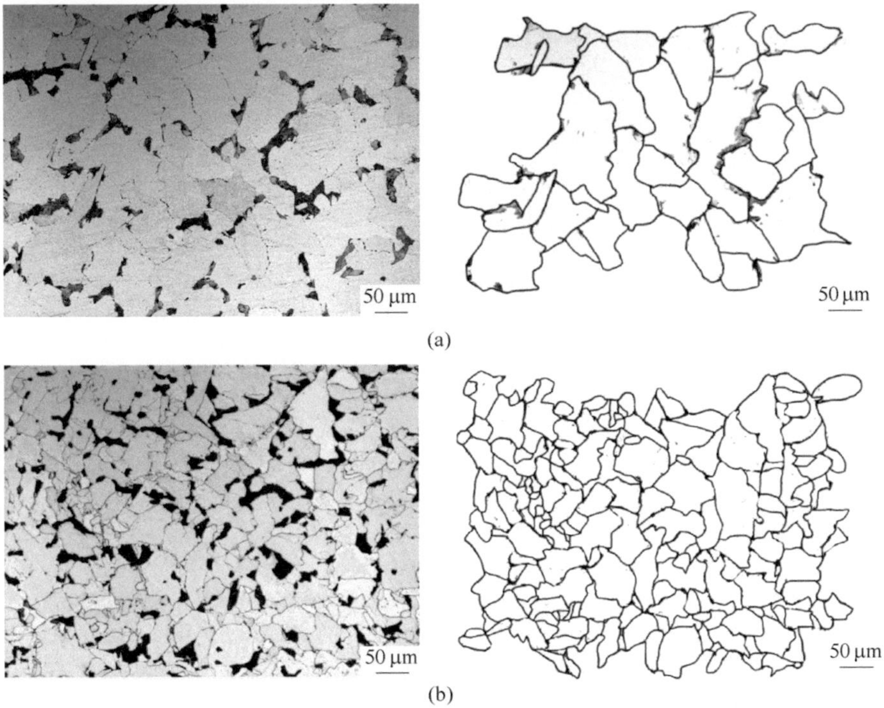

Fig. 2.79 Comparison of metallographic structures in the position of 3 mm below the slab corner by using the different mold. **a** Traditional flat-plate mold; **b** Novel convex-structure mold

2.6 Development of Slab Corner Grain Refine Technology and Its Application

The previous chapters of this book mentioned that the size of the austenite grains on the slab surface is one of the key indexes determining whether it forms crack. If the slab corner of the micro-alloyed steel can form fine austenite grains during the solidification process, on the one hand, the fine solidification structure will significantly increase the hot ductility of the steel; On the other hand, the atomic arrangement of the fine austenite grain boundaries can intensify the complexity of its dislocation and other structures. When a crack breaks through the grain boundary to another grain, the fine grain structure can effectively hinder the expansion of the crack, thereby significantly reducing the crack sensitivity of the slab corner. At the same time, if we want to refine the grain structure of the slab corner, it must be accompanied by rapid cooling behavior during its high temperature solidification process, which can also disperse the precipitation of microalloy carbonitrides in the grain and at grain boundaries, further strengthening the grain boundary of the slab corner structure, thereby producing a slab corner structure with high crack resistance. Therefore, researching and developing control process and device technology for refining the grain structure

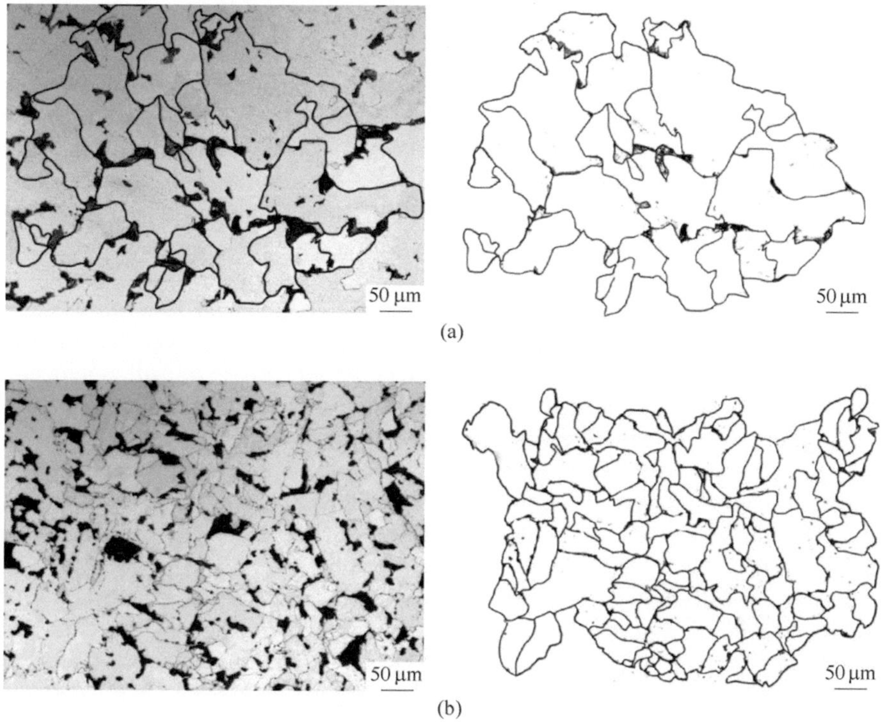

Fig. 2.80 Comparison of metallographic structures in the position of 5 mm below the slab corner by using different mold. **a** Traditional flat-plate mold; **b** Novel convex-structure mold

of the slab corner is another key measure to controlling the crack of the slab corner of the micro-alloyed steel.

However, under the current slab continuous casting process conditions, due to the lack of dynamic recrystallization behavior during the reduction process with large deformation, the refinement of the surface grain structure of the slab mainly depends on the refinement of the initial solidification grains (or preventing high temperature coarsening) and reorganization refinement of phase transition grain. At present, in the research on the refinement of the initial solidification grains of the steel structure, it is mainly focus on the initial solidification cooling rate and the addition of microalloy components. Among them, the research on the mechanism of refining the initial solidification structure grains of steel by rapid cooling has been relatively systematic [77–79]. Practice has also proved that significantly increasing the supercooling of steel solidification can effectively achieve the refinement of the initial austenite grains on the surface of the slab. However, limited by the complex dynamic thermo-mechanical behavior of the slab corner in the mold, even under the efficient heat transfer conditions of the novel convex-structure mold, the refinement of the grain structure of the slab corner is relatively limited, especially the depth of the grain structure refinement is only concentrated at about 0–8 mm below the surface of the slab corner.

In terms of research on the refinement of steel structure grains (or preventing high temperature coarsening) by adding microalloy elements, it is main focus on the precipitation behavior of dual-phase particles after the solidification of the steel structure to prevent the high temperature coarsening behavior of the casting structure [80, 81], and the control of the structure during the rolling and heat treatment process [82, 83]. In the research on phase transition grain refinement, Grange et al. [84] first used the phase transition method of cyclic rapid heating and cooling to achieve the refinement of the casting steel structure grains under laboratory conditions in 1966; Sagaradze [85] based on this control idea, prepared ultrafine grain steel structure; Lee et al. [86] also used this idea to control the phase transition temperature of the steel structure heated to austenization during the heat treatment process of hypopearlitic steel, achieving its surface structure refinement. Domestic scholars have also researched on grain refinement behavior of dual phase transition. However, apart from the reported application of related technologies by Japanese companies such as Kashima Steel, there is no stable application of process and device technology in China yet.

Section 2.6 will study the temperature and cooling rate conditions for the ultrafine grain control cooling of the micro-alloyed steel casting structure, by detection and analysis of the phase transition behavior of the slab at high temperature. Based on this, combining the corner temperature field evolution law of the high temperature solidification process of slab, a new process and device technology based on the ultrafine grain control cooling of the secondary cooling high temperature zone will be developed and implemented.

2.6.1 Determination of Slab Corner Dual Phase Transition Cooling Process

The grain size of the micro-structure and the change of high temperature RA are important indexes to measure the high temperature ductility of steel. In practical applications, the detection of RA is the most commonly used. In the actual detection process, the high temperature RA of the steel structure changes with the controlled cooling process. Specifically, for the current typical composition of micro-alloyed steel structure high temperature $\gamma \rightarrow \alpha \rightarrow \gamma$ dual phase transition ultrafine grain controlled cooling, detection shows that the higher the cooling rate and reheating rate, the better the effect of steel structure grain refine. However, in the actual continuous casting production process, the cooling and reheating rate of the slab in the high temperature zone cannot be maximized ideally. At the same time, excessive cooling and reheating rates also increase the occurrence rate of cracks in the slab due to excessive thermal stress. Therefore, in the actual implementation of grain refine technology based on high temperature dual phase transition of steel structure, on the

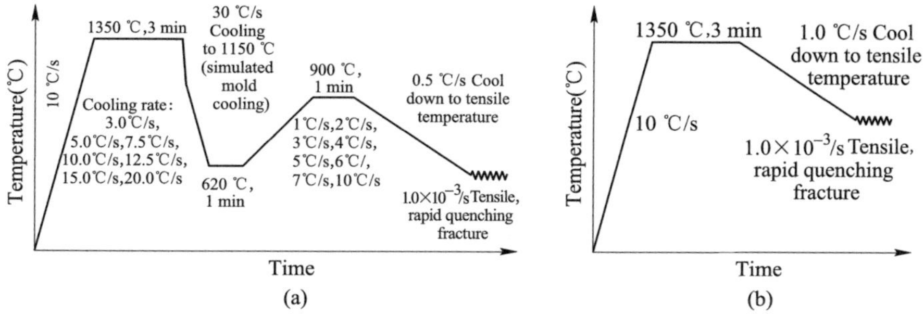

Fig. 2.81 Cooling process of thermal simulation experiment. **a** Dual phase transition cooling; **b** Convention cooling

basis of meeting the effect of grain refine and significantly improving hot ductility, combined with the actual production of continuous casting, the lower cooling rate and reheating rate controlled cooling parameters should be adopted as far as possible.

Therefore, we regard the cooling and reheating rate controlled cooling parameter as the critical controlled cooling process for grain ultrafine. In order to determine the critical controlled cooling process, Gleeble thermal simulation and fracture scanning are adopted to detect the high temperature tensile test of cooling rate of 3.0, 5.0, 7.5, 10.0, 12.5, 15.0, and 20.0 °C/s and reheating rate of 1.0, 2.0, 3.0, 4.0, 5.0, 6.0, 7.0, and 10.0 °C/s. The curve of temperature control is shown in Fig. 2.81a. Therefore, the critical controlled cooling process of grain ultrafine of the above typical Nb-bearing high strength steel composition is determined to be the combination of cooling to 620 °C at a cooling rate of 7.5 °C/s and reheating to above 900 °C from 620 °C at a reheating rate of 5.0 °C/s.

Figure 2.82 shows the evolution of the high temperature RA under the dual phase transition critical cooling process and the traditional process. When the traditional controlled cooling process is adopted, there is a brittle zone where RA is less than 40% at the temperature range of 770–830 °C (generally the 4% shrinkage rate of cross section is considered the critical point for crack formation). When the critical controlled cooling process of dual phase transition is adopted, at the temperature range of 775–850 °C, except for a slight decrease around 825 °C, the high temperature shrinkage of cross section of the steel structure significantly increases and the lowest value is 72.4%, It significantly improves the ability of the slab to resist crack formation during over straightening. The morphology of the tensile fracture of the samples under different tensile processes at the temperature range of 775–900 °C is shown in Fig. 2.83.

Figure 2.84 shows the structural evolution and actual images of the high temperature steel structure in the thermal simulation process. As can be seen, if the high temperature micro-alloyed steel structure can be cooled at a rate of not less than 7.5 °C/s to the complete ferrite temperature, and then reheated at a rate of not less than 5.0 °C/s to the complete austenite temperature, the high temperature cast austenite grain will be refined to about 1/8 of the original austenite grain. According to the hot

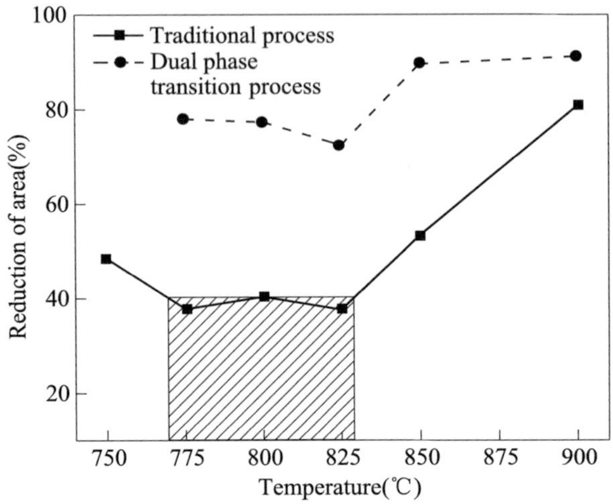

Fig. 2.82 Comparison of RA evolution under different cooling process

ductility curve and metallographic test of the steel, the critical cooling conditions for dual phase transition of typical Nb-bearing high strength steel can be determined: rapid cooling to 620 °C at a rate of 7.5 °C/s, and then rapid reheating to above 900 °C at a rate of 5.0 °C/s, thus achieving ultrafine control of steel structure grain. In addition, the critical cooling conditions of multiple steels containing microalloy components such as Al, V for dual phase transition are concentrated in rapid cooling to 600–620 °C at a rate of 7.0–8.0 °C/s, and then reheating to above 900 °C at a rate of 3.5–5.0 °C/s. Therefore, in the actual continuous casting production, the process of dual phase transition grain ultrafine and controlled cooling of the corner structure in the high temperature zone of the secondary cooling of the slab is implemented, the cooling conditions of the slab corner are uniformly set to be rapidly cooled to 600 °C at a cooling rate of more than 8.0 °C/s and then cooled to above 900 °C at a cooling rate of more than 5 °C/s.

2.6.2 Development of New Cooling Device

As can be known from the above test results, to achieve the grains of the slab corner ultrafine in the high temperature zone of secondary cooling through the dual phase transition, it is necessary to ensure the structure rapid cooling of slab corner to complete the ferrite transition, and then quickly reheat above 900 °C to complete the dual phase transition of $\gamma \rightarrow \alpha \rightarrow \gamma$, thereby achieving the grain refinement. According to the evolution of the slab corner in the secondary cooling casting flow of the continuous caster in Sect. 2.4, practice shows that the slab corner can only be reheated at a rate of 5.0 °C/s or higher under the condition that the shell is thinner in the foot roller and bender of the continuous caster. However, the slab temperature in this area is relatively high. Under the traditional slab continuous caster cooling structure,

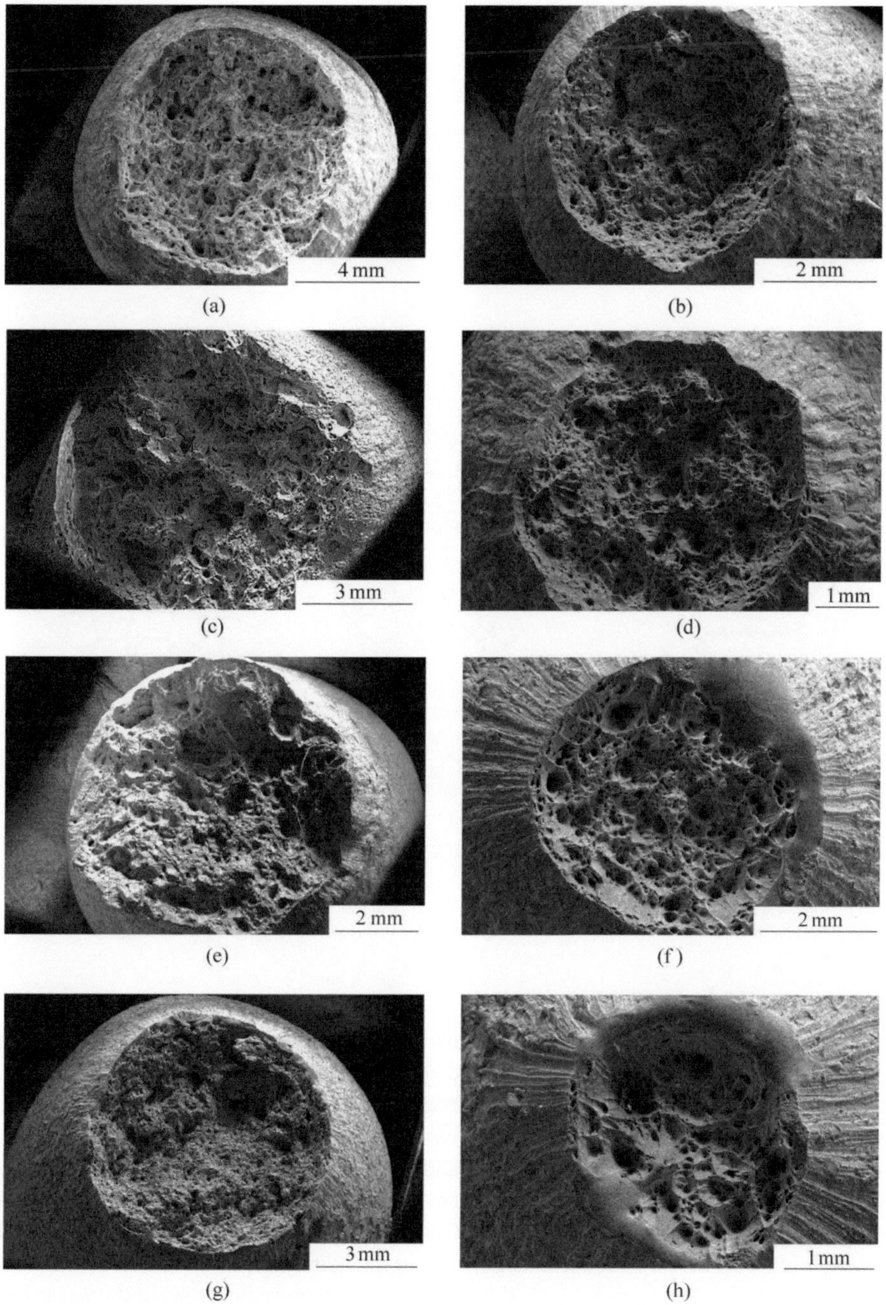

Fig. 2.83 Morphologies of sample fracture under different cooling process. **a** Traditional cooling process at 775 °C; **b** Dual phase transition cooling process at 775 °C; **c** Traditional cooling process at 800 °C; **d** Dual phase transition cooling process at 800 °C; **e** Traditional cooling process at 825 °C; **f** Dual phase transition cooling process at 825 °C; **g** Traditional cooling process at 850 °C; **h** Dual phase transition cooling process at 850 °C; **i** Traditional cooling process at 900 °C; **j** Dual phase transition cooling process at 900 °C

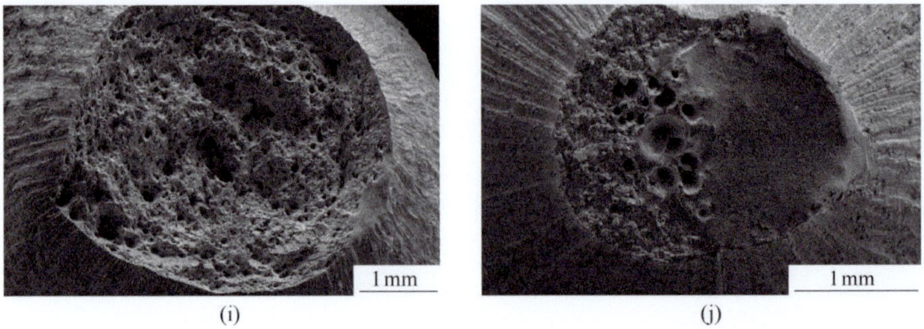

Fig. 2.83 (continued)

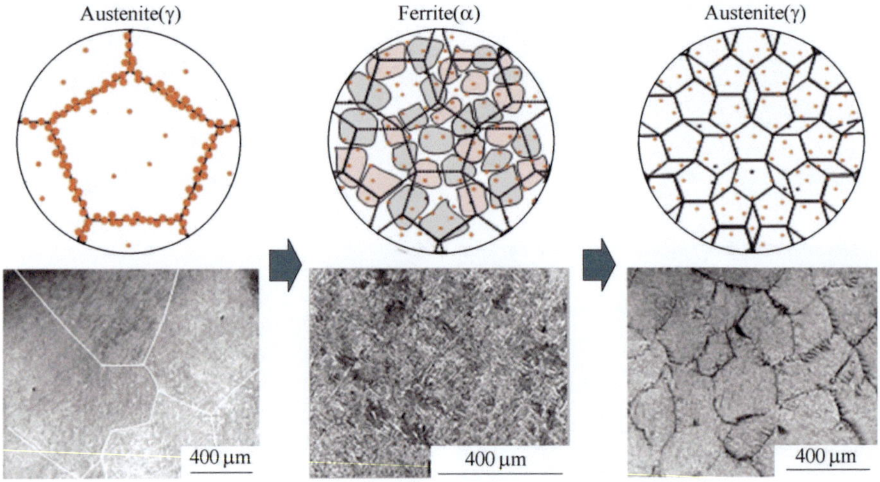

Fig. 2.84 Schematic diagrams of the micro-structure evolution during dual phase transition cooling process

by significantly enhancing the spray intensity of the foot roller zone of the wide face and narrow face, and the second and third zones of the wide face secondary cooling, to quickly reduce the slab corner temperature, on the one hand, because the cooling zone is short and the increasing cooling water has a relatively limited effect on the slab corner, it is difficult to achieve rapid and significant cooling of the slab corner; on the other hand, using the slab surface overall ultra-strong cooling process, practices in Baowu Steel and Pan Steel show that this process is prone to form longitudinal crack defects on the slab surface. Therefore, in the actual implementation of the dual phase transition cooling process of the slab corner structure in the secondary cooling high temperature zone, the ultra-strong cooling control of the slab corner is a limiting link.

In order to ensure rapid cooling and sufficient reheating of the slab corner in the foot roller and bender of the mold, while minimizing the impact of the ultra-strong

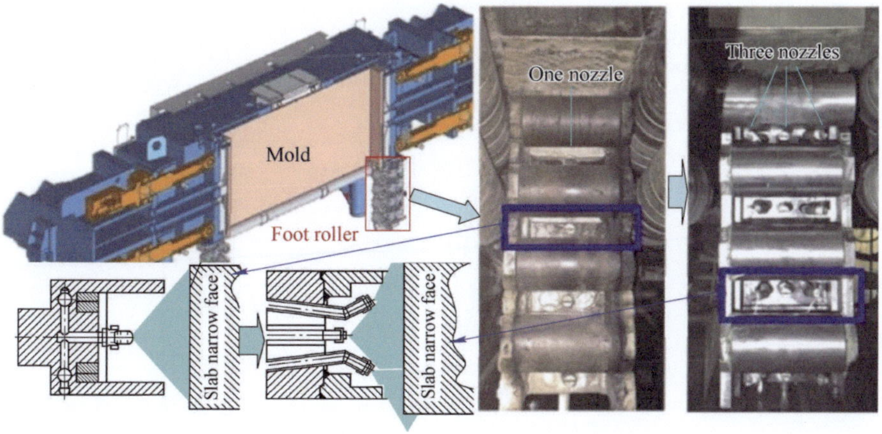

Fig. 2.85 Cooling structure of the slab corner

control cooling in this high temperature area on the secondary cooling temperature field, the shrinkage of the casting flow roller gap, and the reduction process at the end of slab solidification, this book has developed a ultra-strong cooling spray structure for the high temperature area of the slab corner based on the transverse 3 nozzles structure of the narrow face foot roller of the mold, as shown in Fig. 2.85, and equipped with an independent water supply circuit for the edge nozzle. Through this ultra-strong spray structure for the slab corner, and in conjunction with the cooling process design of the slab in the wide face foot roller zone, secondary cooling zone No.2 and No.3, it can conveniently achieve the completion of the $\gamma \rightarrow \alpha \rightarrow \gamma$ dual phase transition of the slab corner before exiting the bender, while ensuring that the slab temperature field, shell solidification growth when entering the bow segments is equivalent to the traditional continuous casting process.

2.6.3 Slab Temperature Field by Using the New Cooling Process

Based on the strong spray structure of the slab corner in the narrow face foot roller zone of the mold mentioned above, using the finite element model of slab secondary cooling solidification heat transfer developed in Sect. 2.4, the secondary cooling temperature field under the $\gamma \rightarrow \alpha \rightarrow \gamma$ dual-phase transition control cooling conditions in the foot roller zone and bender of the slab was simulated. The temperature field evolution of the slab wide face and narrow face center, slab corner is shown in Fig. 2.86. The temperature of the slab corner after exiting the mold continues to decrease rapidly under the forced spray cooling of the additional nozzle in the narrow side foot roller zone. When the slab exits the narrow face foot roller zone of

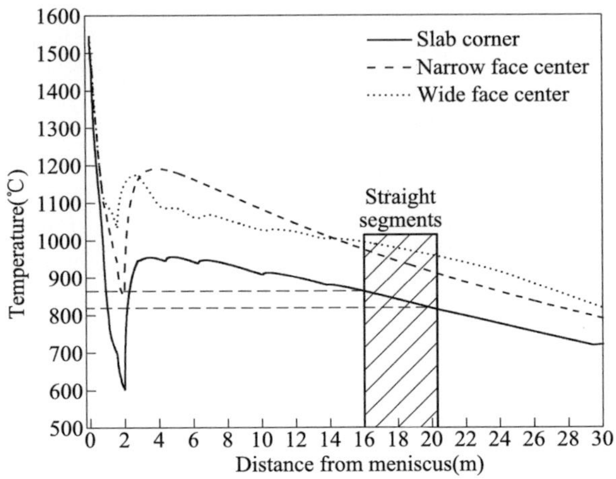

Fig. 2.86 Evolution of slab surface temperature under dual phase transition cooling process

the mold, the corner temperature drops to about 600 °C, with an average cooling rate exceeding 8.0 °C/s, achieving a rapid ferrite transition of the slab corner structure. Subsequently, the slab enters the secondary cooling No.3 zone. Due to the lack of spray cooling effect on the slab narrow face and its corner, and the cooling intensity of No.3 zone is relatively lower than the water supply intensity of the traditional secondary cooling control cooling process, the slab corner quickly reheats. When the slab enters secondary cooling No.4 zone, its corner reheats to the highest point of about 950 °C, with a reheating rate reaching an average of about 8.5 °C/s, the structure is austenitized again, thus completing the $\gamma \rightarrow \alpha \rightarrow \gamma$ dual-phase transition, achieving the ultrafine control of the slab corner structure grain reorganization.

Under the above controlled cooling conditions, for the wide and narrow faces of the slab, after the slab exits the mold, both its wide and narrow faces are subject to more rapid spray cooling than the traditional secondary cooling process. The temperatures of the wide and narrow faces of the slab significantly decrease after the foot roller and the end of the secondary cooling zone. Specifically, after the wide face of the slab exits the secondary cooling zone, the temperature drops to about 1030 °C, which is more than 30 °C lower than the traditional secondary cooling process. For the narrow face of the slab, due to the combined action of the strong spray of the original middle nozzle on narrow face of mold and the ultra-strong spray in the corner zone of the narrow face of the slab, the temperature in the center of the narrow face of the slab decreases rapidly. When the slab exits the narrow face foot roller zone, its temperature drops to about 860 °C, which is 160 °C lower than the traditional secondary cooling process. The rapid cooling and temperature reduction of the narrow face of the slab will significantly increase the strength and shell growth of the narrow face, helping to prevent the narrow face bulging of wide-thick slab and ultra-thick plate.

At the same time, when the slab exits the rapid cooling zone, due to the relatively slow cooling of the wide face in the secondary cooling No.3 zone and the lack of spray cooling on the narrow face, the surface temperature of the wide face of the slab rapid recovers to about 1175 °C. Compared with the traditional secondary cooling process, the surface temperature at the center of the wide face of the slab increases by about 50 °C. However, when the slab moves downward to the entrance of the bow segment, the surface temperature at the center of the wide face of the slab under the aforementioned slab surface structure dual-phase transition cooling process is only about 20 °C higher than the traditional secondary cooling process, so it will not affect the casting flow roller gap shrinkage system in the bow segments and the position of reduction zone at the end slab solidification. As for the narrow face of the slab, due to its rapid cooling effect in the narrow face foot roll zone, the highest recovery temperature of the narrow face is only 1190 °C, which is about 70 °C lower than the traditional secondary cooling process. When the slab enters the straight segment, the temperature of the wide face and corner, which greatly affect the force behavior of the slab, is basically the same as the corresponding temperature under the traditional secondary water distribution process, so it will not have a significant impact on the straightening of the slab.

2.6.4 Effect of Slab Corner Grain Refinement

The mentioned corner dual phase transition grain refinement control cooling process of the slab is implemented in the actual continuous casting process, and samples are taken to corrode and metallographically detect the casting structure at different positions below the slab corner before and after application. The morphology of the structure at different positions of the slab corner under the new process and the original process are compared as shown in Figs. 2.87 and 2.88. Among them, Fig. 2.86 shows the metallographic structure under the corner of Nb-bearing steel slab cast by the traditional secondary cooling weak cooling process at depths of 5, 10, 15, and 20 mm. From the fig., it can be seen that the structure within 0–20 mm below the slab corner cast by the traditional secondary cooling weak cooling process for Nb-bearing steel mainly presents a structure of ferrite film on the original austenite grain boundary, and as the depth below the slab surface increases, the size of the original austenite grains gradually increases, and the width of the ferrite film also gradually increases. According to the corner crack generation mechanism of the micro-alloyed steel continuous casting slab, this “austenite + first proeutectoid ferrite film” structure cause the slab corner to form intergranular cracking, because the ferrite film structure is softer, the bending and straightening stress of the slab is very easy to concentrate on the grain boundary ferrite film. At the same time, because the original austenite structure grains at the slab corner are abnormally coarse, it is also very unfavorable for preventing the expansion of intergranular cracks.

Figure 2.88 shows the metallographic structure within the range of 0–20 mm below the corner of the Nb-bearing steel slab produced by the dual phase transition controlled cooling process. Due to the rapid cooling control of the slab corner structure in the high temperature zone, its corner temperature quickly decreases from nearly 1000 °C to below 600 °C. The high cooling rate, on the one hand, promotes Nb

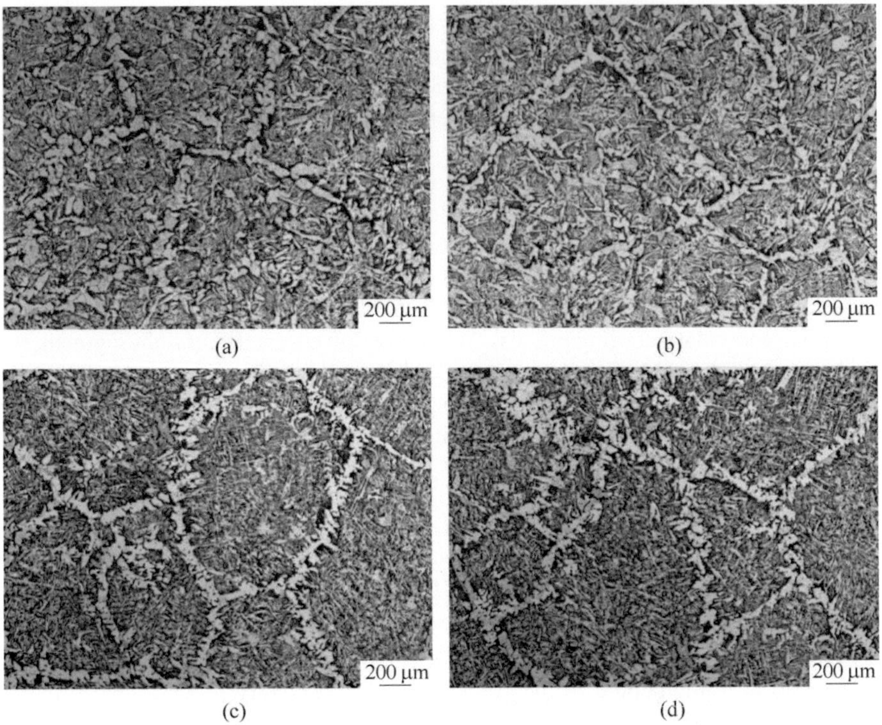

Fig. 2.87 Morphology of the micro-structure in the different position below the Nb-bearing steel slab corner under traditional cooling process. **a** 5 mm; **b** 10 mm; **c** 15 mm; **d** 20 mm

(C, N) and other dual-phase particles to quickly precipitate and disperse in the original austenite grain boundary and within the grain, providing more sufficient nucleation conditions for the ferrite transition and reheating stage austenite transition; On the other hand, during this cooling stage, the rapid cooling control promotes a higher γ-α transition rate, during the reheating stage of the slab corner, the ferrite and Nb (C, N) formed by the above process provide new nucleation sites for the formation and growth of new austenite. Under this effect, the residual austenite in the newly formed and original austenite competes for growth during the slab reheating process, thereby breaking the original austenite structure and achieving grain reorganization and refinement as well as secondary distribution of elements such as C, presenting a "ferrite + pearlite" metallographic structure as shown in Fig. 2.86. At the same time, due to the more dispersed dual-phase particles generated in the original austenite structure during the rapid cooling control stage, it further provides nucleation conditions for the new austenite structure in the subsequent ferritization process, promoting the ultrafine control of the casting structure.

In addition, it can be seen from Fig. 2.88 that the ferrite and other structural grains in the range of 0–20 mm below the slab corner are fine and evenly distributed, and the closer to the slab surface, the higher the rate and degree of phase transition, the

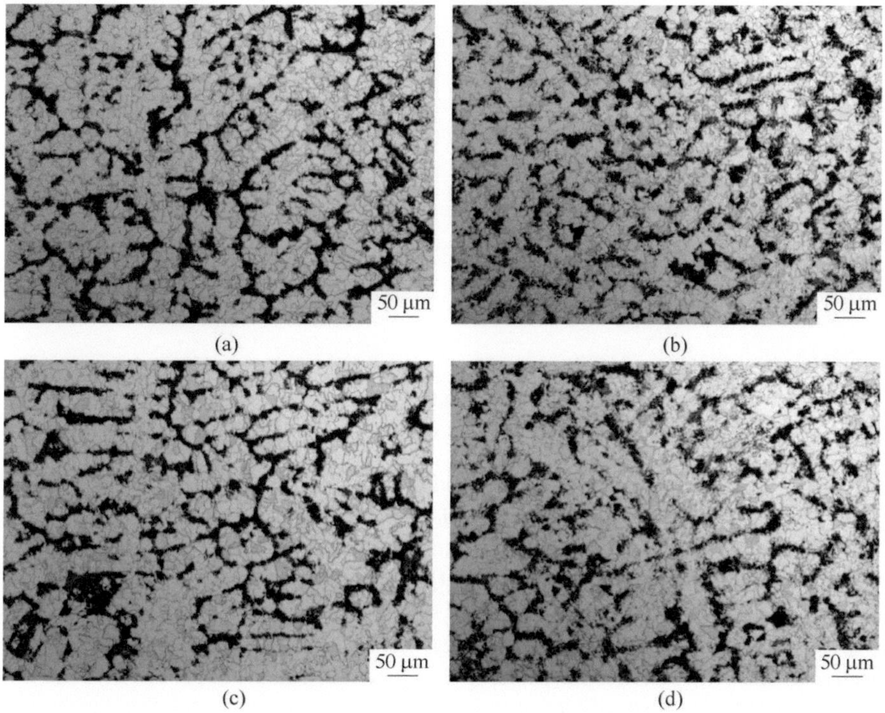

Fig. 2.88 Morphology of the micro-structure in the different position below the Nb-bearing steel slab corner under dual phase transition cooling process. **a** 5 mm; **b** 10 mm; **c** 15 mm; **d** 20 mm

finer the ferrite and other structural grains, thus the structural ductility significantly increases, which meets the deformation ductility requirements of the straightening process. At the same time, the fine and evenly distributed grain structure is also very beneficial to prevent the expansion of intergranular cracks, which can effectively prevent the formation and expansion of crack in the slab corner.

Figure 2.89 shows the precipitation morphology of microalloy carbonitrides at 3 mm and 7 mm below the slab corner of the above Ni-bearing high strength steel. It can be seen that the size of precipitations is very fine and the distribution is more dispersed, which achieves the effect of dispersion distribution. According to the thermodynamics and kinetics of carbonitride precipitation of Ni-bearing microalloy in Sect. 2.2, it can be seen that the corner (especially the structure far from the corner) is still in the precipitation temperature zone after the slab exit the mold, its structure is in the ultra-strong cold ferrite, due to the average cooling rate reaches more than 8.0 °C/s, the slab corner (especially the structure far from the corner of the slab in a certain depth range) continues to be rapidly cooling on the basis of the corner high efficiency heat transfer novel convex-structure mold, so that the structure of the corner of the slab in a larger range achieves the high cooling rate control in the temperature zone of the total precipitation of carbonitride. Thus, the precipitation

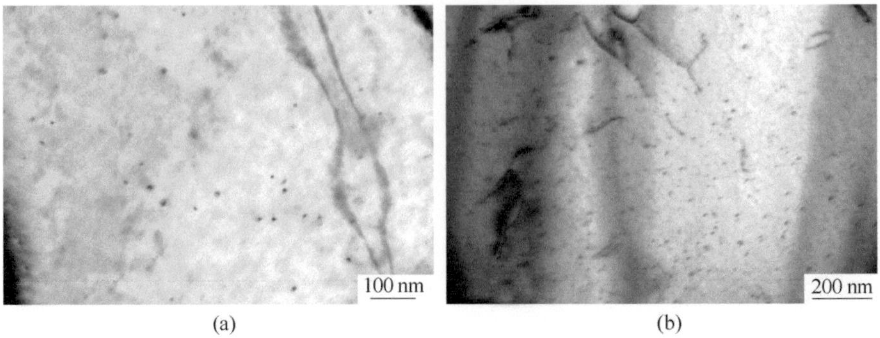

Fig. 2.89 Precipitation morphology of Nb-bearing steel slab corner under dual phase transition cooling process. **a** 3 mm below the corner; **b** 7 mm below the corner

size of microalloy carbonitrides is finer and the distribution is more dispersed, which further improves the hot ductility of the slab corner structure.

Figure 2.90 shows the metallographic structure morphology at 3 and 7 mm depths below the slab corner of Al-bearing steel (Q235B plus Al steel). As can be seen from the fig., under the traditional continuous casting process, there are obvious crack defects in the slab corner structure. The slab corner structure is mainly composed of relatively coarser ferrite (average above 95 μm) and pearlite, but it can be seen that is distributed in the form of "austenite + ferrite film" under high temperature, and the grain size of austenite is relatively coarser, up to 2 mm or more. Affected by this, the structural ductility of the slab under high temperature conditions is low, and the slab has produced crack defects extending along the grain boundary as shown in the fig. during the straightening process.

Figure 2.91 shows the casting structure morphology at 3, 5, 10, 15, and 20 mm below the slab corner after adopting the dual phase transition grain ultrafine controlled

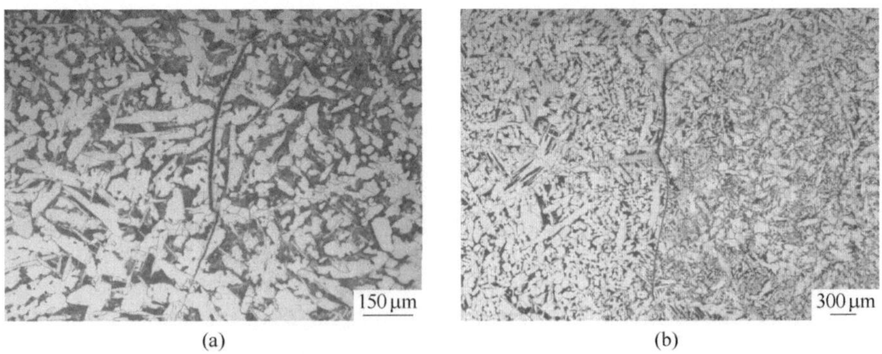

Fig. 2.90 Metallographic structure morphology at different position below the Al-bearing steel slab corner under traditional cooling process. **a** 3 mm; **b** 7 mm

cooling process. It can be seen that the 2 times phase transition nucleation reorganization in the slab corner structure, the structure within the 0–20 mm range below the surface of the slab has significantly refined, and the ferrite grain size is all in the range of 15–20 μm, eliminating the obvious Widmanstatten brittle structure characteristics under the traditional secondary cooling process. All grains are nearly round and evenly distributed, which will greatly improve the ductility and crack resistance of the slab corner structure. In the actual continuous casting production process, using this technology to produce Al-bearing steel grades, the slab corner can achieve free defect production, and all produced slab corners meet the requirements for free cleaning charging for rolling.

2.7 Summary

Through the detection and analysis of the micro-structure and precipitation morphology of the corner of micro-alloyed steel slab, the calculation of the thermodynamics and kinetics of carbonitride precipitation during the solidification process of micro-alloyed steel, the thermo-mechanical evolution of the solidified shell in the slab mold, and the thermal simulation analysis of the high temperature phase transition behavior of the steel structure, the formation mechanism of transverse crack at the corner of the micro-alloyed steel slab is clarified. On this basis, a slab corner crack control technology based on the dispersion of microalloy carbonitrides and grain ultrafine of the structure is proposed, and a novel convex-structure mold and slab secondary cooling high temperature zone corner grain ultrafine controlled cooling process and device technology are developed. The main conclusions are as follows:

(1) The main reason for the transverse cracks in the corner of the micro-alloyed steel slab is that during the high temperature solidification of the slab, microalloy carbonitrides are precipitated in a chain distribution at the austenite grain boundaries of the edge corner surface layer. On the one hand, the carbonitrides embrittle the grain boundaries, and on the other hand, during the subsequent solidification and deformation of the slab, microvoids form around the carbonitrides precipitated at the grain boundaries, significantly reducing the ductility of the corner structure of the slab. At the same time, a low ductility structure of "austenite + proeutectoid ferrite film at the grain boundary" is formed during the subsequent solidification of the slab, forming a significant third brittle temperature zone. The slab cracks along the austenite grain boundaries due to insufficient ductility during deformation processes such as straightening and poorly controlled roller gap, extend to form corner cracks.
(2) In the current main composition of micro-alloyed steel slab continuous casting process, the carbonitrides precipitated at the high temperature containing Nb, B, Al, etc., in the corner of the slab are precipitated in the lower part of the mold and are completely precipitated before the secondary cooling straight

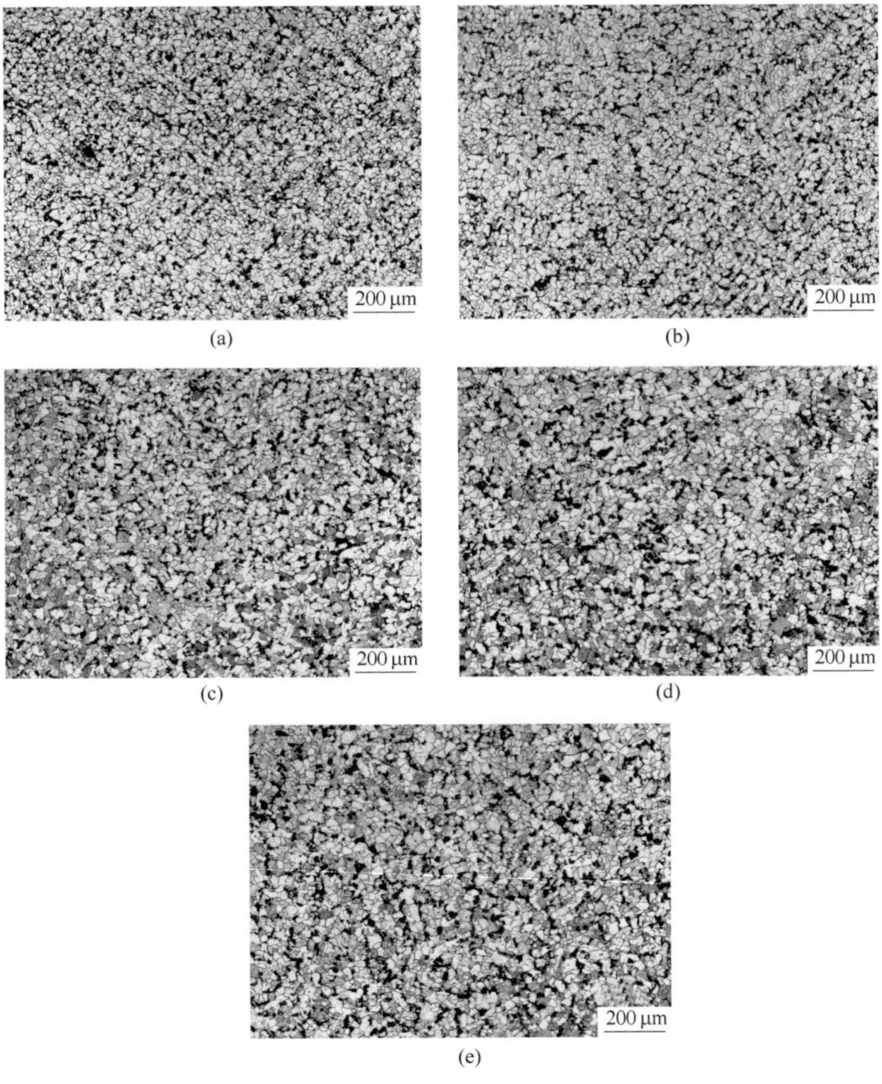

Fig. 2.91 Metallographic structure morphology at different position below the Al-bearing steel slab corner under dual phase transition cooling process. **a** 3 mm; **b** 5 mm; **c** 10 mm; **d** 15 mm; **e** 20 mm

segments. Increasing the cooling rate in the precipitation temperature zone of the corresponding microalloy carbonitrides to above 5.0 °C/s can significantly disperse precipitation and improve the hot ductility of the steel structure.

(3) A 3D thermo-mechanical calculation model of the solidified shell in the mold was developed to consider the dynamic distribution of mold flux film and air gap. It is clear that the traditional flat-plate mold cannot compensate the solidification

shrinkage of the shell in real-time, leading to the concentrated distribution of thick air gap and mold flux film in the corner of the shell, which leads to the low heat transfer efficiency of the slab corner in the lower part of the mold. This is the key reason for the concentrated precipitation of Nb-bearing microalloy carbonitrides along the grain boundary at high temperature in the slab corner. Based on this, a corner high efficiency heat transfer novel convex-structure mold with "fast compensation at the top, slow compensation at the middle and lower parts, and multi-compensation at the corners" was developed to satisfy the solidification shrinkage of the shell. The cooling rate of the middle and lower corner of the mold is greater than 10 °C/s, the precipitation of microalloy carbonitride at the high temperature is dispersed, and the grains of the solidified structure are significantly refined, which improves the hot ductility of the corner structure of the slab from the source.

(4) Under traditional slab continuous casting secondary cooling conditions, the formation of a low ductility solidification structure of "austenite + proeutectoid ferrite film at the grain boundary" in the corner of the slab is an inherent characteristic. Implementing dual-phase transition control cooling in the high temperature zone of the secondary cooling corner of the slab can ultrafine its grains. In response to the situation where the corner of the slab cannot be rapidly cooled to meet the $\gamma \rightarrow \alpha$ transition after the traditional slab continuous casting process, a strong spray structure and an automatic water distribution system for the narrow face foot roller of the continuous caster were developed, it can make the corner of the slab ultra rapid cooling. This achieves uniform ultrafine of the structure within the 0–20 mm range below the corner of the slab during the production process, and changes the traditional coarse "original austenite + grain boundary ferrite film" low ductility structure to a size not larger than 20 μm uniform "ferrite + pearlite" high ductility distribution, improving the ductility of the corner structure of the slab and controlling the generation of transverse cracks.

References

1. Maehara Y, Ohmori Y (1984) The precipitation of AlN and NbC and the hot ductility of low carbon steel. Mater Sci Eng 62(1):109–119
2. Dippenaar R, Moon SC, Szekeres ES (2007) Strand surface cracks-The role of abnormally large prior-austenite grains. AISE Steel Technol 4(7):105–115
3. Seo SC, Son KS, Lee SK et al (2008) Variation of hot ductility behavior in as-cast and remelted steel slab. Met Mater Int 14(5):559–563
4. Dippenaar R, Bernhard C, Schider S et al (2014) Austenite grain growth and the surface quality of continuously cast steel. Metall Mater Trans B 45(4):409–418
5. Rappaz M (2016) Modeling and characterization of grain structures and defects in solidification. Curr Opin Solid State Mater Sci 20:37–45
6. Kong LX, Xu JJ (2018) Research progress on thermodynamic model of non-ferrous metal alloy solution. Technol Innov Appl 12:88–89

7. Xu K, Thomas BG, O'Malley R (2011) Equilibrium model of precipitation in microalloyed steels. Metall Mater Trans A 42(2):524–539
8. Sun XL, Wang F, Chen XC et al (2017) Study on primary carbonitrides in H13 steel based on the two sublattice model. Chin J Eng 39(1):61–67
9. Narita K (1975) Physical chemistry of the groups IV a(Ti, Zr), Va(V, Nb, Ta) and the rare earth elements in steel. Trans Iron Steel Inst Jpn 15:142–152
10. Irvine KJ, Pickering FB, Gladman T (1967) Grain refined C-Mn steels. J Iron Steel Inst 205(2):161–182
11. Xiong L (2015) Study on mechanism of edge cracking of SS400 boron-bearing steel. Wuhan University of Science and Technology
12. Dutta B, Sellars CM (1987) Effect of composition and process variables on Nb(C, N) precipitation in niobium microalloyed austenite. Mater Sci Technol 3(3):197–206
13. Yong QL (2006) Second phases in structural steels. Metallurgical Industry Press, Beijing
14. Christian (1975) The theory of transformation in metals and alloys in equilibrium and general kinetic theory. Pergamon Press, Oxford
15. Savage J, Prichard WH (1954) The problem of rupture of the billet in the continuous casting of steel. J Iron Steel Inst 178:269–277
16. Lait J, Brimacombe JK, Weinberg F (1974) Mathematical modeling of heat flow in the continuous casting of steel. Ironmak Steelmak 2(1):90–98
17. Sawai T, Ueshima Y, Mizoguchi S (1990) Microsegregation and precipitation behavior during solidification in a nickel-base superalloy. ISIJ Int 30(7):520–528
18. Won YM, Kim KH, Yeo TJ et al (2000) A new criterion for internal crack formation in continuously cast steels. Metall Mater Trans B 31(4):779–794
19. Thomas BG, Li G, Moitra A et al (1997) Analysis of thermal and mechanical behavior of copper molds during continuous casting of steel slabs. In: 80th steelmaking conference proceedings, pp 1–19
20. Dippenaar RJ, Samarasekera IV, Brimacombe JK (1985) Mould taper in continuous casting billet machines. In: Electric finance proceedings, pp 103–117
21. Savage J (1962) A theory of heat transfer and air gap formation in continuous casting moulds. J Iron Steel Inst 197:41–47
22. Hills AWD (1965) Simplified theoretical treatment for the transfer of heat in continuous casting machine moulds. J Iron Steel Inst 203:18–26
23. Wimmer F, Thone H, Lindorfer B (1996) Thermomechanically-coupled analysis of the steel solidification process as a basis for the development of a high-speed billet casting mold. In: Proceedings of the international conference-MSMM '96, pp 366–371
24. Perkins A, Irving WR (1975) Two-Dimensional heat transfer model for continuous casting of steel. In: Mathematical process models in iron and steel making, pp 187–199
25. Chen K, Xu BS (1992) Mold dynamic heat flow and its influence on continuous wheelbelt casting. In: Steelmaking conference proceedings, pp 867–872
26. Meng Y, Thomas BG (2003) Heat transfer and solidification model of continuous slab casting: CON1D. Metall Mater Trans B 34(5):685–704
27. Shibata H, Kondo K, Suzuki M et al (1996) Thermal resistance between solidifying steel shell and continuous casting mold with intervening flux film. ISIJ Int 36:S179-187
28. Watanabe K, Suzuki M, Murakami K et al (1997) The effect of crystallization of mold powder on the heat transfer in continuous casting mold. Tetsu-to-Hagané 83(2):115–126
29. Yamauchi A, Sorimachi K, Sakuraya T (1993) Heat transfer between mold and slab through mold flux film in continuous casting of steel. Tetsu-to-Hagané 79(2):37–49
30. Cho JW, Shibata H, Emi T et al (1997) Heat transfer from shell across flux film to continuous casting mold. CAMP-ISIJ 10(2):180–189
31. Cho JW, Shibata H, Emi T et al (1998) Thermal resistance at the interface between mold flux film and mold for continuous casting of steels. ISIJ Int 38(5):440–446
32. Han HN, Lee JE, Yeo TJ et al (1999) A finite element model for 2-Dimensional Slice of cast strand. ISIJ Int 39(5):445–454

33. Zhu LG, Wang SM, Jin ST (1999) Mathematical simulation for state film in C.C mold. J Univ Sci Technol Beijing 21(1):13–16
34. Thomas BG (1989) Application of mathematical models to the continuous slab casting mold. Trans Iron Steel Soc 16(12):53–66
35. Moitra A (1993) Thermo-mechanical model of steel shell behavior in continuous slab casting. University of Illinois at Urbana-Champaign, Illinois
36. Kim K, Han HN, Yeo T et al (1997) Analysis of surface and internal cracks in continuous cast beam blank. Ironmaking Steelmaking 24(3):249–256
37. Cai ZZ, Zhu MY (2011) Simulation of thermal behavior during steel solidification in slab continuous casting mold I. Mathematical model. Acta Metallurgica Sinica 47(6):671–677
38. Ueshima Y, Mizoguchi S, Matsumiya T et al (1986) Analysis of solute distribution in dendrites of carbon steel with δ/γ transformation during solidification. Metall Mater Trans B 17(4):945–859
39. Won YM, Thomas BG (2001) Simple model of microsegregation during solidification of steels. Metall Mater Trans A 32(7):1755–1767
40. Flores F, Tania M, Castillejos E et al (2016) Study of shell-mold thermal resistance: Laboratory measurements, estimation from compact strip production plant data, and observation of simulated flux-mold interface. Metall Mater Trans B 47(4):2509–2523
41. Nakada H, Susa M, Seko Y et al (2008) Mechanism of heat transfer reduction by crystallization of mold flux for continuous casting. ISIJ Int 48(4):446–453
42. Saraswat R, Maijer DM, Lee PD et al (2007) The effect of mould flux properties on thermo-mechanical behaviour during billet continuous casting. ISIJ Int 47(1):95–104
43. Yamauchi A, Emi T, Seetharaman S (2002) A mathematical model for prediction of thickness of mould flux film in continuous casting mould. ISIJ Int 42(10):1084–1093
44. Tsutsumi K, Nagasaka T, Hino M (1999) Surface roughness of solidified mold flux in continuous casting process. ISIJ Int 39(11):1150–1159
45. Cho JW, Emi T, Shibata H et al (1998) Heat transfer across mold flux film in mold during initial solidification in continuous casting of steel. ISIJ Int 38(8):834–842
46. Cai ZZ, Zhu MY (2013) Thermo-mechanical behavior of peritectic steel solidifying in slab continuous casting mold and a new mold taper design. ISIJ Int 53:1818–1827
47. Anand L (1982) Constitutive equations for the rate-dependent deformation of metals at elevated temperatures. Trans ASME 104(1):12–17
48. Brown SB, Kim KH, Anand L (1989) An internal variable constitutive model for hot working metals. Int J Plast 5:95–130
49. Koric S, Thomas BG (2008) Thermo-mechanical models of steel solidification based on two elastic visco-plastic constitutive laws. J Mater Proces Technol 197(1/2/3):408–418
50. Cai ZZ, Zhu MY (2011). Simulation of thermal behavior during steel solidification in slab continuous casting mold II. Model verification and result analysis. Acta Metallurgica Sinica 47(6):678–687
51. Koric S (2006) Efficient thermo-mechanical model for solidification processes and its applications in steel continuous casting. University of Illinois at Urbana-Champaign, Illinois
52. Harste K (1989) Investigation of the shrinkage and the origin of mechanical tension during the solidification and successive cooling of cylindrical bars of Fe-C alloys. Technical University of Clausthal, German
53. Harste K, Suzuki T, Schwerdtfeger K (1992) Thermomechanical properties of steel: viscoplasticity of γ iron and γ Fe-C alloys. Mater Sci Technol 8(1):23–33
54. Harste K, Schwerdtfeger K (1996) Thermomechanical properties of iron: viscoplasticity of ferrite and austenite-ferrite mixtures. Mater Sci Technol 12(5):378–384
55. Jimbo I, Cramb AW (1993) The density of liquid iron-carbon alloys. Metall Mater Trans B 24(1):5–10
56. Thomas BG, Samarasekera IV, Brimacobe JK (1987) Mathematical model of the thermal processing of steel ingots: Part II. Stress model. Metallurg Mater Trans 18(3):131–147
57. Yoo J, Rubinsky B (1983) Numerical computation using finite element for the moving interface in heat transfer problems with phase transformation. Numer Heat Transf Part A Appl 6(2):209–222

58. Voller RV, Cross MA (1981) Accurate solution of moving boundary problems using the enthalpy methods. Int J Heat Mass Transf 24(3):545–556
59. Voller RV, Cross MA (1983) Explicit numerical method to track a moving phase change front. Int J Heat Mass Transf 26(1):147–150
60. Voller RV, Swaminathan CR, Thomas BG (1990) Fixed grid techniques for phase change problems: a review. Int J Numer Meth Eng 30(4):875–898
61. Roltf WD, Bathe K (1982) An efficient algorithm for analysis of nonlinear heat transfer with phase change. Int J Numer Meth Eng 18(1):119–134
62. Tacke K (1985) Discretization of explicit enthalpy method for planar phase change. Int J Numer Meth Eng 21(3):543–554
63. Carnes-Pintaux AM, Nguyen-Lamba M (1986) Finite element enthalpy method for discrete phase change. Numer Heat Transf Part A Appl 9(3):403–417
64. Runnels SR, Carey GF (1991) Finite element simulation of phase change using capacitance methods. Numer Heat Transf Part B Fundam 19(1):13–30
65. Tamma KK, Namburu RR (1990) Recent advances, trends and new perspectives via enthalpy-based finite element formulations for applications to solidification problems. Int J Numer Meth Eng 30(4):803–820
66. Li CS, Thomas BG (2004) Thermomechanical finite-element model of shell behavior in continuous casting of steel. Metall Mater Trans B 35(12):1151–1172
67. Clyne TW, Wolf M, Kurz W (1982) The effect of melt composition on solidification cracking of steel with particular reference to continuous casting. Metall Mater Trans B 13(2):259–266
68. Davies GJ, Shin YK (1979) Solidification technology in the foundry and cast house. The Metal Society, London
69. Kim K, Yeo T, Oh K et al (1996) Effect of carbon and sulfur in continuously cast strand on longitudinal surface cracks. ISIJ Int 36(3):284–289
70. Matsumiya T, Saeki T, Tanaka J et al (1982) Mathematical model analysis on the formation mechanism of longitudinal surface cracks in continuously cast slabs. Tetsu-to-Hagané 68(13):1782–1791
71. Mizukami H, Mizukami K, Miyashita Y (1977) Mechanical properties of continuously cast steel at high temperatures. Tetsu-to-Hagané 63(146):S652
72. Uehara M, Samarasekera IV, Brimacombe JK (1986) Mathematical modeling of unbending of continuously cast steel slabs. Ironmaking Steelmaking 13(3):138–153
73. Rajil S, Peter D, Kenneth C (2008) The effect of mould flux properties on thermo-mechanical behavior during billet continuous casting. ISIJ Int 47(1):95–104
74. Cai ZZ, Zhu MY (2014) Non-uniform heat transfer behavior during shell solidification in a wide and thick slab continuous casting mold. Int J Miner Metall Mater 21:240–250
75. Hanao M, Kawamoto M, Yamanaka A (2009) Growth of solidified shell just below the meniscus in continuous casting mold. ISIJ Int 49(3):365–374
76. Zhu M, Cai Z (2013) Thermo-mechanical behavior and cracking susceptibility of solidifying shell in continuous casting mold. In: The 8th Pacific RIM international congress on advanced materials and processing, pp 3063–3072
77. Ramachandrarao P (1992) Rapid solidification of steels. Bull Mater Sci 15(6):503–513
78. Guan Y, Song B, Mao JH et al (2003) Cleanness undercooling and grain size of low carbon steel. Acta Metall Sin 39(3):283–286
79. Mohsen AP, Shahverdi HR et al (2016) Microstructural evolution and mechanical properties of a novel FeCrNiBSi advanced high-strength steel: slow, accelerated and fast casting cooling rates. Mater Sci Eng A 668:188–200
80. Rezaean A, Zarandi F, Yue S (2008) Mechanism of hot ductility improvement of a peritectic steel containing vanadium using very-high-temperature compression. Metall Mater Trans A 39(11):2635–2644
81. He T, Hu R, Zhang TB (2016) Effect of Nb content on solidification characteristics and microsegregation in cast Ti–48Al–xNb alloys. Acta Metallurgica Sinica (Eng Lett) 29(8):714–721

82. Fu LM, Shan AD, Wang W (2010) Effect of Nb solute drag and Nbc precipitate pinning on the recrystallization grain growth in low carbon Nb-microacloyed steel. Acta Metall Sin 46(7):832–837
83. Chen J, Lv MY, Tang S et al (2014) Microstructure, mechanical properties and interphase precipitation behaviors in V-Ti microalloyed steel. Acta Metall Sin 50(5):524–530
84. Grange RA (1966) Strengthening steel by austenite grain refinement. ASM Trans Quart 59(1):26–48
85. Sagaradze VV (1997) An ultra-fine grain structure formed as a result of cyclic γ–α–γ transformations. Nanostruct Mater 9(8):201–204
86. Lee UH, Park TE, Son KS (2010) Assessment of hot ductility with various thermal histories as an Alternative Method of in situ Solidification. ISIJ Int 50(4):540–545

Chapter 3
Corner Crack of Micro-alloyed Steel Thin Slab and Its Control

3.1 Development and Process Features of Thin Slab Continuous Casting and Rolling Technology

3.1.1 Development of Thin Slab Continuous Casting and Rolling Technology

Thin slab continuous casting and rolling technology is a new short process strip steel production process successfully developed and applied to industrial production in the late 1980s and early 1990s. It has the advantages of significant energy saving, high product qualification rate, process simplification, short production line, and short production cycle, and has received widespread attention from major steel companies domestic and foreign [1]. Since the invention of this technology, major steel producing countries represented by the USA, South Korea, and China have invested a lot of labor, financial, and material resources to conduct systematic research and optimization development [2]. In recent years, with the increasing competition pressure of domestic and foreign steel companies in the market, the advantages of low cost and high efficiency production of continuous casting and rolling of thin slab, especially the endless rolling process of thin slab, have been fully demonstrated, which accelerates the upgrading of existing production lines and the construction of new production lines. The main process of thin slab continuous casting and rolling technology is: the production line layout of thin slab continuous caster, uniform heating furnace (or electromagnetic induction), and compact rolling is adopted. The high temperature molten steel is prepared into a 20–100 mm thin slab through the mold, secondary cooling and other links, and then heated directly by uniform heating furnace or entered directly into hot rolling mill to roll into hot rolled coil. The successful development and application of this process has changed the traditional long process production of strip steel, which is an important part of the current green production process of strip steel at home and abroad.

© Metallurgical Industry Press 2025

Z. Cai and M. Zhu, *The Control Technology for Slab Surface Crack During Continuous Casting of Micro-alloyed Steel*, Key Steel Manufacturing Technologies by Collaborative Innovation, https://doi.org/10.1007/978-981-96-7171-7_3

Currently, the thin slab continuous casting and rolling technology is developing and promoting rapidly [3, 4]. As of 2017, there are 67 thin slab continuous casting and rolling production lines in the world, with a total flow of 102 and a production capacity of 114 million tons per year. Among the completed thin slab continuous casting and rolling production lines, there are a total of 35 CSP lines, accounting for more than half of the thin slab continuous casting and rolling production lines. China is the country with the most construction and production of thin slab continuous casting and rolling production lines and the largest capacity [5], with 19 types of thin slab production lines (31 flows), with a capacity of nearly 40 million tons, accounting for 35% of the total capacity of the global thin slab continuous casting and rolling production line. At the same time, with the gradual stabilization of endless rolling technology in China, Rizhao Steel, Shougang Steel, Tang Steel, and others are constructing or planning to build 5 thin slab endless rolling production lines, with an additional capacity of more than 11 million tons, becoming an important part of strip steel production of China. The distribution of the constructed thin slab continuous casting and rolling production lines and capacity worldwide of 2017 is shown in Table 3.1.

Currently, thin slab continuous casting and rolling production lines of China are distributed in Wuhan Steel, Han Steel, Bao Steel, Magang Steel, Lian Steel, Jiugang Steel, and other companies, these companies are mainly CSP lines. While the thin slab continuous casting and rolling production lines of Tang Steel, Ben Steel, Shougang Steel, and other companies are FTSR. Angang Steel adopts medium and thin slab (ASP) casting and rolling production line [6]. The endless rolling production line, which represents the most advanced technology, is mainly distributed in the Rizhao Steel ESP production line [7], Shougang Steel MC-CR production line [8], and so on. As of 2017, the statistics of thin slab continuous casting and rolling production lines and capacity of China are shown in Table 3.2.

Since 1989, the US Crawfordsville Nucor Corporation using SMS Group CSP technology built and put into production first industrial thin slab continuous casting

Table 3.1 Global thin slab continuous casting and rolling production lines and annual capacity at the end of 2017

Country	Number of production lines									Production capacity (10 kt/a)	Number of caster flow
	CSP	ISP	FTSR	QSP	CON ROLL	TSP	ESP	ASP	Number of casters		
USA	9	0	0	2	1	2	0	0	14	2098	19
China	7	0	3	0	0	0	3	6	19	3946	31
India	5	0	0	0	0	0	0	0	5	800	7
Korea	1	1	1	0	0	0	1	0	4	830	7
Italy	1	1	0	0	0	0	1	0	3	310	3
Other	11	4	5	1	3	0	0	0	24	3300	33
Total	34	6	9	3	4	2	5	3	66	11,284	100

Table 3.2 Thin slab continuous casting and rolling production lines in China

Serial number	Company	Process type	Number of continuous caster flows	Annual production capacity (10 kt)	Date of commissioning
1	Pearl Steel	CSP	2	180	1999
2	Han Steel	CSP	2	247	1999
3	Bao Steel	CSP	2	200	2001
4	Tang Steel	FTSR	2	250	2002
5	Magang Steel	CSP	2	200	2003
6	Lian Steel	CSP	2	220	2004
7	Anshan Steel	ASP	2	250	2000
8	Anshan Steel	ASP	4	500	2005
9	Ben Steel	FTSR	2	280	2004
10	Tonghua Steel	FTSR	2	250	2005
11	Jiuquan Steel	CSP	2	200	2005
12	Jinan Steel	ASP	2	250	2006
13	Wuhan Steel	CSP	2	253	2009
14	Rizhao Steel	ESP	1	222	2014
15	Rizhao Steel	ESP	1	222	2015
16	Rizhao Steel	ESP	1	222	2015
Total			31	3946	

and rolling production line of the world, thin slab continuous casting, and rolling process in nearly 30 years in the USA, Europe, and China has been widely developed. In addition to its continuous application on a large scale, its technology has also made significant progress, showing the following main development characteristics:

(1) Endless rolling technology has gradually been industrialized and stably applied. In 2009, the Italian company Arvedi first developed and industrialized the first endless rolling strip steel production line (ESP) based on the ISP production line at its Cremona plant. The emergence and stable industrial application of this technology, known as the third technological revolution in the steel industry [9], representing the highest technical level of hot rolled strip production in the world today. In May of the same year, the "HIGH MILL" [10] production line of Pohang Steel in Korea was also reformed to realize endless rolling industrial production. In 2014 and 2015, Rizhao Steel in China successively put three ESP production lines introduced by Arvedi into operation and successfully produced hot rolled coils with a minimum thickness of 0.8 mm (currently capable of producing 0.6 mm thick hot rolled coil). The successive commissioning of these advanced endless rolling production lines marks the successful industrial application of efficient, large scale, and low cost ultra-thin strip endless rolling technology.

(2) The casting speed is getting higher and higher [11]. The casting speed of traditional thin slab continuous casting is 3.5–4.3 m/min. In recent years, with the continuous breakthrough of efficient continuous casting technology, the casting speed has been continuously improved. Among the thin slab continuous casters that have been put into operation stably, CSP, FTSR, and other production lines have broken through the 5.0 m/min casting speed to achieve efficient production. The production lines represented by ESP of Arvedi and Rizhao Steels have higher casting speed, which can generally reach 5.5–6.0 m/min, and the maximum steel throughput reaches 6.5 t/min. Some production lines (Pohang, Korea) have achieved ultra-high-speed continuous casting production of 8.0 m/min, greatly improving the production capacity of the thin slab continuous casting and rolling production line.

(3) The coverage of steel grades is getting wider and wider [12–14]. In recent years, with the continuous advancement of thin slab continuous casting, rolling, and post-rolling cooling technologies, the range of steel grades produced by the thin slab continuous casting and rolling production line has gradually expanded. It has evolved from the earliest single mild steel production to gradually include medium carbon steel and even high carbon steel with C content of 0.7–0.8%, achieving high quality production of all steel grades. Especially in recent years, with the rapid development of microalloy and other technologies, the steel grades produced by the thin slab continuous casting and rolling production line have expanded from relatively single ordinary carbon steel to include microalloyed steel such as Ti, Nb, V, B, Al and high grade silicon steel, etc., and the proportion of high grade steel is showing a trend of continuous increase (for example, the proportion of high grade steel in CSP production lines such as Handan Steel has reached more than 30%).

(4) The competitiveness of the production line is getting stronger and stronger [15]. With the deepening of research and development of thin slab continuous casting and rolling technology in various countries, the process and device technology of this type of production line is gradually developing in the direction of thin specifications, multiple varieties, stabilization, unmanned, and intelligent. Especially in recent years, with the emergence of endless rolling technology, it has almost achieved stable continuous production of unmanned casting, the stability of the mechanical properties of the product is high, and the production of 0.6–0.8 mm thin specification products has achieved "hot rolling instead of cold rolling," greatly enhancing the competitiveness of the products. Statistics show that compared with the traditional hot rolled strip steel production line, the thin slab continuous casting and rolling production line can reduce investment by about 42%, reduce energy consumption by about 50%, reduce production costs by about 12%, increase the yield rate by about 1.8%, and reduce device maintenance costs by about 61% [16], which greatly improves the market competitiveness of this type of production line.

3.1.2 Features of Thin Slab Continuous Casting Process

Thin slab continuous casting and rolling production line is rigid, which lacks rhythmic buffer and defective slab cleanup. The stability, casting speed, and slab quality of continuous casting directly determine the stability, production efficiency, and product quality of the whole production line. Therefore, thin slab continuous casting is the core link of the thin slab continuous casting and rolling production line. The thin slab continuous casting process mainly has the following characteristics [17–20]:

(1) The slab thickness is small. At present, the thickness of slab produced by thin slab continuous caster at home and abroad is generally 20–100 mm, and the most common typical thickness includes 50, 70, and 90 mm. Because of the small thickness of thin slab, the arc radius of this type of continuous caster is small and the production line is short, which is helpful to realize the high temperature requirements of slab running out. At the same time, the solidification rate of the small thickness slab is fast, which also provides the basic conditions for the high-speed continuous casting of the thin slab.

(2) The casting speed is fast. High-speed continuous casting is the basic requirement and main feature of the thin slab continuous casting and rolling process. At present, the design casting speed of typical thin slab continuous casters such as CSP and FTSR is higher than 5.0 m/min, and the actual casting speed of Baowu Steel, Han Steel, and Lianyuan Steel is generally in the range of 3.5–5.0 m/min. In recent years, with the invention of endless rolling technology, the casting speed of thin slab continuous casting has further increased, reaching 5.5–6.0 m/min high-speed stable continuous casting production, and the highest casting speed of advanced companies in South Korea has reached 8.0 m/min. The casting speed of conventional and wide-thick slab at home and abroad is mainly concentrated in the range of 0.8–1.8 m/min, and the maximum casting speed is not more than 3.0 m/min, which is significantly lower than the casting speed of thin slab.

(3) The solidification rate of slab is fast. Due to the small thickness and the high cooling strength of the thin slab, the solidification rate of the slab is fast. For the thin slab with a thickness of 50 mm, the total solidification time is about 1–2 min, while the total solidification time of the conventional slab with a thickness of 250 mm is about 23–25 min. Due to the characteristics of rapid solidification of thin slab, the grain size of the solidified structure of the continuous casting slab is significantly finer than that of the conventional and wide-thick slab, and the spherical crystal zone is large, the center segregation is small, and the density of the slab is high.

(4) The slab has a high temperature when it runs out. It is because of the high casting speed of the thin slab continuous caster that the overall temperature of the slab is relatively high. The surface temperature at the solidification end can reach 1100–1150 °C, the corner temperature can also reach 950–970 °C, and the average temperature can reach 1300 °C. The high temperature characteristic of thin slab

continuous casting lays the temperature foundation for the implementation of endless rolling.

(5) The metallurgical length is short. The thickness of the thin slab is small and the solidification rate is fast, which makes its metallurgical length relatively short. At present, the metallurgical length of conventional thin slab production line at home and abroad is generally about 5–8 m, and the metallurgical length of endless rolling production line varies depending on the casting speed and slab thickness and generally does not exceed 12.5 m. The liquid core length of traditional continuous casting slab generally exceeds 20 m.

(6) The specific surface area is large. The specific surface area of the thin slab with the section size of 1500 mm × 50 mm reaches 5.3 m^2/t, while the specific surface area of the conventional slab with the same width and thickness of 250 mm is only 1.2 m^2/t. The large specific surface area increases the heat dissipation rate of the slab during the solidification, thereby accelerating the solidification rate of the slab, but this also aggravates the solidification thermal stress of the slab, thereby increasing the occurrence rate of surface and subsurface crack defects of the slab.

3.1.3 Liquid Core Reduction Process

In order to achieve high-speed continuous casting of thin slab, the thickness of the slab is generally required to be smaller. However, the significant reduction of the slab thickness will directly affect the insertion of the submerged nozzle and the capacity of the molten pool in the mold, restrict the use of the thick submerged nozzle, and aggravate the fluctuation in the mold liquid level during the continuous casting production. To solve this contradiction, the thin slab continuous casting production line introduced the liquid core reduction [21, 22]; by increasing the thickness of the slab in the mold, it is reduced along the thickness direction to thin the slab after it exit the mold, thereby accelerating the solidification of the slab to achieve high-speed continuous casting production.

The liquid core reduction of thin slab continuous caster is also known as soft reduction, as shown in Fig. 3.1. Liquid core reduction refers the reduction of the thin slab in the case of small shell thickness and a large amount of liquid core after it exits the funnel mold. This process can achieve the purpose of reducing slab thickness, increasing casting speed, improving the center segregation and reducing the number of rolling mills or rolling passes [23]. The liquid core reduction of thin slab continuous casting is usually implemented at the No.1 segment of the bow segment of the continuous caster. By applying a reduction of 5–20 mm to the slab, the slab thickness is significantly reduced and then cooled by the secondary cooling segment of continuous casting to accelerate the complete solidification of the slab.

Implementation of the liquid core reduction process for the thin slab continuous casting process has the following multiple advantages [24, 25]:

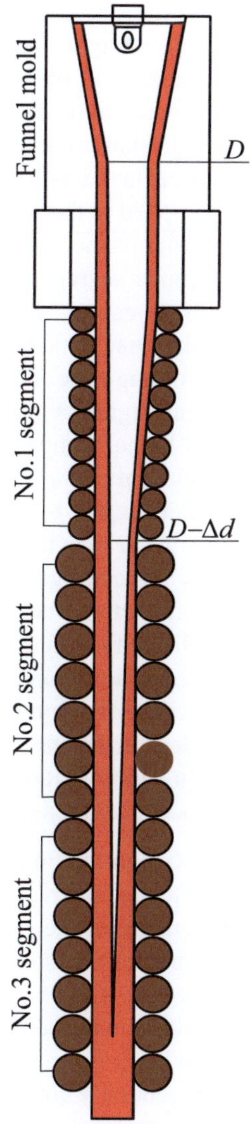

Fig. 3.1 Schematic diagram of liquid core reduction of thin slab

(1) Reduce the thickness of the slab and increase the casting speed of the continuous caster. Among the factors that limit the high-speed continuous casting of thin slab, in addition to the high incidence of surface cracks on the slab caused by high casting speed and the aggravation of solidification inhomogeneity of the slab, the aggravation of the mold level fluctuation caused by the increase of casting speed is also one of the important factors. The capacity of the mold molten pool directly determines the sensitivity of the mold level fluctuation caused by the combined action of the impact of the nozzle flow and the dynamic deformation of the shell in the casting flow. Implementation of the liquid core reduction process

of the slab can effectively reduce the thickness of the continuous casting slab and increase the casting speed of the continuous caster without reducing the capacity of the mold molten pool.

(2) Reduce the center segregation of the slab. The original intention of implementing thin slab liquid core reduction technology is to increase the capacity of molten steel in the mold as much as possible. However, after the liquid core reduction is applied to the thin slab, it helps to break the "crystal bridge" at the solidification front of the slab, compensate the cooling shrinkage of the slab and reduce the center segregation of the slab.

(3) Save energy. Before the thin slab is completely solidified, about 30–40% of the heat is taken away by the cooling effect of the mold and the secondary cooling zone. The implementation of thin slab liquid core reduction technology can significantly improve the casting speed of the slab, so that the slab has a higher temperature, thereby improving the utilization rate of slab heat. At the same time, the solidified shell of the slab at the liquid core reduction is thinner and temperature is higher, the force of the reduction process is much smaller than the solid phase rolling force after the slab is completely solidified, which can significantly reduce the energy consumption of the subsequent rolling production line.

(4) Simplify rolling device and reduce investment costs. The liquid core reduction technology can reduce the burden of the continuous rolling process of thin slab, reduce the number of finishing mill group, and alleviate the burden of rolling device. The flexibility of the device and the service life of the roll is improved. In addition, it can also shorten the length of the production line and reduce the occupied area and investment costs.

Although the liquid core reduction process of thin slab has many advantages, in the actual implementation process, due to the high casting speed and low corner temperature of the slab, the rapid thinning process of the slab will cause the stress concentration at the edge and corner. When the hot ductility of the slab corner structure is low, it is easy to form surface crack defects such as the corners of the slab. Therefore, in the actual implementation of the liquid core reduction process, in addition to limiting the reduction amount of the casting roller in the liquid core reduction zone (generally not exceed 1.5 mm/roller), the temperature at the slab corners should be increased as much as possible to improve the hot ductility of the steel structure, or the morphology of the slab corners should be improved to reduce the deformation resistance of the slab during the liquid core reduction process.

3.2 Cause of the Micro-alloyed Steel Thin Slab Corner Crack

3.2.1 *Morphology of the Cracks*

The frequent occurrence of corner crack defects in micro-alloyed steel thin slab continuous casting is a common technical problem in the continuous casting and rolling process of thin slab [26–30]. The formation of corner cracks of micro-alloyed steel thin slab has the following main common characteristics:

(1) Cracks mainly occur in the troughs of the oscillation marks at the corners of the slab, are mostly transverse cracks. Compared with the conventional and wide-thick slab, the distribution of transverse cracks at the corners of thin slab is relatively continuous and clearly visible. The position of the corner crack of the thin slab is correlated with the trough of the oscillation mark on the slab surface, but the difference between different steel grades is quite large, as shown in Fig. 3.2.

(2) Similar to the conventional and wide-thick slab casting, the transverse cracks at the corners of thin slab also mainly occur in alloy steel containing elements such as Nb, V, B and Al, etc., and the crack defects become more severe with the increase of alloy and N content. The depth of the slab crack is relatively deep, and in severe cases, the opening can exceed 2 mm, and the depth can reach 5–8 mm. After rolling, the slab crack will expand to form the edge crack defect of the coil, as shown in Fig. 3.3.

(a) (b)

Fig. 3.2 Morphology of transverse corner crack of thin slab. **a** Nb-bearing steel; **b** Al-bearing steel

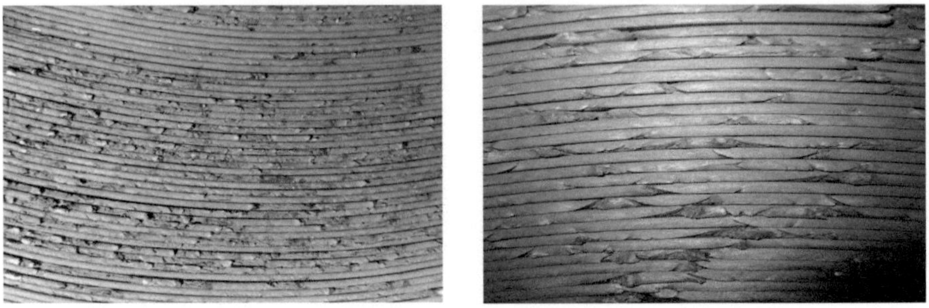

(a) (b)

Fig. 3.3 Edge crack morphology of hot rolled coil caused by transverse corner crack of thin slab. **a** Nb-bearing steel; **b** Al-bearing steel

3.2.2 Cause of the Cracks

During the production process of thin slab continuous casting, due to the rapid heat transfer of the mold and the rapid cooling effect of the secondary cooling zone, the grain size of the surface solidification structure of the slab is significantly refined compared to the conventional and wide-thick slab. Figure 3.4 shows the room temperature metallographic structure of the Qste380TM low carbon Nb-Ti-bearing steel thin slab at 5 and 10 mm below the corner in a certain steel plant. The slab corner microstructure at room temperature consists of fine, uniform ferrite and pearlite, there is no low ductility micro-structure, such as coarse austenite grains in conventional and wide-thick slab corner or coarse austenite + grain boundary ferrite film. It is generally believed that the ductility of the high temperature casting structure of this type of slab is relatively high, and under the normal roller gap accuracy, the slab generally does not form obvious corner crack defects.

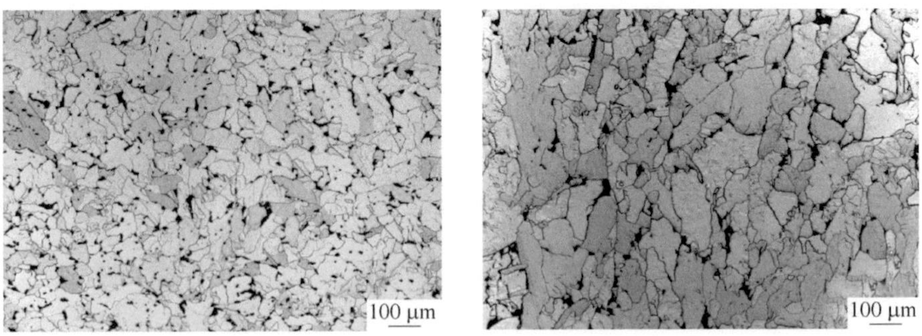

(a) (b)

Fig. 3.4 Micro-structure morphology of Qste380TM low carbon Nb-Ti-bearing steel thin slab corner. **a** 5 mm below slab corner; **b** 10 mm below slab corner

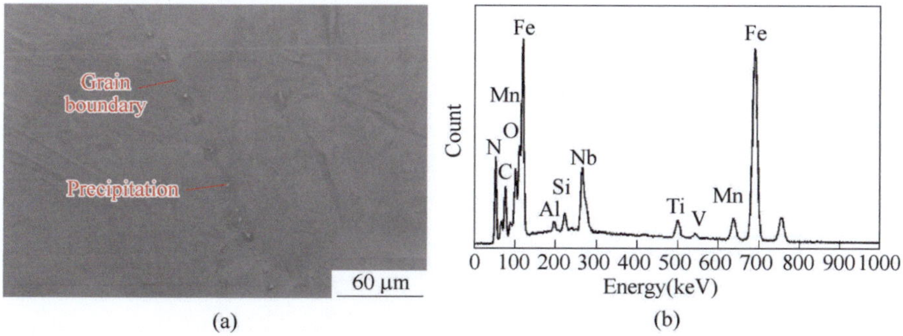

Fig. 3.5 Morphology and composition of carbonitride precipitation at 5 mm below thin slab corner during casting Qste380TM low carbon Nb-Ti-bearing steel. **a** Morphology of carbonitride precipitation; **b** energy spectrum analysis

Figure 3.5 shows the morphology and composition of the precipitation at 5 mm below the slab corner surface of the Qste380TM low carbon Nb-Ti-bearing steel. The carbonitride with the obvious chain structure is distributed on the austenite grain boundary of the corner structure of the slab with a size of about 20 nm. The distribution of carbonitride particles is mainly concentrated on the grain boundaries, and there are few precipitation inside the grains. From the energy spectrum analysis results of Fig. 3.5b, it can be seen that the carbonitrides precipitated at the grain boundary of the corner structure of the slab are mainly Nb-Ti-bearing composite dual-phase particles.

One of the main roles of the precipitation of the microalloy carbonitride is pinning strengthening [31, 32]. However, the cast structure of the steel will be affected by concentrated precipitating at the grain boundary during the solidification process. On the one hand, the pinning effect of the precipitation will limit the grain boundary deformation and slip of the slab during the deformation process, thereby reducing the hot ductility of the slab corner structure. On the other hand, during the deformation process of the liquid core reduction, bending and straightening of continuous casting slab, a large number of microalloy carbonitride precipitated at the grain boundary will easily lead to the formation of continuous distribution of microvoid defects at the grain boundary, which will lead to the stress concentration at the grain boundary microvoid during the further deformation process of the steel structure, thus causing the grain boundary of the slab to crack and expand to form cracks [33–35].

Figure 3.6 shows the evolution curve of the RA of Qste380TM low carbon Nb-Ti-bearing steel thin slab from a certain steel plant under the different deformation rate conditions, and the test scheme is shown in Fig. 3.7a. This steel has a RA of over 60% under the high temperature conditions above 900 °C, showing good ductility. However, when the temperature is in the 800–900 °C temperature range, the ductility of the structure significantly decreases. Especially under the large deformation rate conditions, the RA of the sample drops to below 40% (at a deformation rate of 5.0×10^{-2}/s, the RA of the sample drops to the lowest about 35% at 850 °C), forming a noticeable third brittle temperature zone. When the sample temperature further drops to below 800 °C, the ductility of the structure rises again. It can be seen that although the grain size of this steel is fine under the conditions of thin slab continuous casting

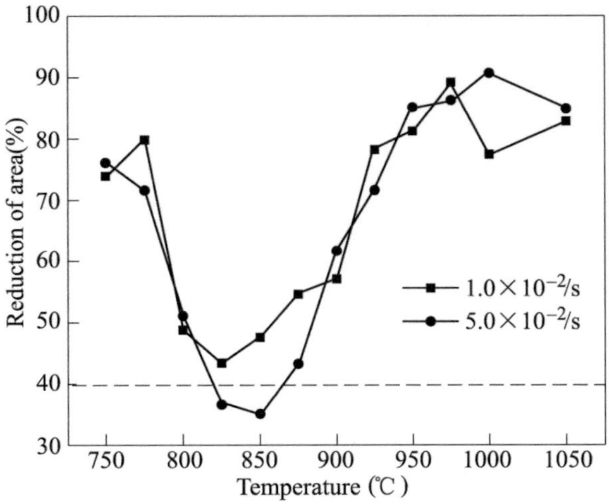

Fig. 3.6 Evolution of RA with temperature of Qste380TM low carbon Nb-Ti-bearing steel under the different deformation rates

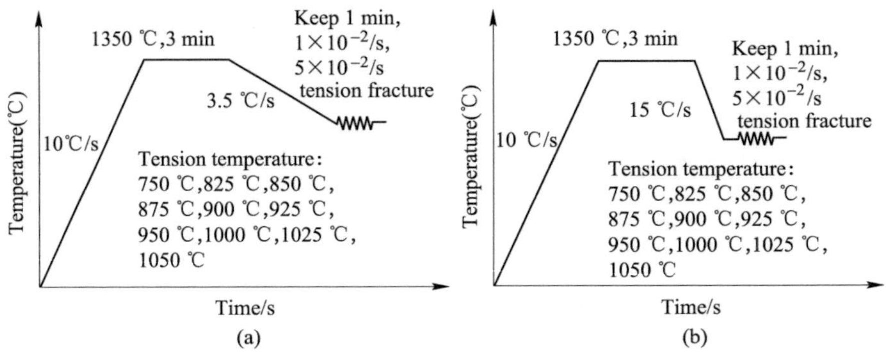

Fig. 3.7 Test scheme for RA of Qste380TM low carbon Nb-Ti-bearing steel under the different cooling condition. **a** Conventional cooling; **b** rapid cooling

production, it still forms a noticeable third brittle temperature zone in the 800–900 °C temperature range.

Moreover, under the higher deformation rate conditions, the RA of Qste380TM low carbon Nb-Ti-bearing steel in the temperature range of 800–925 °C is slightly lower than that under the relatively low deformation rate. This phenomenon indicates that the thin slab with the same solidification structure is more prone to form slab crack defects under the rapid deformation conditions such as the liquid core reduction, top bending, and straightening during the continuous casting process.

In order to further clarify the cause of corner cracks induced by the continuous casting process of thin slab, this book, based on the control idea of the strategy microalloy carbonitride dispersion precipitation in Chap. 2, tested the evolution of RA of Qste380TM low carbon Nb-Ti-bearing steel under the rapid cooling conditions. The test scheme is shown in Fig. 3.7b, and the curve of the RA with the temperature evolution is shown in Fig. 3.8. As can be seen from the figure, the difference of RA of tensile samples at temperatures above 925 °C under the different deformation rates and the corresponding tensile samples at a cooling rate of 3.5 °C/s is small for the 15 °C/s rapid cooling test samples. When the tensile temperature is below 925 °C, the RA of the 15 °C/s rapid cooling samples is significantly higher that the RA of the samples at a cooling rate of 3.5 °C/s. Although the steel structure still has a certain degree of brittleness in the temperature range of 775–900 °C, the minimum RA under the different deformation rates is all above 60% (compared to the minimum RA at a lower cooling rate of 3.5 °C/s, an increase of about 26%), meeting the high ductility requirements of the slab structure under the large deformation.

Figure 3.9 shows the metallographic structure and carbonitride precipitation transmission morphology of the sample cooled to 850 °C at the cooling rate of 15 °C/s (without fracture) and then air cooling to room temperature after preservation with 1 min. As can be seen from the fig., the structure of the sample air cooling to room temperature after the rapid cooling is comparable to the structure of the thin slab under the actual continuous casting production conditions, both of which are fine ferrite and pearlite structures. However, the precipitation of Nb and Ti is dispersed. Therefore, it can be inferred that the main reason for the lack of ductility of Qste380TM low carbon Nb-Ti-bearing steel in the third brittle temperature zone during the continuous casting production process of thin slab is the precipitation of the microalloy carbonitrides along the grain boundaries, causing the grain boundaries to be brittle.

Therefore, based on the above Gleeble thermal simulation tensile test results, it can be inferred that the fundamental cause of the edge and corner crack defects of the micro-alloyed steel thin slab is the thin slab continuous casting process. During this process, the microalloy carbonitrides precipitate in a chain distribution at the grain boundaries of the slab corner, leading to a significant decrease in the hot ductility. Under the high casting speed and liquid core reduction, and during the process of excessive straightening and rapid deformation, the aggravated ductility of the slab,

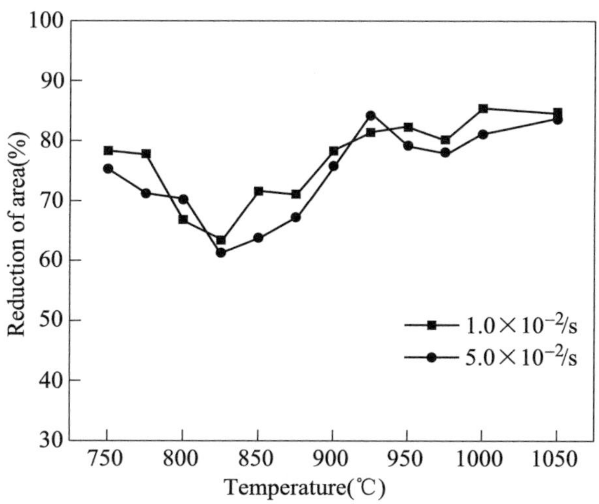

Fig. 3.8 Evolution of RA with temperature of Qste380TM low carbon Nb-Ti-bearing steel under different deformation rate with rapid cooling process

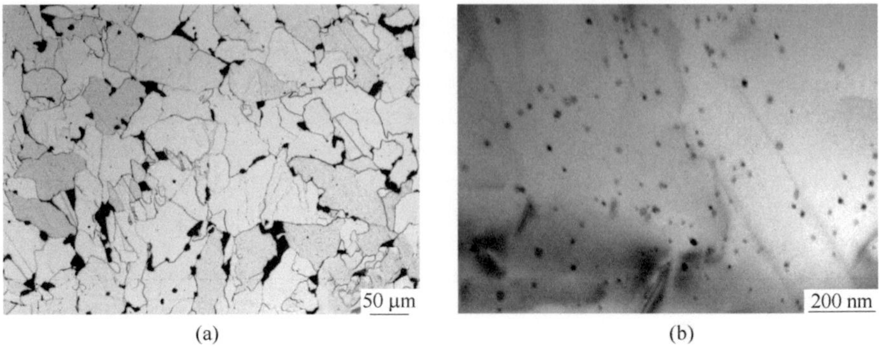

Fig. 3.9 Micro-structure and carbonitride precipitation of the sample cooled to 850 °C with rapid cooling process. **a** Micro-structure morphology; **b** carbonitride precipitation morphology

especially at the inner arc and other corner areas, is subject to the significant stress, causing cracks along the grain boundaries, which then expand to form the edge and corner crack defects of the thin slab.

3.3 Control Strategy of the Cracks

From the analysis of the formation mechanism of corner cracks of micro-alloyed steel thin slab, it is known that the key to eliminating the corner crack is to control the precipitation and distribution of carbonitride along the grain boundary during the solidification process of the slab corner. It is necessary to inhibit the chain precipitation of carbonitrides at the grain boundaries and reduce the stress of the edge corner of the thin slab in the continuous casting production process. Currently, the micro-alloyed steel produced by the main thin slab continuous casting and rolling lines at home and abroad, with its main microalloy elements and their contents similar to those of the conventional and wide-thick slab continuous casting lines, mainly includes the strong C and N binding precipitating strengthening elements such as Ti, Nb, Al, B, V. Therefore, during the production of the micro-alloyed steel thin slab, if steel companies can appropriately reduce the addition of microalloy in the steel and minimize the N content in the steel while increasing the temperature of the slab corner deformation process, it can reduce the occurrence of corner cracks of the thin slab.

However, in the actual production of the micro-alloyed steel, the amount of alloy added directly determines the mechanical properties of the steel, and it is impossible to significantly reduce the alloy content in the steel during the actual production. The control of N content in the molten steel has always been a difficulty in steel smelting. The N content in steel produced by the "converter or electric furnace + LF refining" smelting process is generally between 0.0035 and 0.0070%, making it difficult to significantly reduce the N content in the steel. Especially in the current large scrap steel ratio smelting process, the control of N content in molten steel is a problem. The crack control method based on increasing the temperature control of the slab corners is difficult to implement due to the high casting speed of thin slab continuous casting, the high cooling intensity of the secondary cooling zone, and the fact that in the thin slab continuous casters with the vertical bending structures like CSP, the secondary cooling water in the vertical zone will flow to the straight segment, making it difficult to achieve ultra-high temperature control at the slab corner. Therefore, in the current production of the micro-alloyed steel thin slab by steel companies, the control of corner cracks of the slab can only be achieved under the premise of narrow composition control of microalloy and N content in the steel, by changing the precipitation form, size and distribution of microalloy carbonitride in the solidification structure of the slab, dispersing the microalloy carbonitrides in the matrix of the solidification structure of the slab, improving the ductility of the solidification structure at the slab corner, and reducing the stress of the slab during the deformation process.

As can be seen from the thermodynamic calculation results of the microalloy carbonitride precipitation in Chap. 2 of this book, the precipitation temperature of TiN during the solidification process of Ti-bearing micro-alloyed steel is relatively high, which precipitates first during the solidification process of the slab, thus helping to fixate N. In the actual thin slab continuous casting process of the micro-alloyed steel containing trace amounts of Ti (generally less than 0.035% Ti), the slab produces few edge and corner crack defects. However, for carbonitrides of micro-alloyed elements such as Nb, B, and Al, the distribution of their precipitation has a significant effect on the hot ductility of the steel, causing the largest amount of edge and corner crack defects of thin slab (with a general occurrence rate of over 2.0%, and in some cases, the corner crack rate of thin slab of micro-alloyed steel with high Nb, B content can even reach over 30%), severely affecting the quality of hot rolled coil.

Based on the microalloy carbonitride precipitation thermodynamic model in Chap. 2 of this book, typical micro-alloyed steel containing Nb, Al, V, and B elements such as Qste series, J55, SS400 were selected as research objects, and their carbonitride precipitation thermodynamics were calculated. The changes in the carbonitride precipitation process with temperature are shown in Fig. 3.10.

As can be seen, the starting precipitation temperatures of carbonitrides in the continuous casting process of micro-alloyed steel under the different types of alloys and their contents vary greatly. Among them, the starting precipitation temperature of carbonitride containing Nb, Al, and B elements is higher, which is above 1050 °C, that is, it has begun to precipitate in the mold, and the ending temperature of precipitation is about 800 °C. Combined with the temperature evolution of the slab corner during the thin slab continuous casting process, it can be seen that the carbonitride precipitation of the slab corner during the continuous casting process of Nb, Al, B micro-alloyed steel breaks through the mold and the entire secondary cooling zone. For carbonitrides containing V, the starting precipitation temperature is only about 880 °C, and the ending precipitation temperature has dropped to about 600 °C. Due to the high corner temperature in the actual thin slab continuous casting process, the precipitation amount of V-bearing carbonitride in the corner structure before the slab

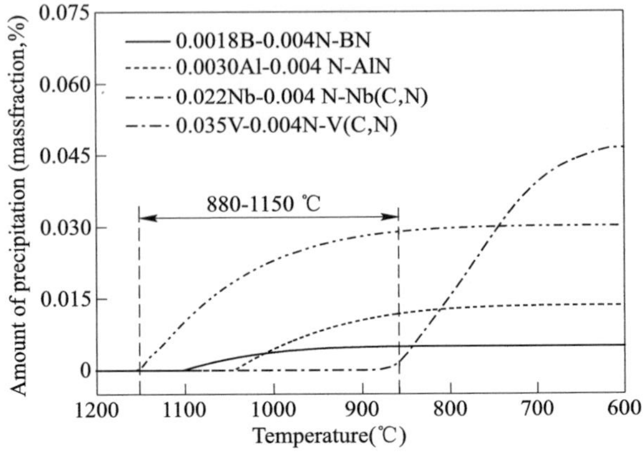

Fig. 3.10 Carbonitride precipitation with temperature of typical composition of micro-alloyed steel produced by thin slab continuous caster

straightening is limited. Actual production statistics show that the edge crack ratio of hot rolled coils from the continuous casting and rolling of thin slab of V-bearing micro-alloyed steel is relatively low. Therefore, the main control objects for corner cracks of thin slab should be Nb, Al, and B microalloy, and the main control link should also be the mold and secondary cooling.

Similarly, according to the results of the precipitation distribution of microalloy carbonitrides with typical composition contents at different cooling rates in Chap. 2 of this book, it is necessary to ensure that the cooling rate of the slab corner in the corresponding carbonitride precipitation temperature zone is higher than 5 °C/s. Combined with the cause of corner cracks of thin slab, it is necessary to strengthen the corner cooling of the slab in the area before the liquid core reduction segment from the mold and disperse the microalloy carbonitride to precipitate within its structure.

3.4 Thermo-mechanical Behavior of Shell Solidification in Thin Slab Mold

According to the analysis of the control strategy for corner cracks in micro-alloyed steel thin slab, ensuring that the cooling rate at the corner of the shell in the mold exceeds 5 °C/s to achieving high temperature dispersion precipitation of carbonitrides in the of the slab corner. This is a key factor to controlling corner crack in Nb, Al, and B micro-alloyed steel thin slab. For this, it is essential to first investigate the heat transfer behavior of the solidification shell within the thin slab mold.

In the production process of thin slab continuous casting, the high casting speed results in a short solidification time for molten steel in the mold. Consequently, the cooling efficiency of the thin slab mold should be significantly higher than that of the conventional slab mold and wide-thick slab mold. Therefore, the cooling structure of the thin slab mold is significantly different from that of the conventional slab mold and wide-thick slab mold. Specifically, the wide face copper of the thin slab mold adopts the U-shaped water slot, which increases the width of the water slot. Additionally, to enhance the uniformity of heat transfer of the copper, a circular water slot is installed at the upper part of the mold and at the connection point of the back plate bolt, as illustrated in Fig. 3.11. In the design of the U-shaped water slot, an inserted water slot structure is adopted on the backplate to increase water flow speed and reducing the thickness of the cooling water in the U-shaped water slot to approximately 4.0 mm. This significantly improves the velocity of the cooling water, while effectively supporting the copper plate and reducing the deformation of the copper plate. The structure of the narrow face copper plate water slot is a circular. Therefore, when simulating the thermo-mechanical behavior of the shell solidification in the thin slab mold, the impact of the rapid cooling water in the mold water slot on the heat transfer through the copper should be considered.

In addition, the mold currently adopted by thin slab continuous casters is predominantly funnel mold. The slab begins to solidify from the meniscus in the mold and subsequently moves downward at a certain casting speed. During this process, the solidified shell in the funnel zone undergoes a large deformation, which in turn affects the overall heat transfer of the shell solidification. Traditional 2D thermo-mechanical model of the slab fails to accurately reflect the slab solidification process as the inner cavity structure of the mold changes. Therefore, to simulate the thermo-mechanical behavior of the thin slab solidification in the mold, a 3D mathematical model must be developed to comprehensively describe both the heat transfer and stress evolution during the shell solidification.

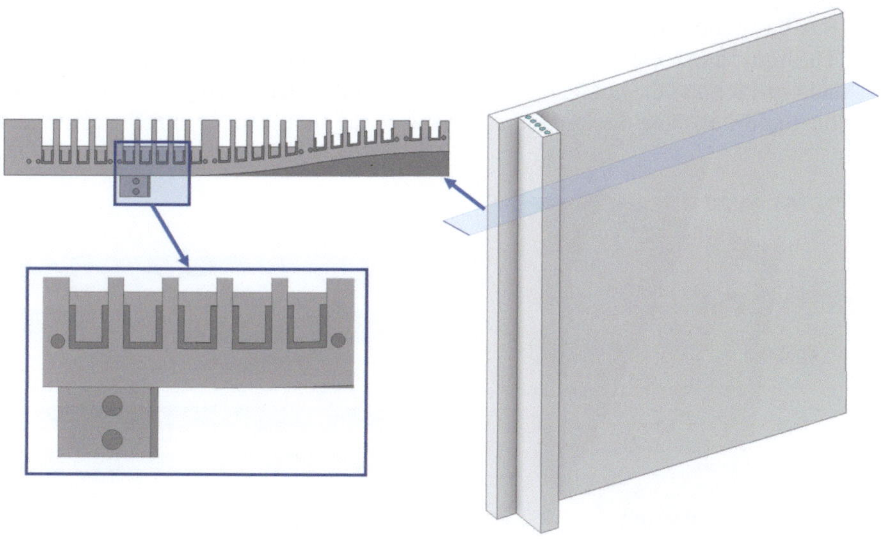

Fig. 3.11 Structure of the thin slab mold copper plate

3.4.1 Mathematical Model

Based on the structure of a CSP funnel mold in a steel plant (as shown in Fig. 3.12), the system comprising the slab (molten steel) and copper plate in the mold is defined as the finite element calculation domain. To comprehensively account for the influence of the mold cooling water on the heat transfer of the "slab mold" system, the MpCCI fluid-solid coupling interface software is employed to couple the calculation of the cooling water flow and heat transfer behavior in the mold copper plate water slot, the heat transfer between the slab and the copper plate, and the solidification and deformation behavior of the shell, to achieve the 3D thermo-mechanical behavior coupling analysis of the "slab-copper plate-cooling water" system in the mold, accurately describing the complex thermo-mechanical behavior during the shell solidification process in the thin slab mold.

3.4.1.1 Model Assumptions

To effectively develop the complex heat transfer and stress behavior of the thin slab mold and the solidification process of the shell, this book has made the following assumptions during the model development:

(1) The composition of the glass phase and crystalline phase in the solid flux layer in the flux channel. The mold flux flowing into the flux channel is only used to distinguish between the solid flux layer and the liquid flux layer, and the liquid flux has good fluidity.
(2) The liquid and solid state of the mold flux is determined by the temperature of the slab surface and the hot face of the mold, as well as the melting characteristics of the mold flux. The thickness distribution of the flux layer is jointly determined

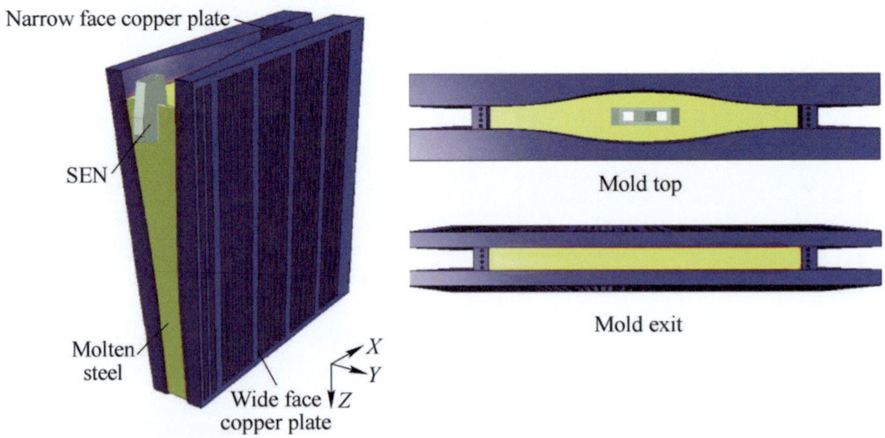

Fig. 3.12 CSP funnel mold

by the gap at the interface between shell and cooper plate and the state of the mold flux.

(3) Ignoring the shrinkage of the flux film, the thermal resistance caused by the solidification shrinkage of the mold flux is addressed by introducing the thermal resistance of the interface between mold flux and copper plate.
(4) The thermal conductivity of the mold copper plate is assumed to be isotropic, as well as the density and specific heat are considered as constants.
(5) Ignoring the deformation of the mold, the contact between the mold copper plate and the shell is set as rigid-flexible contact.
(6) Ignoring the friction between the slab and the mold, and the sliding contact between the shell and the mold is considered as smooth contact.
(7) The heat transfer and stress of the solidified shell and the mold are assumed to be symmetrical.
(8) Ignoring the influence of mold oscillation on the heat transfer and force of copper plate and shell solidification.

3.4.1.2 Model Development

A. Heat Transfer Control Equation between Shell and Copper Plate.

Based on the above symmetry assumption, a 1/4 cross section of the thin slab and its mold copper plate is taken as the calculation domain. Since the heat transfer of the slab in the mold is a unsteady heat transfer process with an internal heat source, the heat transfer differential equations of the two are selected as formulas (3.1) and (3.2):

$$\frac{\partial H(T)}{\partial t} = \frac{\partial}{\partial x}\left(\lambda_s(T)\frac{\partial T}{\partial x}\right) + \frac{\partial}{\partial y}\left(\lambda_s(T)\frac{\partial T}{\partial y}\right) + \frac{\partial}{\partial z}\left(\lambda_s(T)\frac{\partial T}{\partial z}\right) \tag{3.1}$$

$$\rho c\frac{\partial T}{\partial t} = \frac{\partial}{\partial x}\left(\lambda_m\frac{\partial T}{\partial x}\right) + \frac{\partial}{\partial y}\left(\lambda_m\frac{\partial T}{\partial y}\right) + \frac{\partial}{\partial z}\left(\lambda_m\frac{\partial T}{\partial z}\right). \tag{3.2}$$

In the formula $H(T)$, $\lambda_s(T)$ represent the enthalpy and thermal conductivity of the steel, which are to be calculated as functions of temperature;

ρ, c, λ_m denote the density, heat capacity, and thermal conductivity of the copper substrate and nickel deposits layer, respectively.

B. Control Equation for the Water Flow in Water Slot.

As discussing earlier in this book, the heat transfer efficiency of thin slab mold is significantly higher than that of conventional slab and wide-thick slab mold. This enhanced efficiency can be attributed not only to the optimized heat transfer structure design of the copper plate but also to the flow and heat transfer characteristics of the cooling water within the water slots of copper plate, which directly influence the heat transfer of the mold. To investigate the influence of cooling water flow and heat transfer on both the heat transfer and mechanical properties of the solidified shell, this book has developed the following model for cooling water flow and heat transfer in the mold copper plate water slot.

According to the laws of mass, momentum, and energy conservation, the following control equations are used for the flow and heat transfer of cooling water in the mold water slot.

Continuity equation:

$$\frac{\partial u_j}{\partial x_j} = 0, \tag{3.3}$$

where

u_j The flow rate of cooling water in the j direction, m/s;
x_j The coordinate in the j direction, m.

Momentum equation:

$$\frac{\partial}{\partial x_j}\rho_w u_i u_j = -\frac{\partial p}{\partial x_i} + \frac{\partial}{\partial x_j}\mu_{eff}\left(\frac{\partial u_j}{\partial x_i} + \frac{\partial u_i}{\partial x_j}\right) + \rho_w g_i, \tag{3.4}$$

where

ρ_w The density of water, kg/m^3;
u_i The water flow rate in the i direction, m/s;
p Pressure, Pa;
g Gravitational acceleration, m/s^2;
μ_{eff} Effective viscosity, kg/(m s), which is the combined viscosity of laminar and turbulent flow, $\mu_{eff} = \mu_1 + \mu_t$, μ_1 is the laminar molar viscosity, μ_t is the turbulent viscosity.

Energy equation:

$$\frac{\partial}{\partial x_i}\left[u_i(p_w H + p)\right] = \frac{\partial}{\partial x_j}\left(\lambda_{eff}\frac{\partial T_w}{\partial x_j}\right)$$
$$\lambda_{eff} = \lambda_w + \frac{c_w \mu_t}{\mathrm{Pr_t}}, \tag{3.5}$$

where

H Total energy, J;
λ_{eff} Effective thermal conductivity, W/(m K);
λ_w, c_w, Tw The thermal conductivity, thermal capacity and temperature of the water, with corresponding units of W/(m K), J/(kg K) and K;
Pr_t Turbulent Prandtl constant, the value taken in this calculation is 0.85.

Given the high-speed water in the mold cooling water slot and the large Reynolds number, the effect of cooling water turbulence on the heat transfer at the cooling interface between water and slot needs to be considered. Therefore, the standard k-ε model is adopted for the flow model. In this model, the turbulent kinetic energy k (m^2/s^2) and dissipation rate ε (m^2/s^3) are calculated by the transportation formula (3.6)–(3.8):

$$\frac{\partial}{\partial x_i}(\rho_w k u_i) = \frac{\partial}{\partial x_j}\left[\left(\mu_l + \frac{\mu_t}{\sigma_k}\right)\frac{\partial k}{\partial x_j}\right] + G - \rho_w \varepsilon \tag{3.6}$$

$$\frac{\partial}{\partial x_i}(\rho_w \varepsilon u_i) = \frac{\partial}{\partial x_j}\left[\left(\mu_l + \frac{\mu_t}{\sigma_\varepsilon}\right)\frac{\partial \varepsilon}{\partial x_j}\right] + C_1\frac{\varepsilon}{k}G - C_2\rho_w\frac{\varepsilon^2}{k} \tag{3.7}$$

$$\mu_t = \rho_w C_\mu \frac{k^2}{\varepsilon}. \tag{3.8}$$

In the formula, $G = \mu_t S^2$, $S = \sqrt{2S_{ij}S_{ij}}$, $S_{ij} = \frac{1}{2}\left(\frac{\partial u_i}{\partial x_j} + \frac{\partial u_j}{\partial x_i}\right)$, $C_1 = 1.44$, $C_2 = 1.92$, $C_\mu = 0.09$, $\sigma_k = 1.0$, $\sigma_\varepsilon = 1.3$.

C. Stress Control Equation of Shell

Compared with the conventional slab and wide-thick slab continuous casting process, the solidification process of the thin slab in the mold produces significant ductile deformation and creep phenomena when passing through the funnel area. Therefore, the mechanical control equation of the thin slab uses the Anand rate-related constitutive equation proposed by Anand and Brown [38, 39]:

$$\dot{\overline{\varepsilon}}_{ie} = A\exp\left(-\frac{Q_A}{T}\right)\left[\sinh\left(\xi\frac{\overline{\sigma}}{s}\right)\right]^{\frac{1}{m}}, \tag{3.9}$$

where the evolution equation for s is:

$$\dot{s} = \left(h_0\left|1 - \frac{s}{\tilde{s}\left[\frac{\dot{\overline{\varepsilon}}_{ie}}{A}\exp\left(\frac{Q_A}{T}\right)\right]^n}\right|^{\alpha}\operatorname{sign}\left(1 - \frac{s}{\tilde{s}\left[\frac{\dot{\overline{\varepsilon}}_{ie}}{A}\exp\left(\frac{Q_A}{T}\right)\right]^n}\right)\right)^{\frac{\dot{\overline{\varepsilon}}_{ie}}{A}} \tag{3.10}$$

The meanings and corresponding values of each parameter in the formula can be found in Table 2.6 of Chap. 2.

D. Model Initial and Boundary Conditions

(a) Initial Conditions

The initial heat transfer conditions of the model mainly include the temperature of the meniscus, the thickness of the mold flux, and the temperature of the mold copper plate. Among them, the initial temperature of the meniscus is the casting temperature of the molten steel. The initial thickness of the mold flux film is calculated based on the average flux consumption and the cross-sectional size of the slab. The initial temperature of the copper plate is obtained by initializing the pretreatment program shown in Fig. 2.33 of Chap. 2 of this book.

In the mechanical initial conditions, the initial stress of the slab in the calculation domain is set to 0.

(b) Thermal Boundary Conditions

In this heat transfer calculation model, the setting of the heat transfer boundary conditions of the slab and the mold is similar to the shell solidification heat transfer in the traditional slab and wide-thick slab mold in Sect. 2.3.1.

The convective heat transfer boundary is still used between the mold copper plate and the cooling water, which is determined by the following formula:

$$q = h_{\mathrm{w}}(T - T_{\mathrm{w}}), \tag{3.11}$$

where

T, T_{w} The temperature of the mold water slot surface and the cooling water temperature, K;
h_{w} The convective heat transfer coefficient between the water slot and the cooling water, W/(m^2 K).

However, in this model, the heat transfer coefficient of the interface between cooling water and water slot is calculated according to the near-wall heat transfer law, and the calculation formula is shown in formula (3.12) [40]:

$$\begin{aligned}
T_{\mathrm{w}}^* &= \frac{T_{\mathrm{m}} - T_{\mathrm{w}}}{q}\rho_{\mathrm{w}}c_{\mathrm{w}}C_{\mu}^{1/4}k^{1/2} = \begin{cases} \mathrm{Pr}\, y^* & y^* < y_{\mathrm{T}}^* \\ \mathrm{Pr_t}\left[\frac{1}{\kappa}\ln(Ey^*) + P\right] & y^* > y_{\mathrm{T}}^* \end{cases} \\
y^* &= \frac{\rho_{\mathrm{w}}y_{\mathrm{w}}}{\mu_{\mathrm{l}}}C_{\mu}^{1/4}k^{1/2} \\
P &= 3.15\,\mathrm{Pr}^{0.695}\left(\frac{1}{E'} - \frac{1}{E}\right)^{0.359} \\
&\quad + 9.24\left(\frac{E'}{E}\right)^{0.6}\left[\left(\frac{Pr}{\mathrm{Pr_t}}\right)^{\frac{3}{4}} - 1\right]\left(1 + 0.28\mathrm{e}^{-0.007\,\mathrm{Pr}/\mathrm{Pr_t}}\right),
\end{aligned} \tag{3.12}$$

where,

T_{w}^* Dimensionless temperature;
T_{m} Wall temperature of the water slot, K;
T_{w} Temperature of the cooling water film adjacent to the water slot wall, K;
Pr Molar Prandtl constant;
$y*$ Dimensionless distance from water slot wall;

y_w Distance from the geometric center of the cooling water grid adjacent to the water slot wall, m;

y_T^* —Dimensionless heat transfer boundary layer thickness, determined by the value of y^* at the intersection of the linear law and the logarithmic law;

κ Von Karman constant;

E Empirical constant, $E = 9.793$;

E' Rough wall surface correction function.

The value of E' is obtained through the following formula:

$$\frac{1}{\kappa}\ln\frac{E}{E'} = \begin{cases} 0 & K_s^+ \le 2.25 \\ \frac{\sin\left[0.4258\left(\ln K_s^+ - 0.811\right)\right]}{\kappa}\ln\left(\frac{K_s^+ - 2.25}{87.75} + C_s K_s^+\right) & 2.25 < K_s^+ \le 90 \\ \frac{1}{\kappa}\ln\left(1 + C_s K_s^+\right) & 90 < K_s^+, \end{cases} \tag{3.13}$$

where

K_s^+ Dimensionless roughness height $\left(K_s^+ = \rho_w K_s C_\mu^{1/4} k^{1/2}/\mu_l\right)$;

K_s Physical roughness height, m;

C_s Roughness constant, C_s=0.5.

The water slot exit is set as pressure -outlet. Other boundary conditions, such as the temperature and speed at the inlet, and the roughness height of the water slot wall, are set according to the actual roughness. The physical parameters of the cooling water are detailed in Table 3.3.

(c) Mechanical Boundary Conditions

In the actual thin slab continuous casting process, both the shell and the mold undergo complex mechanical behavior. According to the above assumption of this book, the mechanical boundary conditions of the shell solidification and the mold mainly include displacement boundaries and force boundaries. For the mold displacement constraint, since the deformation of the mold copper plate is ignored, the displacement constraints of the mold wide face and narrow face copper plates are all set to 0. The displacement of the slab symmetry plane is also set to 0, and the ferrostatic pressure of the molten steel at the solidification front is also applied to the shell element corresponding to the solid phase fraction $f_s = 0.884$. The mechanical constraint between the slab surface and the hot face of the mold copper plate is rigid-soft contact, and then a 3D thermo-mechanical calculation analysis is performed on the slab, mold and cooling water.

Table 3.3 Physical parameters of cooling water for thin slab caster

Parameter	Value	Unit
Density	998.2	kg/m^3
Specific heat	4182	J/(kg K)
Thermal conductivity	0.6	W/(m K)
Viscosity	0.0006	kg/(m s)

3.4.2 *Thermo-mechanical Behavior of Shell Solidification in Mold*

3.4.2.1 Shell Deformation

Figure 3.13 shows the 3D deformation distribution of the shell solidification in the width and thickness direction in the funnel mold during the production of Qste380TM low carbon Nb-Ti-bearing steel by a CSP continuous casting caster in a certain steel plant. The continuous casting process parameters are shown in Table 3.4. For the deformation in the width direction of the slab, because the solidification process of the shell within a certain height range at the upper part of the mold is the main process of eliminating the superheat of the molten steel, especially in the mold area transitioning from the funnel area to the straight area, the solidification of the shell significantly lags behind, and the shrinkage and other deformations of the shell are not significant. Therefore, in the thin slab mold, the shrinkage deformation of the shell solidification in the width direction is mainly concentrated in the middle and lower part of the mold, and it is most significant at the corners of the slab, as shown in Fig. 3.13a.

The deformation of the shell solidification along the thickness direction is due to the funnel structure of the wide face copper plate of the thin slab caster, which gradually transitions from the upper part to the lower part (as shown in Fig. 3.14). The slab is affected by the extruding and thinning effect of the copper plate in the funnel area on its thickness direction during the solidification process. The deformation of the solidified shell along the thickness direction is mainly concentrated in the wide face funnel area of the mold and also mainly concentrated in the middle and lower part of the mold, as shown in Fig. 3.13b.

Figure 3.15a shows the overall deformation of the slab and its shrinkage appearance in different height directions caused by the combined effect of solidification shrinkage and copper plate extruding during the solidification process of the Qste380TM low carbon Ni–Ti-bearing steel thin slab in the funnel mold (the deformation amount of the solidified shell in the figure is magnified 5 times). The shrinkage deformation of the thin slab in the mold is mainly concentrated in the slab corner, and from the height of 300 mm below the meniscus of the mold, the shrinkage deformation of the wide and narrow face corner of the solidified shell increases gradually as it moves downward.

Specifically for the wide face corner of the slab, within the range of 50 mm height below the meniscus, due to the high temperature and thinness of the solidified shell, the shrinkage of the slab in the thickness direction is small, and the solidified shell is closely attached to the wide face copper plate for rapid heat transfer, as shown in Fig. 3.15b. However, as the slab solidifies and moves downward, the temperature of the narrow shell continues to decrease, thus the shell continues to shrink in the thickness direction and drives the deformation of the wide face corner of the slab and the shell in off-corner toward the center of the slab thickness. In this process, due to the lack of effective compensation for the shrinkage of the slab in the thickness

Fig. 3.13 3D deformation behavior of the shell in funnel mold. **a** Wide face direction; **b** thickness direction

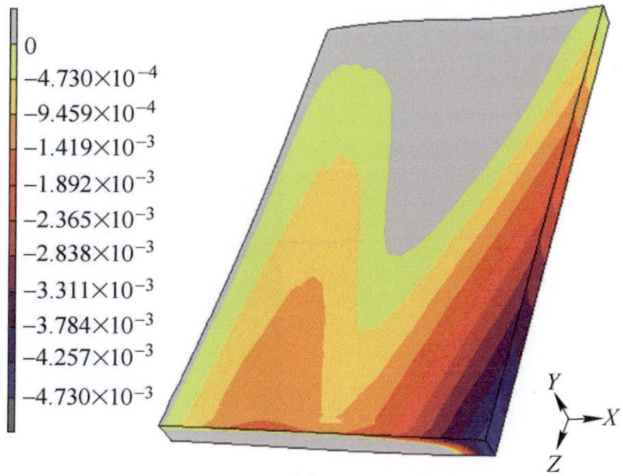

(a)

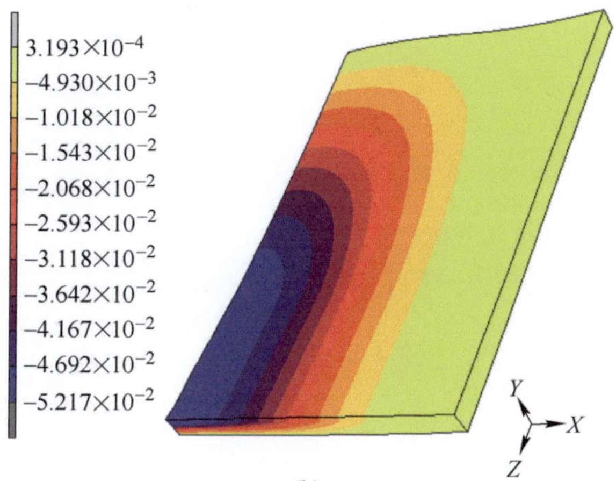

(b)

Table 3.4 Casting process of Qste380 low carbon Nb-Ti-bearing steel thin slab

Casting speed (m/min)	Cross-sectional size (mm × mm)	Casting temperature (°C)	Wide face water volume (L min^{-1})	Narrow face water volume (L min^{-1})	Water temperature difference (°C)	Liquidus of steel (°C)
4.0	1250 × 90	1550	6677	310	4.5	1525

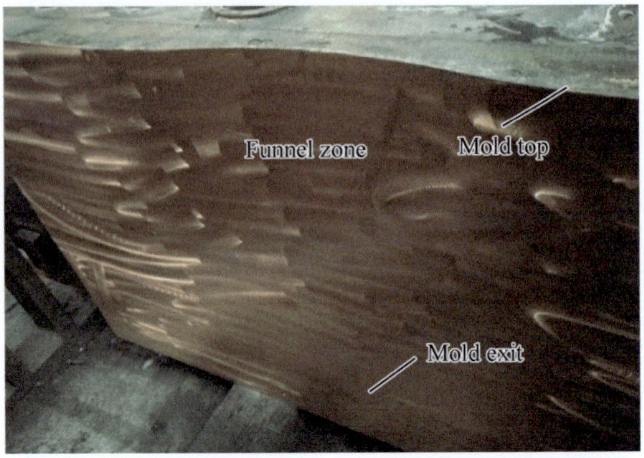

Fig. 3.14 Wide face copper plate of the CSP funnel mold

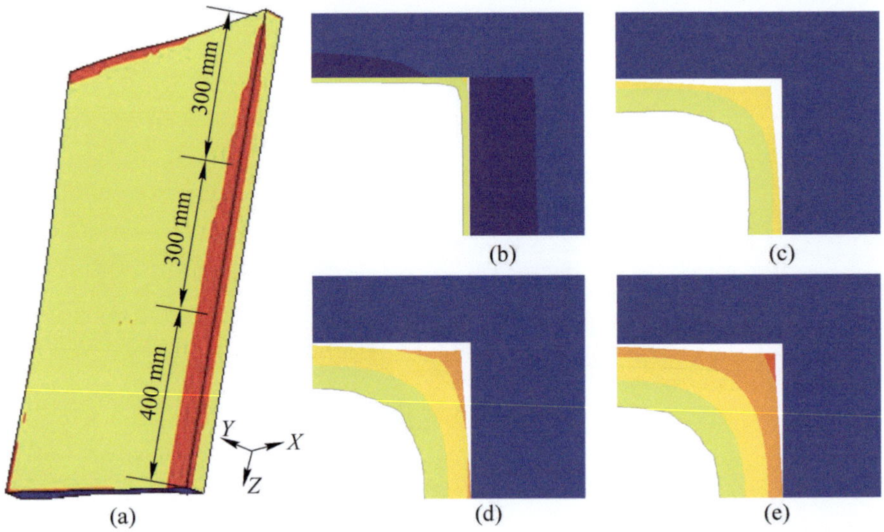

Fig. 3.15 Deformation and shrinkage of solidified shell in funnel mold. **a** Overall deformation; **b** 50 mm below meniscus; **c** 300 mm below meniscus; **d** 600 mm below meniscus; **e** at mold exit

direction by the parallel area of the wide face copper plate of the thin slab mold (the width difference between the upper and lower parts of the narrow face copper plate of the CSP mold in a certain steel plant, the taper compensation amount of the parallel area along the slab thickness direction is only 0.2 mm), it causes the solidified shell in the wide face corner and the off-corner to detach from the wide face copper plate earlier and form a large interfacial gap, as shown in Fig. 3.15c–e.

For the narrow face corner of the slab, the solidification shrinkage deformation characteristics of the shell are similar to those of the wide face corner. The shrinkage deformation within the 50 mm height range below the meniscus is not significant, and the shell is closely attached to the narrow face copper plate. As the slab continues to cool down, the shell gradually separates from the copper plate to form a large interfacial gap. However, in this process, due to the continuous linear taper compensation of the solidification shrinkage of the slab in the direction of the wide face center by the narrow face copper plate, the gap between the shell and the narrow face copper plate of the mold shows a trend of increasing and then slightly decreasing along the height direction of the mold.

3.4.2.2 Mold Flux Film Distribution

Due to the combined effects of the shrinkage gap of the solidified shell in the wide and narrow face corners of the thin slab mold and the evolution of the slab surface temperature, a significant phenomenon of concentrated distribution of mold flux film and air gap in the slab corner has formed. Figure 3.16 shows the calculated results of the distribution of the thickness of the mold flux film in the flux channel in the slab corner at different heights inside the thin slab mold. The distribution of the thickness of the mold flux film in the wide and narrow face corners of the thin slab is generally similar, all showing that the thickness is thinner in the upper part of the mold and mainly concentrated in the slab corner. As the solidified shell moves downward, it gradually concentrates in the off-corner of the wide and narrow faces. When the solidified shell moves downward from 100 to 400 mm below the meniscus, the thickness of the mold flux film in the off-corner of the wide and narrow faces of the slab increases rapidly, with the increment reaching 0.27 and 0.30 mm, respectively, as shown in Figs.3.16a, b. This phenomenon will cause "hot spots" to appear in the off-corner of the wide and narrow faces of the thin slab, which will then cause longitudinal crack or steel leakage accident in the off-corner of the thin slab (most significant on the wide face). Afterward, as the shrinkage deformation of the wide and narrow face corners and the off-corner of the slab is gradually stable, the distribution of the thickness of the mold flux film in the wide and narrow face corners of the slab and its off-corner also tends to be stable.

3.4.2.3 Air Gap Distribution

Figure 3.17 shows the distribution of air gap thickness in the wide and narrow face corner of the slab in the mold under the typical continuous casting process of a certain steel plant. As can be seen from Fig. 3.17a, the air gap at the corner of the wide face of the slab first forms at about 100 mm below the meniscus and then continues to increase in thickness as the slab moves downward. The maximum value of the air gap on the wide face of the slab appears at the corner of the mold exit and is about 0.33 mm. The phenomenon of the increasing air gap in the wide face corner of the thin slab mold as the slab moves downward is mainly due to the increasing shrinkage of the solidified shell in the thickness direction, and the mold at 400 mm below the meniscus has completely solidified. However, in this process, the shrinkage of the solidified

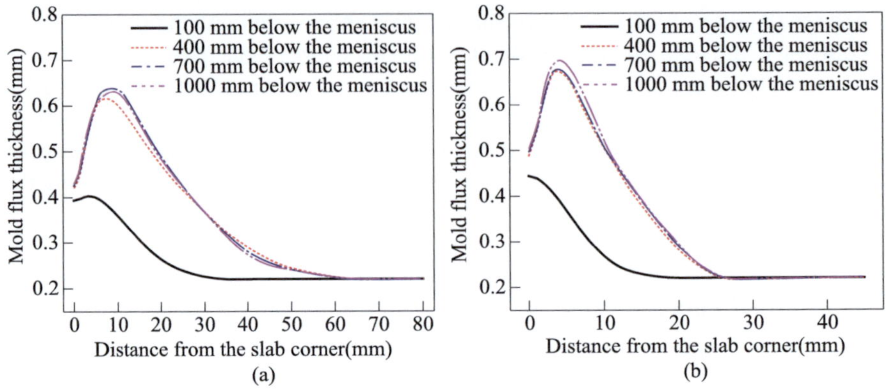

Fig. 3.16 Distribution of solid flux film thickness, **a** Wide face; **b** narrow face

shell in the corner along the thickness direction lacks effective compensation from the taper of the wide face copper plate (from the meniscus to the mold exit, the effective compensation of the copper plate taper along the slab thickness direction is only about 0.18 mm), which causes the interfacial gap between the wide face corner of the slab and the copper plate to gradually increase. In this process, since the thickness of the mold flux film in the wide face corner of the slab is constant in the middle and lower part of the mold (as shown in Fig. 3.16a), the increased interfacial gap is converted into the thickness increase of the air gap, forming a trend of gradually increasing air gap thickness along the height of the mold in the wide face corner of the mold.

In addition, the air gap near the wide face corner of the slab rapidly decreases from the corner to its off-corner, and the thick air gap is mainly distributed within 0–20 mm

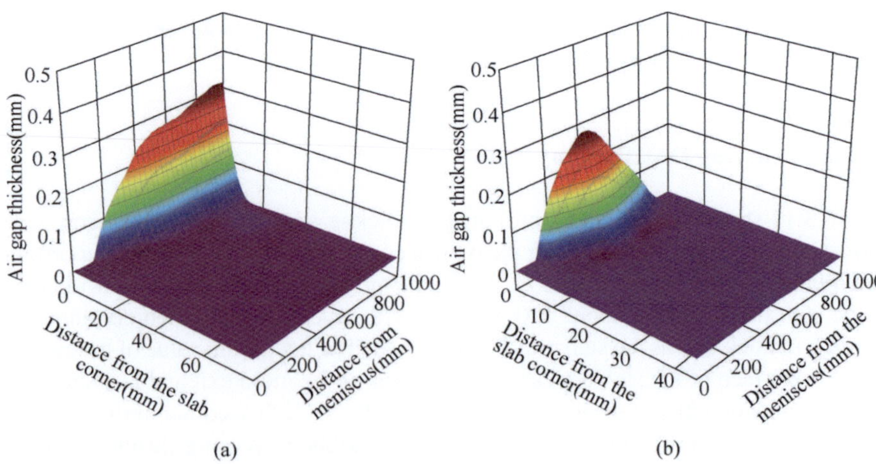

Fig. 3.17 Air gap distribution around slab corners. **a** Wide face; **b** narrow face

from the wide face corner of the slab. The main reason for this distribution of the air gap is that during the solidification process of the thin slab in the mold, the wide face of the slab within 0–20 mm from the solidified shell corner cools rapidly to below the solidification temperature of the mold flux, while the surface temperature of the slab in the off-corner far from the corner is higher, there is still a liquid flux layer inside the interface between shell and mold. Under the action of mold oscillation, the fluid flux film fills the gap formed by the solidification shrinkage of the shell along the thickness direction, as shown in Fig. 3.16a. This forms a trend of rapid decrease in the air gap from the wide face corner of the slab to the off-corner.

Figure 3.17b shows the distribution of air gap in the narrow face corner of the slab and its vicinity in the thin slab mold. It can be seen that the air gap in the narrow face corner of the slab and its vicinity in the thin slab mold is mainly concentrated in the middle part of the mold height ranging from 100 to 820 mm below the meniscus, showing a trend of rapid increase in the upper part of the mold height and rapid decrease in the lower part. The maximum thickness of the air gap is located at about 400 mm below the meniscus and is about 0.28 mm. The main reason for this distribution trend of air gaps in the narrow face corner of the thin slab is: at the middle height of the mold, the heat transfer rate of the slab is high, and the thermal shrinkage in the direction of the wide face center also increases significantly. However, the narrow face copper plate of the traditional thin slab mold is a linear plane structure, and the taper compensation method adopted cannot effectively attach the shrinkage of the shell, resulting in a large gap between the narrow face shell corner in the middle part of the mold and the copper plate, leading to the formation and rapid increase of air gap. As the slab continues to move downward, the shrinkage of the solidified shell toward the wide face center gradually slows down. Under the continuous compensation of the narrow face copper plate taper, the gap between the narrow face corner of the slab and the copper plate gradually decreases and finally completely disappears at about 820 mm below the meniscus.

In addition, similar to the lateral distribution of air gap in the wide face corner of the slab, the air gap in the narrow face of the thin slab mold is also mainly concentrated in the area within 0–10 mm from the solidified shell corner. The reason for this phenomenon is similar to that of the corner of wide face.

3.4.2.4 Shell Temperature Distribution

Figure 3.18 shows the 3D temperature field distribution of the solidified shell under the heat transfer conditions at interface between shell and copper plate in the above thin slab mold. It can be seen that during the solidification process of the thin slab in the caster, due to its high heat transfer rate in the area near the meniscus, the temperature gradient of the solidified shell in the casting direction is large. However, in the range of about 100 mm below the meniscus, the heat transfer medium between the solidified shell and the copper plate is mainly the liquid flux layer and the solid flux layer, and the distribution is relatively uniform, so the temperature distribution of the solidified shell on its wide and narrow faces is relatively uniform.

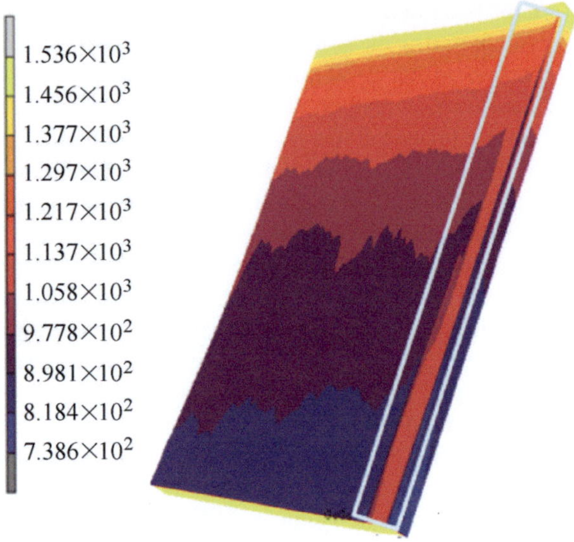

Fig. 3.18 Shell temperature distribution in the mold

However, as the slab moves downward, the shrinkage gap in the off-corner of the solidified shell gradually fills with a thick flux film and air gap, reducing the heat transfer rate in this area, causing the off-corner of the slab to gradually form a "hot spot," as marked in the box in Fig. 3.18. As a result, in the actual production process, the grains of the solidified structure on the surface of the slab in this area will be relatively coarse, thereby reducing the hot ductility of the structure in this area. In the actual production process, the hot rolled coil produced by the CSP line of this steel plant often forms a warping defect within 100–200 mm from its edge. Apart from the off-corner of the slab, the transverse temperature field distribution of the solidified shell is generally uniform on the wide and narrow face. In addition, during the solidification process of the continuous casting slab, as the heat transfer rate from the solidified shell to the mold gradually decreases, the surface temperature evolution of the shell slows down, that is, the temperature gradient along the casting direction gradually decreases.

Specifically for the corner of the thin slab, as shown in Fig. 3.19, within the range of 0–100 mm below the meniscus, because the corner of the slab has not yet formed an air gap and a thick flux film, it is affected by its 2D heat transfer, the heat transfer rate of the solidified shell is fast, and the maximum cooling rate exceeds 100 °C/s. Under the effect of this rapid cooling, the grains of the initial solidified structure formed at the corner of the slab will be significantly refined, and since the temperature of the corner of the slab has dropped to nearly 1100 °C during this process, the precipitation size of the Ti-bearing microalloy carbonitrides will be relatively fine, and the distribution will also be dispersed. Therefore, the ductility of the structure formed by solidification at the corner of the slab will be high, which is beneficial for crack control at the corner of the slab.

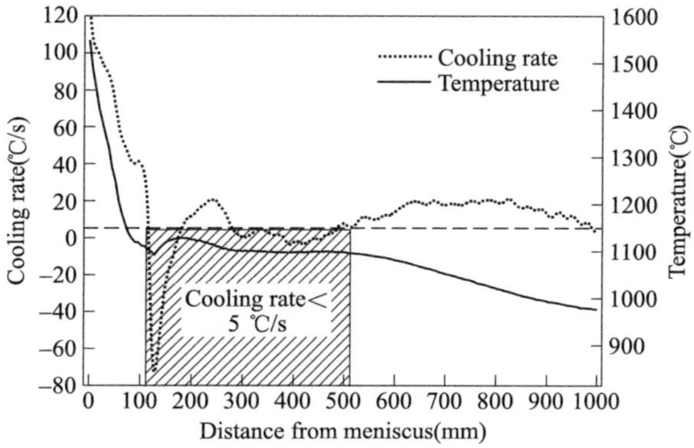

Fig. 3.19 Temperature distribution at the center of wide face and narrow face and slab corner

However, as the slab continues to move downward, the corner of the solidified shell is affected by the concentrated distribution of thick air gap and flux film, and the heat transfer rate decreases rapidly. When the slab moves downward to about 110–515 mm below meniscus, the cooling rate of the corner of the shell drops to less than 5 °C/s. What is more noteworthy is that due to the rapid growth of the air gap at the corners of the wide and narrow face of slab, a reheating phenomenon occurs at the corner of the slab in the range of about 125–160 mm below meniscus. The slow heat transfer characteristic of the slab corner in the middle of the thin slab mold will cause the fine grains initially solidified on the slab surface to coarse, which is not conducive to the high ductility control of the slab corner structure. At the same time, because the temperature of the slab corner at this position is in the high temperature precipitation zone of carbonitride of microalloy elements such as Nb, Al, B, the slow heat transfer of the slab corner will promote the precipitation of its microalloy carbonitrides at austenite grain boundaries, thereby embrittling the grain boundaries and reducing the hot ductility of the micro-alloyed steel slab corner structure.

As the slab continues to move downward, due to the slowing contraction of the slab in the width direction and the continuous compensation effect of the narrow face copper plate taper, the air gap at the corner of the solidified shell gradually decreases, and the heat transfer rate at the corner of the slab increases again to above 5 °C/s. Therefore, in the upper part of the mold, the heat transfer behavior of the corner of the solidified shell is not suitable for the high ductility control of the structure of the thin slab of Nb, B, and Al micro-alloyed steel.

3.5 Thermo-mechanical Behavior of Thin Slab in Secondary Cooling

During the CSP continuous casting production process, the slab is subject to significant stress not only during the bending and straightening processes but also due to the reduction of liquid core. The stress behavior of the slab is related not only to the structure and deformation process of the continuous caster but also to the temperature field distribution within the casting flow. Understanding the evolution of the temperature field within the secondary cooling casting flow under the different continuous casting processes is the basis for improving the stress during the production process of thin slab and improving their surface quality. For this purpose, this book first uses the FLIR A40 infrared thermometer to perform infrared temperature measurement on the slab surface during the production process of Qste380TM Nb-Ti low carbon micro-alloyed steel by CSP continuous caster in a steel plant. Based on this, by constructing a 3D heat transfer and mechanical evolution coupled calculation model of thin slab secondary cooling, the secondary cooling solidification heat transfer and mechanical evolution rules of typical micro-alloyed steel continuous casting production process are studied and analyzed.

3.5.1 Temperature Measurement of Thin Slab in Secondary Cooling

The structure of the CSP continuous caster is very compact, and it is difficult to measure the surface temperature of the slab at all casting flow positions during the actual production. Combining the actual situation of a certain steel plant, this book only measures the infrared temperature of the narrow face and inner arc corner of the slab at the key position of the continuous casting straight segments, that is, 0.7 m before the No.1 withdrawal and straightening roller, the entrance of the No.2 withdrawal and straightening roller, the entrance of the No.3 withdrawal and straightening roller, the entrance of the No.4 withdrawal and straightening roller and 0.7 m after the No.4 withdrawal and straightening roller. The steel grade for the temperature measurement is Qste380TM low carbon Nb-Ti-bearing micro-alloyed steel, the slab cross-sectional size is 1250 mm × 90 mm, the casting speed is 4.0 m/min, and the superheat of the molten steel is 25 °C. The FLIR infrared thermal imager (Thermal CAM) is used for measurement, as shown in Fig. 3.20.

Figure 3.21 shows the infrared thermal images of the slab surface at the key positions such as 0.7 m before the No.1 withdrawal and straightening machine, the entrance of the No.2 withdrawal and straightening roller, the entrance of the No.3 withdrawal and straightening roller, the entrance of the No.1 withdrawal and straightening roller, 0.7 m after the No.4 withdrawal and straightening roller and the temperature change over time at the slab corner (SP01 point in Fig. 3.21) under the above temperature measurement conditions. As can be seen from Fig. 3.21, during the temperature measurement period, the temperatures at the corners of the slab at the corresponding key positions are approximately 911, 905, 901, 893, and 884 °C, respectively (according to the principle of infrared temperature measurement, the highest temperature during the temperature measurement period is closest to the actual temperature of the continuous casting slab surface).

Fig. 3.20 Temperature measurement of slab surface during CSP casting

From the measured temperature results, it can be seen that the corner temperature of the slab during the straightening process is relatively high. According to Fig. 3.6, the RA of Qste380TM low carbon Nb-Ti-bearing steel under the different deformation rates changes with temperature. When the temperature of the slab is above 890 °C, its RA exceeds 56%, and the structure has high ductility. Based on the device accuracy meeting the process conditions, the slab production process generally does not produce surface crack defects (existing research shows [41], when the RA of high temperature steel structure is higher than 40%, the slab will not produce crack defects if the device meets the accuracy).

3.5.2 Evolution of Temperature and Stress of Thin Slab in Secondary Cooling

The secondary cooling of CSP continuous casting has high cooling intensity, and it is accompanied by liquid core reduction, bending, and straightening, so the solidification thermo-mechanical behavior of slab is very complicated. At the same time, during the CSP thin slab continuous casting process, the cooling intensity of the secondary cooling zone is large, and there is a large temperature gradient along the casting direction of the slab, which causes the obvious heat transfer along the casting direction. Based on the structure of CSP continuous caster in a steel plant and the

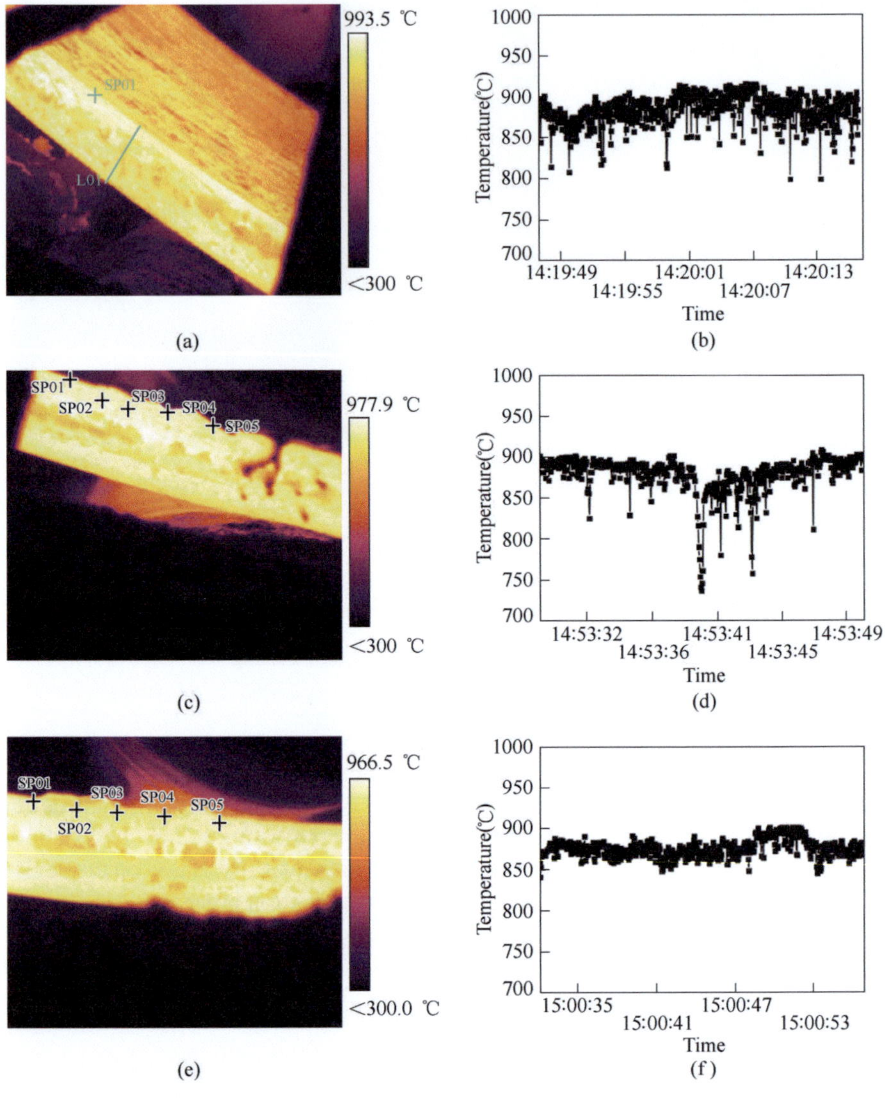

Fig. 3.21 Slab surface temperature measured at different positions of secondary cooling. **a** Thermal imaging map at 0.7 m before No.1 withdrawal and straightening roller; **b** slab corner temperature at 0.7 m before No.1 withdrawal and straightening roller; **c** thermal imaging map at the entrance of No.2 withdrawal and straightening roller; **d** slab corner temperature at the entrance of No.2 withdrawal and straightening roller; **e** thermal imaging map at the entrance of No.3 withdrawal and straightening roller; **f** slab corner temperature at the entrance of No.3 withdrawal and straightening roller; **g** thermal imaging map at the entrance of No.4 withdrawal and straightening roller; **h** slab corner temperature at the entrance of No.4 withdrawal and straightening roller; **i** thermal imaging map at 0.7 m after No.4 withdrawal and straightening roller; **j** slab corner temperature at 0.7 m after No.4 withdrawal and straightening roller

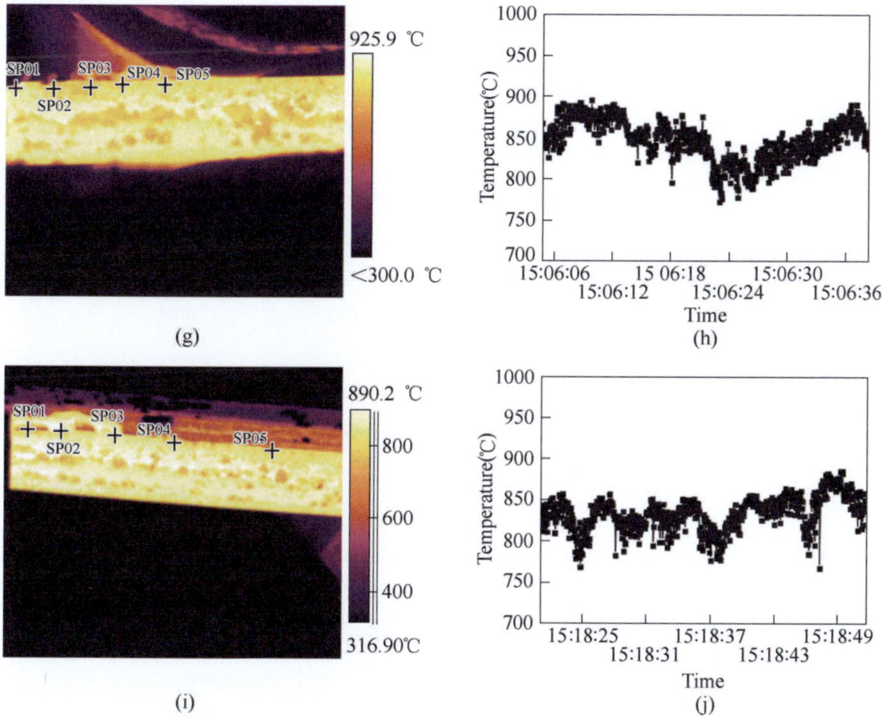

Fig. 3.21 (continued)

Qste380TM low carbon Nb-Ti-bearing micro-alloyed steel production process, this book constructs a 3D thermo-mechanical coupling finite element calculation model of the secondary cooling zone of the slab to explore the solidification and heat transfer as well as mechanical evolution behavior of the slab in the secondary cooling zone during the continuous casting of thin slab of micro-alloyed steel.

According to the secondary cooling zone of CSP continuous caster in a steel plant, the secondary cooling zone includes six cooling zones, which correspond to the foot roller, grid segment, No.1 bow segment, No.2 bow segment, No.3 bow segment, and No.4 bow segment, with a total length of about 8.705 m. The length and spray area of each zone are shown in Table 3.5, and the water volume of each zone of Qste380TM steel under the condition of 4.0 m/min casting speed is shown in Table 3.6.

3.5.2.1 Mathematical Model

In the process of the thin slab continuous casting, it is generally believed that the solidification heat transfer and stress along the width direction is symmetrical. Therefore, in the process of developing the mathematical model, the solid model only needs to develop the transverse 1/2 model of the slab, and the size of the slab model is

Table 3.5 Division of the secondary cooling zone

Cooling zone	Segment	Length of cooling zone (m)	Distance below Meniscus (m)	Cooling area of cooling zone (m^2)
1	Foot roller segment	0.06	1.060	0.20
2	Grid segment	0.49	1.550	1.65
3	No.1 bow segment	1.27	2.820	4.27
4	No.2 bow segment	1.98	4.795	6.64
5	No.3 bow segment	2.45	7.245	8.23
6	No. 4 bow segment	2.45	9.705	8.27

Table 3.6 Water volume of in secondary cooling zone during casting Qste380TM steel at a casting speed of 4.0 m/min

Cooling zone	Included segment	Corresponding control loop	Cooling water volume(L/min)
1	Foot roller segment	1	330.0
2	Grid segment	2	1720.5
3	No.1 bow segment	3.0	1128.4
		3.1	386.5
		3.2	254.2
4	No.2 bow segment	4.0	990.5
		4.1	340.6
		4.2	215.5
5	No.3 bow segment	5.0	665.5
		5.1	244.3
		5.2	149.5
6	No.4 bow segment	6.0	345.5
		6.1	100.5
		6.2	145.0

625 mm × 90 mm. At the same time, in order to fully consider the mechanical effect of the casting roller of the continuous caster bow segment on the slab during the process of liquid core reduction and clamping, based on the actual roller array structure of the CSP continuous caster in a steel plant, a 3D finite element model including the roller array and the slab system is developed, as shown in Fig. 3.22. It is believed that the No.1 bow segment of the CSP continuous caster of a certain steel plant is the liquid core reduction segment, with a total of 9 pairs of rollers.

The reduction amount of Qste380TM low carbon Nb-Ti-bearing micro-alloyed steel continuous casting production process is 13.5 mm. The casting slab is meshed by regular hexahedron unit, and the roller array of bow segment is set as rigid body structure.

A. Model Assumptions

To simplify the calculation model, the following assumptions and simplifications are assumed for the solidification heat transfer and stress process of the thin slab in the secondary cooling casting flow:

(1) The behaviors of heat transfer and stress of the slab solidification process along the width and thickness directions are symmetrical.
(2) The high temperature physical properties of the slab during the solidification process are isotropic, and its heat transfer and mechanical properties are only functions of temperature.
(3) The surface of the slab in the same cooling zone is uniformly cooled.
(4) CSP is a typical straight arc continuous caster, and its solidification is basically completed in the vertical zone, assuming that the heat transfer conditions of the inner arc and outer arc of the slab during the solidification process are consistent.
(5) The heat carried away by the heat conduction of the casting roller is converted to the comprehensive heat transfer coefficient of each zone of the secondary cooling.
(6) Ignore the friction between the slab and the bow segment.
(7) Ignore the deformation of the casting roller in the process of liquid core reduction, the contact between slab and casting roller in the bow segment is rigid-soft contact.

B. Heat Transfer Equation

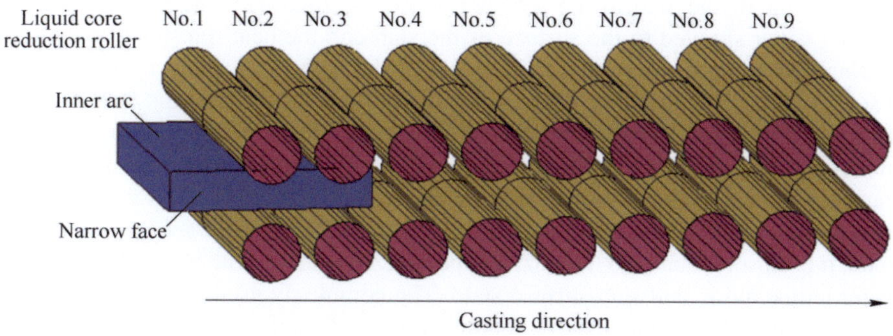

Fig. 3.22 Schematic diagram of the thermo-mechanical calculation model of thin slab continuous casting secondary cooling

Based on the above assumptions, considering that the solidification process of the slab in the secondary cooling is an unsteady heat transfer process, the control equation for the solidification heat transfer of the slab is selected as Eq. (3.14):

$$\rho c \frac{\partial t}{\partial \tau} = \frac{\partial}{\partial y}\left(k_{\text{eff}} \frac{\partial t}{\partial y}\right) + \frac{\partial}{\partial x}\left(k_{\text{eff}} \frac{\partial t}{\partial x}\right) + S_{\text{o}}, \tag{3.14}$$

t Temperature, °C;
ρ Density, kg/m^3;
c Specific heat capacity, kJ/(kg °C);
k_{eff} Thermal conductivity, W/(m °C);
S_{o} Internal heat source term, W/m^3.

The mechanical control equations are the constitutive equations shown in formulas (3.9) and (3.10).

C. Initial Conditions of the Model

Since the restart technology is adopted in the construction process of this calculation model, the initial temperature field and stress of the slab inherit the results of the thermo-mechanical coupling calculation model of the shell solidification in the mold of Sect. 3.4, the initial conditions of the heat transfer and stress of the slab in this calculation model are the temperature and stress state of the slab when it exits from mold.

D. Boundary Conditions

(a) Heat Transfer Boundary in Secondary Cooling Zone

During the thin slab continuous casting process, there are mainly three forms of heat transfer of the slab in the secondary cooling zone: (1) The heat transfer of cooling water sprayed to the surface of the slab and the heat transfer between the casting roller and the surface of the slab; (2) the heat taken away by the air and water evaporation caused by the mist spray cooling; (3) the heat dissipation from the surface of high temperature slab to the air is radiation heat transfer. In the actual model development process, it is difficult to quantitatively describe each form of heat transfer. The comprehensive heat transfer coefficient of the secondary cooling zone is an important parameter to describe the heat transfer effect of the secondary cooling zone. It is closely related to the nozzle type, nozzle arrangement, water flow density, water temperature, and slab surface state. The commonly used empirical formulas for calculating the comprehensive heat transfer coefficient of the secondary cooling zone under the different continuous casting conditions can be found in Sect. 2.4.1.2 of Chap. 2 on the slab heat transfer and mechanical models.

Considering the actual situation of high water flow density in each secondary cooling zone during the thin slab continuous casting process, this book chooses the empirical heat transfer coefficient formula measured by Cai et al. [45]:

$$\begin{aligned} h &= 0.61W^{0.597}\left(3 < W < 10\ \mathrm{L}/\left(\mathrm{m}^2\ \mathrm{s}\right), t_s = 800\ ^\circ\mathrm{C}\right) \\ h &= 0.59W^{0.385}\left(3 < W < 20\ \mathrm{L}/\left(\mathrm{m}^2\ \mathrm{s}\right), t_s = 900\ ^\circ\mathrm{C}\right) \\ h &= 0.42W^{0.351}\left(3 < W < 12\mathrm{L}/\left(\mathrm{m}^2\ \mathrm{s}\right), t_s = 1000\ ^\circ\mathrm{C}\right), \end{aligned} \tag{3.15}$$

h Comprehensive heat transfer coefficient, kW/(m^2 °C);
W Water spray density, L/(m^2 s);
t_s Surface temperature of the slab, °C;

(b) Heat Transfer Boundary in Air Cooling Zone

In the CSP continuous casting process of a steel plant, the slab enters the air cooling zone after exiting the No.4 bow segment. The heat transfer mode of the slab is mainly radiation heat transfer to the surrounding environment. Therefore, the radiation heat transfer in the air cooling zone is set as the radiation heat transfer calculation formula shown in Eq. (3.16):

$$\begin{aligned} q &= \sigma\varepsilon\left((t+273)^4 - (t_{amb}+273)^4\right) = h_{rad}(t - t_{amb}) \\ h_{rad} &= \sigma\varepsilon\left((t+273)^2 + (t_{amb}+273)^2\right)((t+273) + (t_{amb}+273)) \end{aligned} \tag{3.16}$$

σ Stefan–Boltzmann constant, 5.67 × 10^{-8} W/(m^2 K^4);
ε Emissivity of the object, the value is 0.8;
t Surface temperature of the slab, °C;
t_{amb} Environment temperature, °C;
h_{rad} Equivalent heat transfer coefficient of the environment, W/(m^2 °C);

(c) Boundary of the Symmetry Plane

According to the assumption of heat transfer and stress symmetry in the wide face direction of the slab, the heat transfer of the symmetry plane of the slab is set to be adiabatic, and the displacement constraint is also 0, as shown in Eq. (3.17) and Eq. (3.18):

$$-\mathrm{k}_{\mathrm{eff}}\frac{\partial t}{\partial x}|B = 0 \tag{3.17}$$

$$\text{Displacement}_X = 0. \tag{3.18}$$

(d) Ferrostatic Pressure

In the CSP continuous casting process, the vertical height from the meniscus of the mold to the complete solidification point of the slab is 8–9 m, forming a huge ferrostatic pressure of molten steel, which can easily lead to bulge deformation of the shell. In the process of applying the ferrostatic pressure of molten steel, the same treatment method as in Sect. 3.4.1 is adopted, the ferrostatic pressure of molten steel is applied to the solidification front unit with a solid fraction of 0.88. The ferrostatic pressure expression is shown in Eq. (3.19).

$$P = \rho g h, \tag{3.19}$$

P Ferrostatic pressure on the solidification front of the slab, Pa;
ρ Density of molten steel, kg/m^3;
g Gravity acceleration, 9.8 m/s^2;
h Vertical height from the meniscus of the mold, m;
e Contact between Roller and Thin Slab during Liquid Core Reduction.

The inner and outer arc surface units of the slab and the casting roller are defined as rigid-soft contact pairs, and the contact analysis is automatically applied by the finite element software Marc during the calculation process.

3.5.2.2 Model Verification

Figure 3.23 is a comparison of the calculated temperature of the narrow face and corner of the slab in the production process of Qste380TM low carbon Nb-Ti-bearing micro-alloyed steel thin slab in a certain steel plant with the actual measured temperature. The temperature of the center and corner of the slab narrow face at the end of the bender and the entire straight segment calculated by the model is slightly higher than the measured temperatures at the corresponding position. The main reason is mainly due to the fact that the narrow and corner faces of the continuous casting slab in the temperature measurement zone are covered with iron oxide scale to a certain extent (as shown in Fig. 3.21), which reduces the measured temperature of the slab face. Table 3.7 is a quantitative comparison between the simulated and measured temperature values of the center and corner of the slab narrow face at the temperature measurement point. The relative error between the temperature values of the narrow face and corner of the slab calculated by model and the temperature values of the narrow face and corner of the slab measured in the field is only about 0.40% to 1.61%, which is within the allowable error range, the model calculation results can be considered effective.

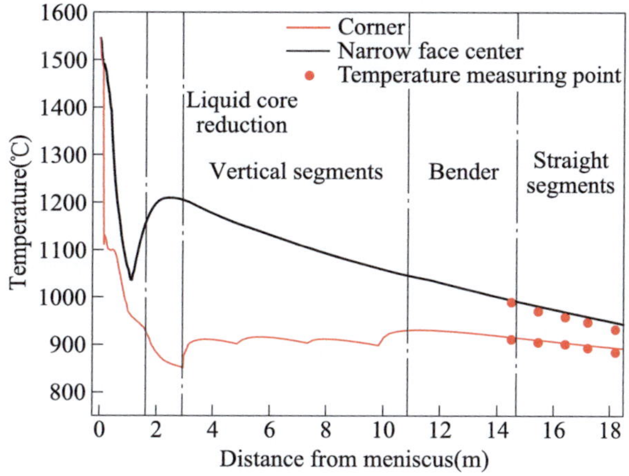

Fig. 3.23 Comparison of the calculated temperature and the measured temperature of thin slab surface in secondary cooling

Table 3.7 Comparison of predict temperature and measured temperature of the slab surface at different position of secondary cooling

Temperature measurement position		Measured value (°C)	Predicted value (°C)	Relative error (%)
0.7 m before No.1 withdrawal and straightening roller	Narrow face center	989	993	0.40
	Corner	911	916	0.55
Entrance of No.2 withdrawal and straightening roller	Narrow face center	970	980	1.03
	Corner	905	909	0.44
Entrance of No.3 withdrawal and straightening roller	Narrow face center	958	968	1.04
	Corner	901	904	0.33
Entrance of No.4 withdrawal and straightening roller	Narrow face center	932	947	1.61
	Corner	893	899	0.67
0.7 m behind the 4th withdrawal and straightening roller	Narrow face center	932	947	1.61
	Corner	884	894	0.13

3.5.2.3 Thin Slab Temperature in Secondary Cooling

Figure 3.24 shows the temperature field contour of Qste380TM Ni–Ti-bearing low carbon micro-alloyed steel at the end of No.1–4 segment, the entrance and end of the straight segment under the typical thin slab continuous casting process in a certain steel plant. It can be seen from the figure, in the zone above the No.3 bow segment of the CSP continuous caster (as shown in Figs. 3.24a–c), due to the high cooling intensity of the secondary cooling zone of the slab wide face, and the narrow face has no spray cooling effect after the narrow face foot roller zone (only two sets of nozzles on each side), the temperature of the slab wide face is significantly lower than the narrow face. When the slab solidifies to the No.4 bow segment and below, the temperature of the slab wide face quickly rebounds under the combined effect of the significant reduction in the cooling intensity of the bow No.4 segment, the air cooling of the top bending and the straight segment, the temperature begins to be higher than the narrow face temperature. Especially after the slab enters the straight segment, the temperature distribution of the slab wide face is very uniform after the air cooling zone is homogenized, and the temperature rises to more than 1100 °C, which meet the temperature requirements of the slab wide face defect-free straightening continuous casting production.

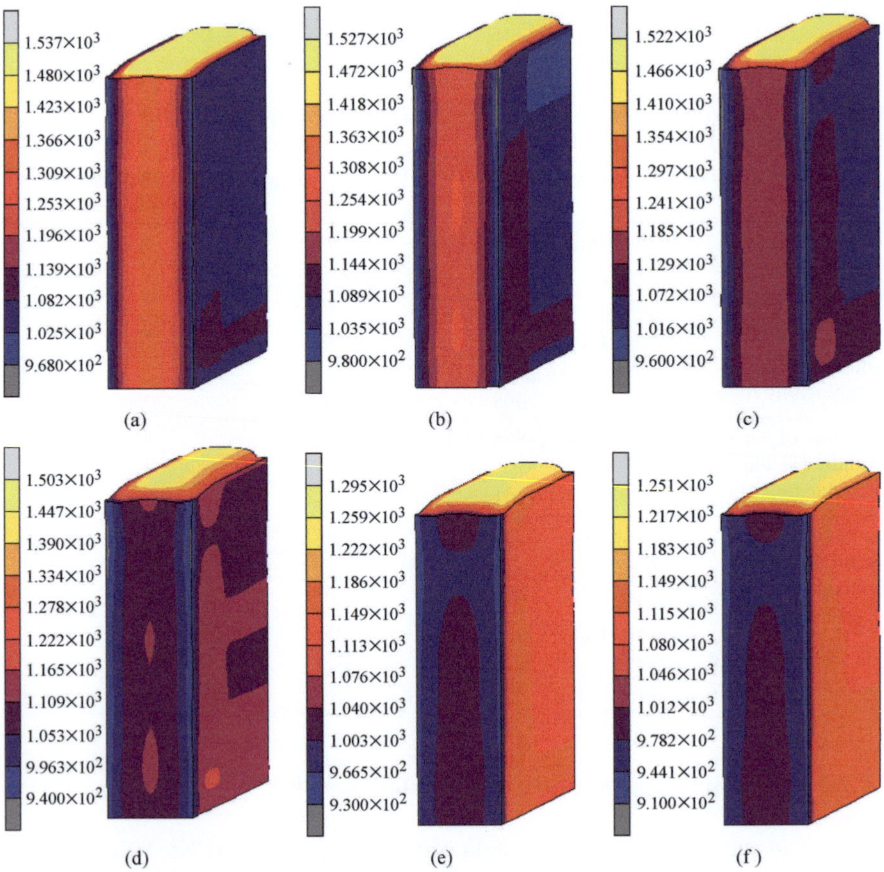

Fig. 3.24 Slab temperature field at different positions of the casting flow. **a** End of No.1 segment; **b** end of No.1 segment; **c** end of No.3 segment; **d** end of No.4 segment; **e** entrance of the straight segments; **f** end of the straight segments

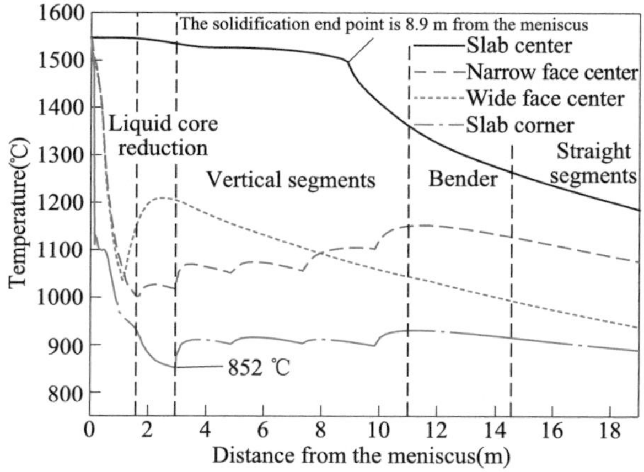

Fig. 3.25 Evolution of thin slab during continuous casting under the traditional process

Figure 3.25 shows the temperature history curve of the center, corner, wide face centerline, and narrow face centerline of the thin slab during the continuous casting process of Qste380TM steel. The temperature evolution of the slab center is relatively flat before the complete solidification point, and it shows a relatively rapid downward trend after the complete solidification point. Due to the small thickness of the thin slab and the high cooling intensity in the secondary cooling zone, the liquid core length of the slab is short. The liquid core length of Qste380TM low carbon Nb-Ti-bearing micro-alloyed steel in a certain steel plant is only about 8.9 m under the typical continuous casting process.

Compared with the center of the slab, the surface temperature fluctuation of the thin slab continuous casting process is more intense. Specific to the wide face of the slab, after exiting the mold, and the surface temperature continues its rapid cooling trend in the mold due to strong spray cooling effect of the foot roller and grid zone. After entering the bow No.1 segment (liquid core reduction segment), due to the decrease of cooling intensity in this segment relative to the foot roller and grid zone, the wide face of the slab begins to slightly reheat, overall change is in the range of about 1000–1025 °C. The test results, RA of Qste380TM low carbon Nb-Ti-bearing micro-alloyed steel with temperature evolution under the different deformation rates are shown in Fig. 3.6. The RA of the steel in this temperature range can reach over 80%, and the slab should not produce crack defects during the liquid core reduction. When the slab enters No.2 and No.3 bow segment, the cooling intensity of the wide face of the slab further decreases, and its surface temperature overall slightly increases. When the slab enters No.4 and below bow segment, the overall temperature of the wide face of the slab increases greatly first and then decreases slowly near the entrance of the top bending segment until it exits the continuous caster. During the straightening process, the overall temperature of the wide face of the slab is higher than 1080 °C, which meets the temperature requirement of high temperature over straight.

For the narrow face of the slab, the solidified shell after exiting the mold is cooled by two sets of foot roller nozzles in the narrow face foot roller zone, and the temperature at the center of the narrow face of the slab rapidly decreases with the temperature at the centerline of the wide face. However, when the slab exits the narrow face foot roller zone, overall temperature rapidly increases due to the lack of spray cooling effect on the narrow face of the shell and reaches the highest temperature of about 1210 °C in the liquid core reduction segment. Afterward, with the continuous increase of the thickness of the solidified shell and the effect of thermal radiation, the surface temperature of the slab begins to gradually decrease. When the slab enters the straight segments, the temperature at the center of the narrow face dropped reduced to about 990 °C.

For the corner of the slab, their overall temperature is lower than the temperature at the centerline of the wide and narrow faces of the slab. After the slab exits the mold, the temperature at the slab corner rapidly decreases due to the combined rapid cooling effect of the narrow face foot roller zone, wide face foot roller and grid zone. It is worth mentioning that when the slab enters No.1 bow segment (liquid core reduction segment), due to the thicker solidified shell at the corners of the slab and the strong spray cooling effect on the wide face, the temperature at the corners of the slab in this zone continues to decrease to a minimum value of about 852 °C (average cooling rate is about 3.9 °C/s). As can be seen from Fig. 3.6, the RA of Qste380TM low carbon Nb-Ti-bearing steel with temperature changes under the different deformation rates, the RA of the slab at this temperature is only about 36% under the conditions of high deformation rate, and the corner of the slab is easy to crack and expand into crack defects due to the deformation force of the liquid core reduction.

When the slab exits the liquid core reduction segment, the cooling water volume of the slab wide face begins to gradually decrease, and the corner of the slab shows a slight fluctuating upward trend in each cooling zone. At the end of each segment of the No.2–No.4 bow segment, the temperature of the slab corner is about 901, 903, and 899 °C, respectively, and the corner temperature is maintained above 895 °C during the whole straight segment. Similarly, it can be seen from Fig. 3.6 that the RA of Qste380TM low carbon Nb-Ti-bearing steel varies with temperature under different deformation rate, the RA of the steel at this temperature is greater than 57%, and the hot ductility of the corner structure of the slab is better. Therefore, under the premise that the accuracy of the device meets the requirements, it can be considered that the straightening process of thin slab continuous casting could not form obvious corner transverse crack defects.

3.5.2.4 Thin Slab Mechanical Behavior During Liquid Core Reduction

The liquid core reduction is a key technology in the thin slab continuous casting and rolling process. It is an important guarantee for accelerating slab solidification, increasing casting speed, matching the thickness of the slab between continuous casting and rolling and improving the center segregation of the slab. However, during the liquid core reduction of the high-speed thin slab, the bow segments will strongly extrude the slab with a liquid core, so that the slab (especially the inner arc) will cause large deformation and stress. As analyzed from above section, the ductility of the corner structure of the slab in the liquid core reduction segment during the production of micro-alloyed steel by the thin slab continuous caster is low, so it is necessary to explore the mechanical behavior characteristics of the slab during the liquid core reduction.

A. Deformation during the Liquid Core Reduction

Figure 3.26 shows the deformation contour along the thickness direction of the slab at the entrance, middle, and end of the reduction segment under the condition of 13.5 mm liquid core reduction for the Qste380TM low carbon Nb-Ti-bearing micro-alloyed steel continuous casting process (note: negative values in the figure represent the reduction deformation amount of the slab along the thickness direction). It is worth noting that the 9 pairs of reduction rollers in the liquid core reduction segment of CSP continuous caster in a certain steel plant adopt the same reduction increment during the actual reduction process, the reduction increment of each roller is 1.5 mm. At the entrance of the liquid core reduction segments, the narrow face of the slab is rapidly cooled by the foot roller and the grid segments, and its thermal shrinkage along the thickness direction is greater than the reduction, while the shell temperature of the wide face is generally high, and the slab does not form obvious deformation due to reduction. However, when the slab enters the middle of the reduction segment, the reduction of the reduction roller exceeds the thermal shrinkage of the slab, and the slab begins to obviously deform along the thickness direction, especially the maximum thinning amount of the slab corner reaches about 7.3 mm. When the slab solidifies to the end of the reduction segments, the reduction deformation amount of the reduction roller on the slab along the thickness direction further increases. At the last roller of the reduction segments, the thinning amount of the slab corner along the thickness direction reaches nearly 14.0 mm.

Figure 3.27 shows the displacement contour of the slab in the width expansion direction at the entrance, middle, and exit of the liquid core reduction segments (note: negative values in the figure represent the width expansion deformation amount relative to the initial size of the slab at the meniscus, mm). During the production process of Qste380TM low carbon Nb-Ti-bearing micro-alloyed steel under the typical CSP continuous casting process in a certain steel plant, the total thermal shrinkage of the slab in the width expansion direction reached 14.30 mm from the mold solidification to the entrance of the liquid core reduction. Subsequently, the slab enters the liquid core reduction segments and is strongly extruded by the reduction roller, causing a certain degree of deformation in the direction of casting and width expansion. When the slab solidifies to the exit of the liquid core reduction segments, the width expansion of slab caused by the liquid core reduction reaches about 7.74 mm.

B. Evolution of Equivalent Stress

Figure 3.28 shows the evolution of equivalent stress in the inner arc wide face and inner arc narrow face of the slab during the continuous casting production process

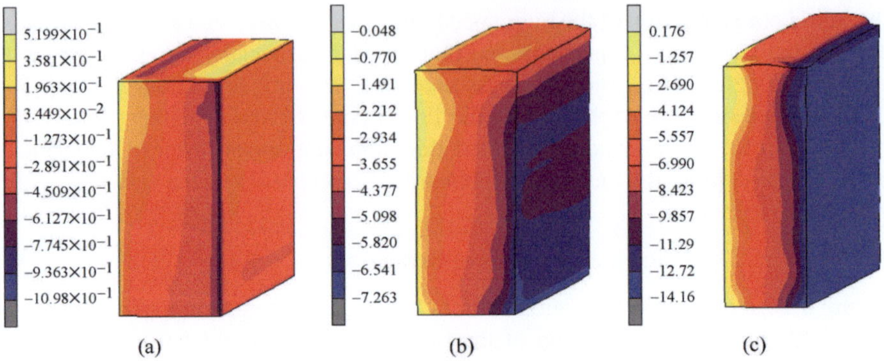

Fig. 3.26 Displacement contour of the thin slab along the thickness direction during liquid core reduction. **a** Entrance of reduction segment; **b** middle of reduction segment; **c** end of reduction segment

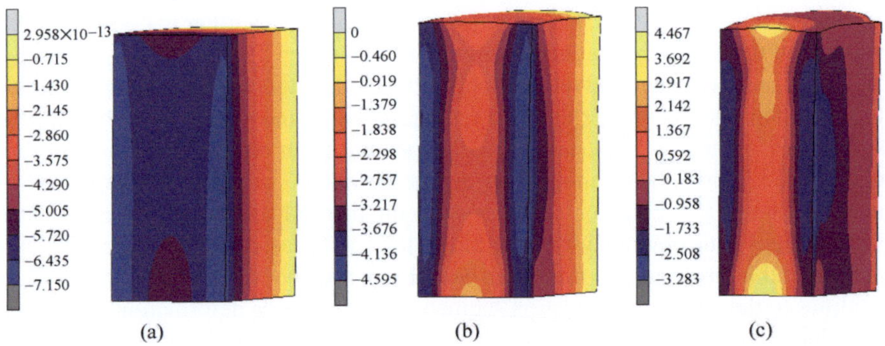

Fig. 3.27 Displacement contour of the thin slab along the with direction during liquid core reduction. **a** Entrance of the reduction segment; **b** middle of the reduction segment; **c** end of the reduction segment

of Qste380TM low carbon Nb-Ti-bearing micro-alloyed steel. As can be seen from the figure, in the liquid core reduction segments, due to the periodic extruding of the casting roller, the stress of the wide and narrow face of the slab fluctuates periodically when the slab moves downward. For the inner arc wide face, under the action of the casting roller reduction stress, the stress at the centerline is concentrated in the range of 40–45 MPa, while the stress at the corner significantly increases, mainly concentrated in the range of 60–65 MPa. Compared with the wide face, the equivalent stress of the centerline of the slab narrow face significantly decreases, mainly concentrated in the range of 4–5 MPa. Therefore, the corner of the slab is subject to the greatest stress in the liquid core reduction segments. As can be seen from Fig. 3.6, RA of Qste380TM low carbon Nb-Ti-bearing steel changes with evolution of temperature under the high deformation rate, the strong stress and low ductility of the slab in the liquid core reduction segments easily cause transverse crack defects in the inner arc

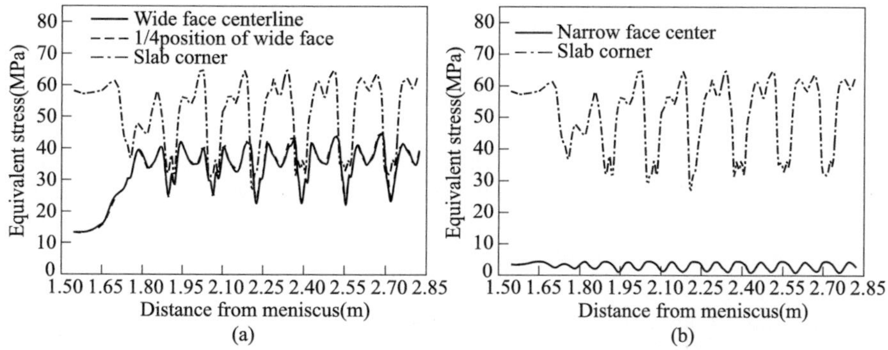

Fig. 3.28 Evolution of slab equivalent stress during liquid core reduction under traditional process. **a** Wide face; **b** narrow face

corner of the slab. Therefore, it further confirms that the concentrated stress on the slab during the liquid core reduction and the low ductility of the slab corner structure in this segment are the key reasons for the corner cracks of the micro-alloyed steel thin slab.

3.6 Thin Slab Corner Carbonitride Precipitation Control

From the analysis of the solidification thermo-mechanical behavior of the thin slab in the traditional mold under the secondary cooling water distribution process in Sect. 3.4, it can be seen that the average cooling rate in the middle part and exit of the mold to the exit of the No.1 bow segment is low (the middle part of the mold in height direction has a corner dynamic reheating, and the average cooling rate of the corner of the area of the mold exit to the No.1 bow segment exit is only about 3.9 °C/s), causing a large amount of precipitation of microalloy carbonitrides along grain boundaries at the slab corner, which decreases its hot ductility. From the results of the precipitation control of microalloy carbonitrides in Chap. 2 and the evolution of the RA of Qste380TM steel under the different cooling rates in Sect. 3.2, it is known that rapidly cooling the micro-alloyed steel structure at a cooling rate greater than 5 °C/s to disperse its carbonitride precipitation is the key to improving the hot ductility of micro-alloyed steel. Therefore, this section will focus on the development of the thin slab efficient heat transfer curved surface mold and foot roller zone rapid cooling technology, to disperse the precipitation of carbonitrides in the corner of the micro-alloyed steel thin slab before entering the liquid core reduction segments, ensuring the high ductility of the slab during the liquid core reduction.

3.6.1 Development of Efficient Heat Transfer Mold

3.6.1.1 Design of the Curved Surface Mold

From the calculation results of the solidification thermo-mechanical behavior evolution of Qste380TM low carbon Nb-Ti-bearing steel thin slab in the mold in Sect. 3.4, it can be seen that during the solidification process in the traditional narrow face flat-plate mold, the shell solidification process in the upper part of the mold lacks efficiency compensation in the width direction of the slab, resulting in a large shrinkage gap in the area near the corner of the solidified shell narrow face, forming a phenomenon of concentrated distribution of thick flux film and air gap, which greatly reduces the heat transfer rate at the slab corner. It is not conducive to the growth of the slab corner grain and the control of carbonitride dispersion.

For the actual thin slab continuous casting mold, the middle part of its wide face is a funnel structure with a wide top and a narrow bottom, and the edge is a parallel structure. Due to the restriction of the large deformation process of the solidified shell in the funnel area, the compensation in the direction of its thickness in the parallel area of the mold edge is very small (according to the different thicknesses of the slab, the compensation in the direction of thickness in the parallel area is usually set to 0–0.2 mm). Therefore, it is difficult to increase the taper compensation of the mold wide face copper plate in the direction of the slab thickness. Therefore, to accelerate the heat transfer rate of the shell corner in the mold, the effective way is to reduce the interface gap size at the corner of the slab narrow face, to reduce the air gap and flux film thickness at the slab narrow face corner in the middle part of the mold.

Based on the design concept of novel convex-structure mold of the conventional slab in Chap. 2 and the structural characteristics of the actual thin slab mold, this book proposes a narrow face structure copper plate with a curved surface structure characteristics of “fast compensation at the upper part, slow compensation at the middle and lower part,” which efficiently compensates for the shrinkage of the slab in the upper part of the mold in the width direction. The structure of the curved-structure mold copper plate is shown in Fig. 3.29. In this design, due to the small thickness of the thin slab and the high temperature of the solidified shell in the upper part of the mold, a large taper compensation is implemented for the slab in the upper part of the mold, which can effectively eliminate the shrinkage gap of the slab corner in the width direction, so there is no need to make a “larger corner compensation” structure design similar to the mold for the corner of the narrow face copper plate. Figure 3.30 shows the distribution of the curved face compensation of the narrow face copper plate of a CSP mold in a certain steel plant along the height direction of the mold. As can be seen from the figure, the maximum compensation of the curved face in the height direction is located about 360 mm height from the upper part of the mold. In the height area of 0–360 mm below the upper part of the mold, the curved face compensation of the narrow face copper plate increases rapidly from 0 mm to the maximum value of about 1.1 mm. Then, as the height of the mold increases,

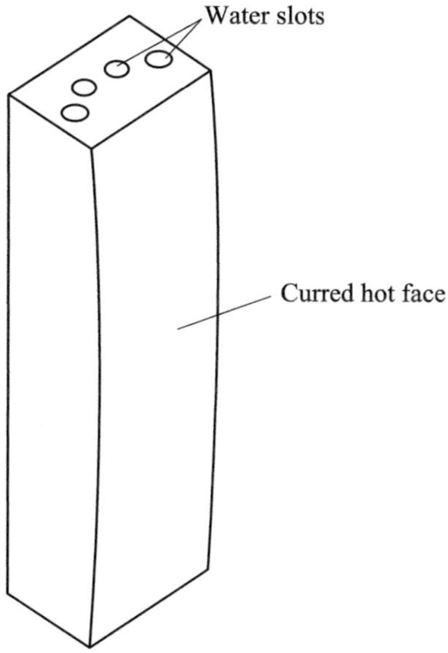

Fig. 3.29 Diagram of the curved narrow face copper plate

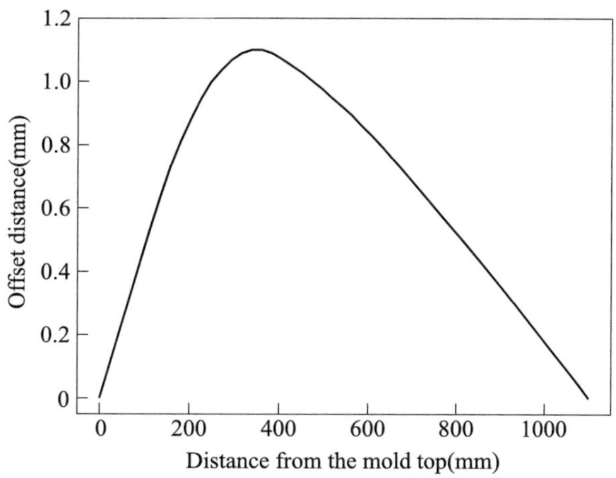

Fig. 3.30 Curve of the working face of narrow face copper plate of new mold

the compensation gradually decreases until the lower part of the mold, when the compensation of the curved face drops to 0 mm.

3.6.1.2 Solidification Behavior of Shell in the New Mold

Figure 3.31 shows the distribution of the thickness of the flux film and air gap in the corner and vicinity of the wide and narrow faces of the slab under the condition of the narrow face curved-structure thin slab mold at different heights of the mold. In the corner and off-corner of the wide face of the mold, the distribution of the flux film in different height directions of the mold is similar to that of the flux film distribution of the traditional narrow face flat-plate mold, both of which are thin in the upper part of the mold and mainly concentrated in the corner, and gradually concentrated in the off-corner of the wide face as the slab moves downward. Under the new curved surface structure mold, when the slab moves downward to 100–400 mm below the meniscus, the thickness of the flux film in the off-corner of its wide face increases sharply, reaching a maximum of about 0.52 mm, and then the distribution of the thickness of the flux film in the corner and its off-corner tends to be stable. However, compared with the distribution of the flux film of the traditional narrow face flat-plate mold, due to the rapid increase of the taper compensation of the new curved narrow face copper plate in the upper part of 0–360 mm height of the mold, the narrow face copper plate supports the solidification of the narrow face shell better to restrict the shrinkage deformation of the corner and off-corner of wide face toward the center of the narrow face. The maximum thickness of the flux film in the corner and off-corner of the solidified shell is reduced by about 10 and 20% respectively, which will improve the heat transfer in the wide face corner and its vicinity of the slab.

For the corner and off-corner at the narrow face of the slab, the thickness of the flux film under the conditions of the new curved-structure mold is significantly less than that in the traditional flat-plate mold. The maximum thickness of the flux film at the narrow face corner of the slab in the new mold has dropped from about 0.45 mm to about 0.25 mm, and the thickness of the flux film in the narrow face off-corner has dropped from about 0.7 mm to about 0.3 mm, with the reductions of 45 and 57% respectively, greatly improving the heat transfer of the corner and off-corner at narrow face of the slab.

Figure 3.32 shows the distribution of the corner air gap of wide and narrow faces along the height direction of the mold under the new curved-structure mold. As can be seen from the figure, the corner air gap of wide and narrow faces in the new mold are significantly reduced compared to the corner air gap of traditional flat-plate mold,

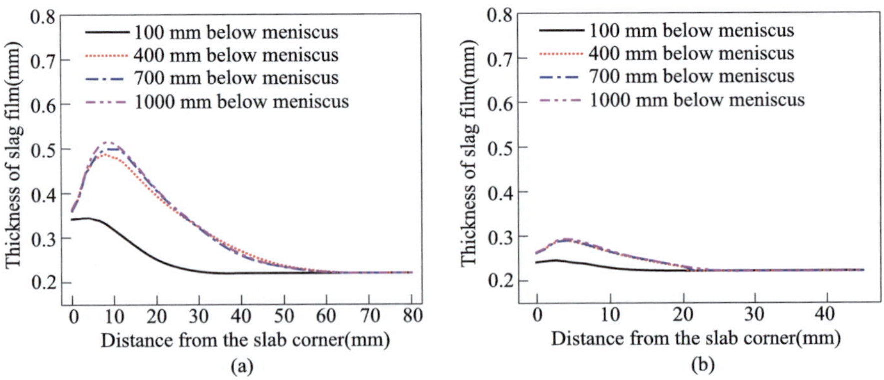

Fig. 3.31 Distribution of solid flux film around thin slab corner. **a** Wide face; **b** narrow face

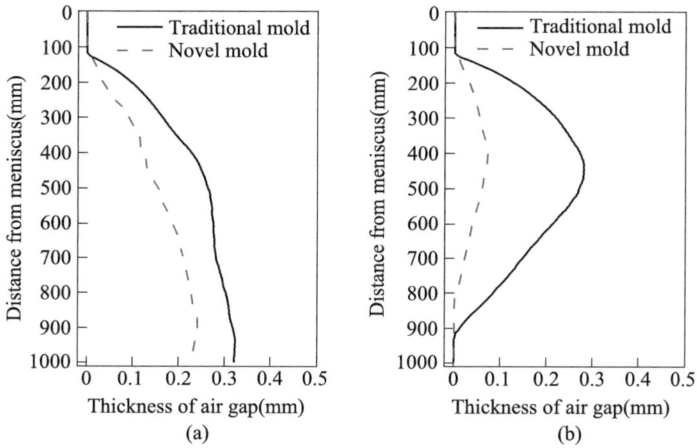

Fig. 3.32 Thin slab corner solid flux film distribution in different mold. **a** Wide face; **b** narrow face

with the most significant reduction in the middle of the height at narrow face, the maximum air gap has dropped from about 0.28 mm to about 0.05 mm. It means the air gap almost completely disappearing. For the wide face corner of the slab, the air gap of the wide face corner of the slab under the new mold also decreases by nearly 0.1 mm on average. Due to the significant reduction of the air gap in the corner of the wide and narrow faces of the solidified shell, the heat transfer efficiency of the slab in the mold will be further improved.

Figure 3.33 shows the temperature and cooling rate evolution curve of the thin slab corner along the height direction of the new mold. The cooling rate of the slab corner at its upper part is very fast, and the temperature quickly drops from the casting temperature to about 1150 °C in the height range of 0–100 mm below the meniscus. During this cooling process, the average cooling rate of the slab corner reaches more than 40 °C/s. Then, as the slab continues to solidify and descend, although the cooling rate of the slab corner decreases, the average cooling rate throughout the process is higher than 10 °C/s. That means it meets the cooling rate requirements for the dispersion precipitation of the carbonitrides of Nb, Al, B, and other microalloy elements, which helps to improve the hot ductility of the slab corner structure.

3.6.2 Development of Thin Slab Corner Rapid Cooling Technology in Secondary Cooling

3.6.2.1 New Cooling Structure for Narrow Face Foot Roller

As mentioned above, there are only two sets of cooling nozzles in the narrow face foot roller zone of a typical CSP continuous caster, as shown in the structure and

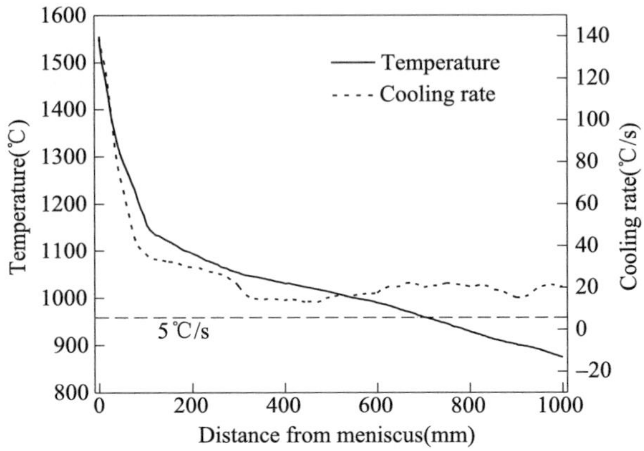

Fig. 3.33 Evolution of temperature and cooling rate of slab corner in new mold

physical diagram of Fig. 3.34. According to the calculation results of the secondary cooling temperature field of Qste380TM low carbon Ni–Ti-bearing micro-alloyed steel thin slab continuous casting in Sect. 3.5.2 of a steel plant, it can be known that the corner temperature of the thin slab in the cooling zone decreases to about 945 °C and the average cooling rate is only about 4.0 °C/s under the combined cooling effect of the narrow face foot roller zone and the wide face foot roller zone. When the thin slab exits the narrow face foot roller zone, there is no spray cooling effect on the narrow face, and the average cooling rate of the slab corner in the grid and bow No.1 segment is further reduced to about 3.8 °C/s. According to the PPT curve of typical Nb, Al, B, and other microalloy carbonitride precipitation in Chap. 2 of this book, 945 °C is often in the carbonitride precipitation "nose point" temperature range of micro-alloyed steel produced by main steel plants. However, the cooling rate of the slab corner under the current typical CSP continuous casting process cannot meet the requirements for the dispersion precipitation of microalloy carbonitride. Therefore, based on the above high efficiency heat transfer new curved-structure mold, the corner cooling of the thin slab after exiting the mold should be further strengthened, so that the corner cooling of slab can fully meet the requirements of current typical microalloy carbonitride dispersion precipitation.

The online infinite width adjustment of mold is the basic requirement of thin slab continuous casting production. To adapt to the infinite width adjustment production of the mold, and ensure that the corner of the thin slab production process at all widths are quickly cooled in the precipitation temperature zone, this book introduces a new spray structure for large flow strong spray cooling at the inner and outer arc of slab corner under the narrow face of the thin slab mold. The spray frame shown in Fig. 3.34 is modified into 6 sets of large flow nozzles shown in Fig. 3.35 (3 sets facing the inner arc side of the slab, 3 sets facing the outer arc side of the slab), a large flow automatic control water supply pipeline is laid for the new spray cooling, and the inner and outer arc corners of the slab exiting the mold are forced to be cooled.

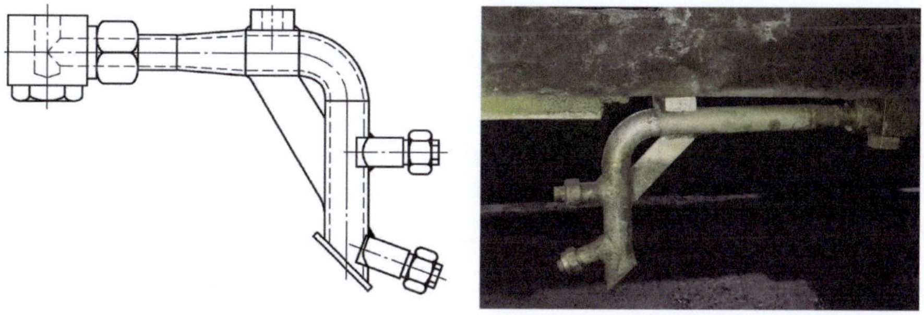

(a) (b)

Fig. 3.34 Cooling structure of the narrow face foot roller of the CSP continuous caster. **a** Structure diagram; **b** physical map

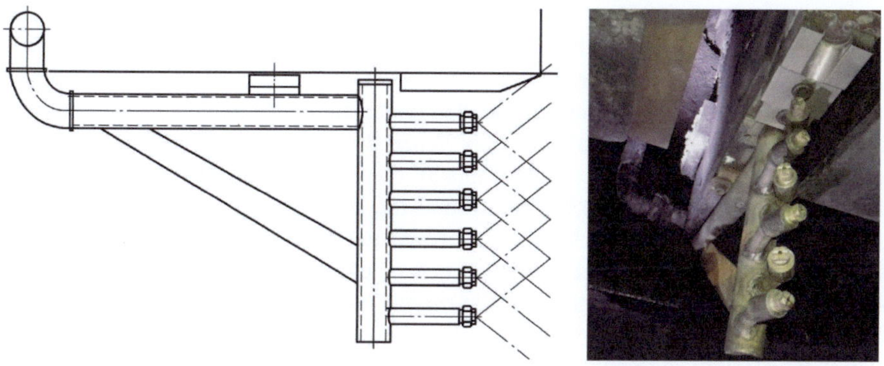

(a) (b)

Fig. 3.35 New cooling structure of narrow face foot roller. **a** Structure diagram; **b** physical map

3.6.2.2 Slab Temperature in New Narrow Face Foot Roller Zone Under Rapid Cooling Process

In the actual implementation process of the narrow face foot roller rapid cooling water distribution in the thin slab continuous caster, while significantly increasing the cooling water volume in the narrow face foot roller zone, the water volume in the wide face foot roller zone also appropriately increases. However, the water volume in the grid segment and No.1 bow segment (liquid core reduction segment) decreases to a certain extent to ensure a rapid reheat in the corner after the slab exits the rapid cooling zone of the narrow face foot roller. On this basis, to ensure that the position of the liquid core of the slab remains basically constant, the cooling water volume in No.2 and No.3 bow segments of the slab is also appropriately increased.

Figure 3.36 shows the evolution of the center and surface temperatures of the thin slab under the rapid cooling water distribution in the new narrow face foot roller zone. As can be seen, after implementing the new cooling technology, due to the enhanced cooling effect in the wide face foot roller zone, the wide face temperature at the entrance of the grid zone of the slab has dropped about 30 °C compared to the secondary cooling process of traditional thin slab. Subsequently, when the slab enters the grid zone, the temperature at the centerline of the wide face of the slab gradually reheat due to the reduced cooling effect on the wide face of the slab. When the slab solidifies at the entrance of the liquid core reduction segment, the temperature reaches 1055 °C, which is basically same as the temperature at the entrance of the liquid core reduction segment in the traditional secondary cooling process. When the slab enters the liquid core reduction segment, as the cooling water volume on the wide face further decreases, the temperature of the wide face of the slab significantly increases. When the slab solidifies to the middle of the liquid core reduction segment, the temperature reheats to the maximum of about 1115 °C, which is about 90 °C higher than the wide face temperature of the slab under continuous casting secondary cooling process of the traditional thin slab. Subsequently, the center temperature of the wide face of the slab begins to gradually decrease. When the slab solidifies to the exit of the liquid core reduction segment, the temperature at the centerline of the wide face drops to about 1085 °C, which is still about 65 °C higher than the wide face temperature of the slab under the continuous casting secondary cooling process of traditional thin slab. When the slab enters No.2 and No.3 bow segment, due to the higher cooling intensity compared to the traditional secondary cooling process, the surface temperature at the centerline of the wide face of the slab gradually presents a fluctuating downward trend, until the slab enters No.4 bow segment, the temperature reheats with a certain degree. When the thin slab enters the straight segment, the temperature at the centerline of the wide face of the slab is in the range of 1080–1125 °C, which is equivalent to the wide face temperature of the slab in the straight segment in the traditional secondary water distribution process.

For the narrow face of the slab, its temperature rapidly decreases due to the large flow spray cooling in the narrow face foot roller zone after exiting the mold. When the slab solidifies to the rapid cooling zone of the narrow face, the temperature of

Fig. 3.36 Evolution of thin slab under the new cooling process of the narrow face foot roller

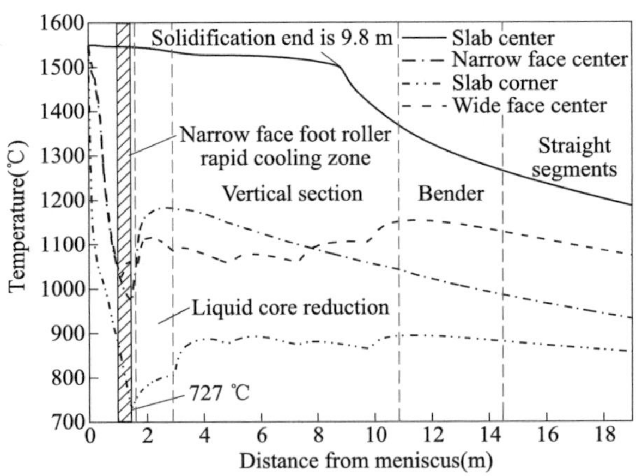

the centerline of the narrow face drops to the minimum of about 970 °C. Compared with the traditional narrow roller zone, the temperature of only two sets of nozzles and their secondary cooling water distribution process decreased by about 40 °C. Afterward, without the spray cooling effect, the temperature of the narrow face of the slab quickly reheats again. When the slab solidifies at the entrance of the liquid core reduction segment, the center temperature of the narrow face reheats to about 1040 °C, but it is about 100 °C lower than the temperature under the traditional thin slab continuous casting process. As a result, the deformation resistance stress of the slab at the entrance of the liquid core reduction segment will increase. When the slab solidifies to the middle and later stages of the liquid core reduction segment, the center maximum temperature of the narrow face reheats to about 1183 °C, which is only about 26 °C lower than the traditional thin slab secondary cooling process. Afterward, under the effect of radiation heat transfer, the temperature of the narrow face of the slab continues to slowly decrease. After the slab enters the straight segments, the temperature at the centerline of the narrow face drops from about 985 °C to about 935 °C, meeting the requirements of the slab high temperature through the straight segment.

For the corner of the slab, its temperature rapidly decreases under the rapid cooling effect of the narrow face and wide face foot roller zone after exiting the mold, mainly due to the rapid cooling effect of the narrow face foot roller zone (the cooling intensity of the grid segment decreases, but the temperature of the slab corner continues to rapidly decrease). When the slab solidifies to the exit of the rapid cooling zone of the narrow face foot roller, the temperature of the slab corner drops to the minimum, about 727 °C. During this rapid cooling process, the cooling rate of the slab corner and its subsurface depth of 10 mm is greater than 15°C/s, as shown in Fig. 3.37. As can be known from the thermodynamics and kinetics of Nb, Al, B, and other microalloy carbonitrides precipitation in Chap. 2, the temperature of the slab corner solidifying to the exit of the rapid cooling zone of the narrow face foot roller covers the entire precipitation temperature zone of the corresponding carbonitrides, and the whole process cooling rate is higher than 10 °C/s (especially the cooling rate in the rapid cooling zone of the narrow face foot roller is higher than 15 °C/s), which will significantly disperse the precipitation of microalloy carbonitrides in the slab corner and overall improve the ductility of the slab corner structure (as shown in Fig. 3.8). In addition, it can be seen that the corner temperature of the slab in the liquid core reduction segment is in the range of 727–810 °C, which is significantly lower than the temperature range of 930–852 °C under the traditional thin slab continuous casting secondary cooling process. As a result, the stress of the slab corner during the liquid core reduction will significantly increase. However, when the slab enters the No.2 bow segment, the corner temperature quickly rises to above 900°C, and it is in the range of 860–880 °C in the straight segment. Similarly, the RA test results of Qste380TM low carbon Nb-Ti-bearing micro-alloyed steel under rapid cooling in Fig. 3.8 show that the RA of the slab corner straightening in this temperature range is 71.6–78.3%, which meets the requirements of high ductility straightening of steel structure. That is, under the premise that the accuracy of the continuous caster

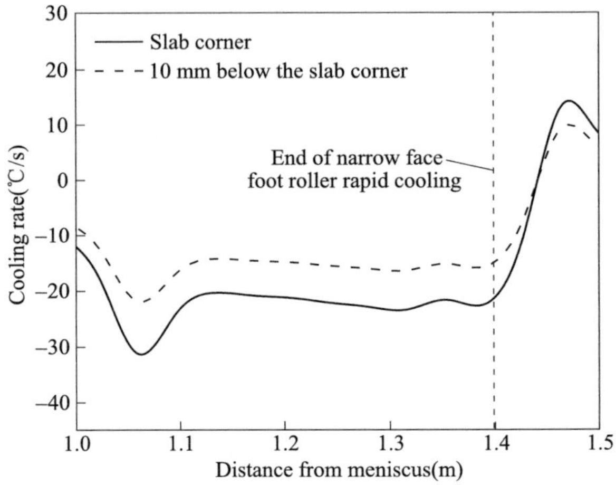

Fig. 3.37 Cooling rate of corner and the position of 10 mm below the corner under the new cooling process of the narrow face foot roller

device meets the requirements, the Nb-Al-B-bearing micro-alloyed steel under the new process will not form slab corner crack defects.

3.6.2.3 Stress Evolution in Narrow Face Foot Roller Zone Under Rapid Cooling Process

Figure 3.38 shows the stress evolution of the slab under the 13.5 mm liquid core reduction in a No.1 bow segment (with a reduction increment of 1.5 mm) under the above narrow face foot roller strong cooling process. Under the narrow face foot roller rapid cooling process, the stress in the center of the wide face of slab still periodically fluctuates with the peak stress of about 44.1 MPa under the action of each foot roller. Compared with the stress maximum of 44.8 MPa under the traditional secondary cooling process of thin slab continuous casting, the maximal stress of the new process of wide surface liquid core is equivalent, and the overall evolution trend is similar. For the narrow face of the slab, due to the strong spray cooling effect of the rapid cooling zone of the narrow face foot roller, the temperature of the narrow face entering the liquid core reduction segments are reduced as a whole. The central stress of the narrow face of the slab in the liquid core reduction segments under the new process is significantly higher than that of the narrow face of the slab under the traditional thin slab continuous casting secondary cooling process, and the range of stress evolution increases significantly during the action and release of the reduction roller.

For the corner of the slab, the stress evolution under the new process is significantly different from the traditional continuous casting process. The stress maximum of the slab corner under the action of the liquid core reduction segments entrance and the second reduction roller is the highest, with the stress maximum increasing by about

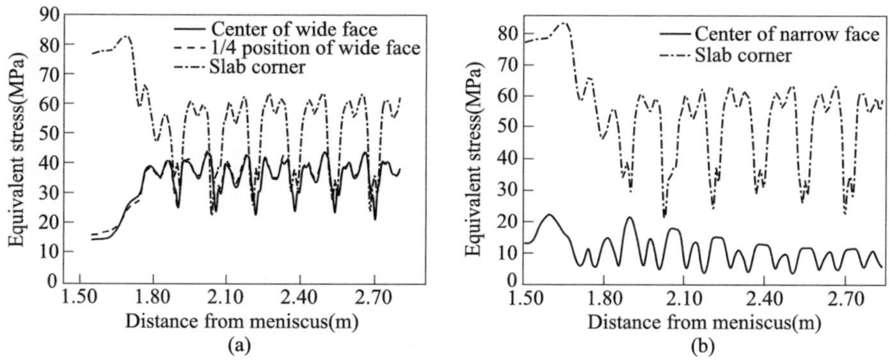

Fig. 3.38 Evolution of equivalent stress of slab under the new cooling process of the narrow face foot roller. **a** Wide face; **b** narrow face

84.2%, reaching 82.5 MPa. Then, as the temperature of the slab corner quickly increase, the stress maximum significantly reduces by about 64 MPa, reaching a level equivalent to the stress maximum of the slab corner under the traditional thin slab continuous casting secondary cooling process. It can be seen that although the ductility of the slab corner structure under the narrow face foot roller rapid cooling process significantly increases, the reduction stress of the slab corner in the upper part of the liquid core reduction segments also significantly increases, which will increase the wear of the reduction roller surface and is also not conducive to the control of slab corner cracks.

3.6.3 Development of Gaussian Concave Mold Technology

Although the rapid cooling process of narrow face foot roller can highly ductile the slab corner structure, it also significantly increases the corner reduction stress of the slab in the upper part of the liquid core reduction segment, which is not conducive to the control of slab corner cracks. Therefore, based on the mold with high ductility control of thin slab corner structure and secondary cooling control technology, this book has developed a mold with narrow face of Gaussian concave surface technology. That is, based on the above new curved-structure mold with the curved narrow surface copper plate, the working face of the narrow face copper plate of the mold is horizontally designed as a Gaussian concave structure (as shown in Fig. 3.39), which promotes the metal of the slab narrow face to flow along the width direction during the liquid core reduction, thereby reducing the corner stress of the slab during the liquid core reduction.

In the actual design process, the distance between the hot face of the narrow face copper plate of the mold and the root of the water slot, the taper compensation curve of the mold along the height direction, and the rapid cooling process of narrow face

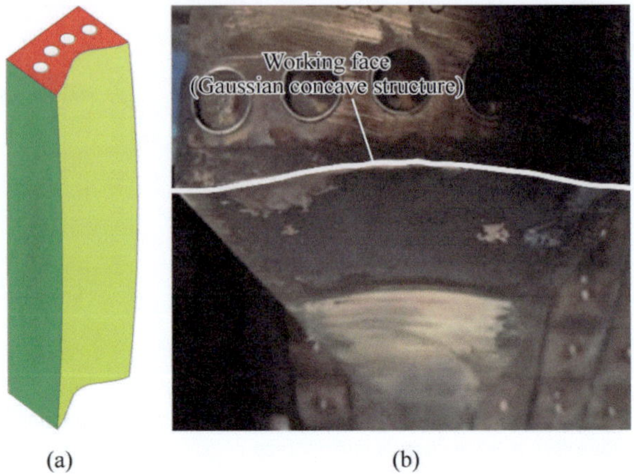

Fig. 3.39 Mold with narrow face of Gaussian concave surface. **a** Schematic diagram; **b** physical diagram

foot roller remain constant, the cooling conditions of the slab from the mold to the exit of the No.1 bow segment of the secondary cooling zone remain constant. Hence, the temperature field evolution of the slab solidification process is very similar to the solidification behavior under the above conditions of traditional flat-plate mold, so the temperature field evolution of the slab under the novel structure copper plate in the mold and the secondary cooling zone is no longer described in this section.

Figure 3.40 shows the stress evolution of the slab under the mold with narrow face of Gaussian concave surface and the rapid cooling process of narrow face foot roller. Under the same process of secondary cooling water distribution of mold and narrow face foot roller rapid cooling, due to the metal of narrow face Gaussian convex slab (the physical slab is shown in Fig. 3.41) is easier to flow toward the width during the liquid core reduction, the stress at the centerline and 1/4 of the slab wide face is slightly smaller than the narrow face transverse straight convex mold. For the narrow face of the slab, the stress of the centerline of the slab narrow face increases to a certain extent due to the greater deformation during the expanding process, but the stress peak is less than 30 MPa, which should not form defects such as cracks on the narrow face of the slab.

For the corner of the slab, under the effect of the mold with narrow face of Gaussian concave surface, the corner stress of the liquid core reduction is significantly reduced. However, due to the temperature of the slab at the entrance of the liquid core reduction segment is less than 730 °C, the upper part of the liquid core reduction segment still has a stress peak. However, the maximum stress is only about 65 MPa, which is equivalent to the stress peak at corner reduction of the slab under the secondary cooling process of traditional thin slab continuous casting. When the slab enters the middle and lower parts of the liquid core reduction segment, the peak value of the liquid core reduction stress is reduced to about 51–55 MPa due to the rapid reheat of the slab corner, compared with secondary cooling process of the traditional thin slab continuous casting and the rapid cooling water distribution process of narrow face foot roller, the peak of the reduction stress is reduced by nearly 20%. Therefore,

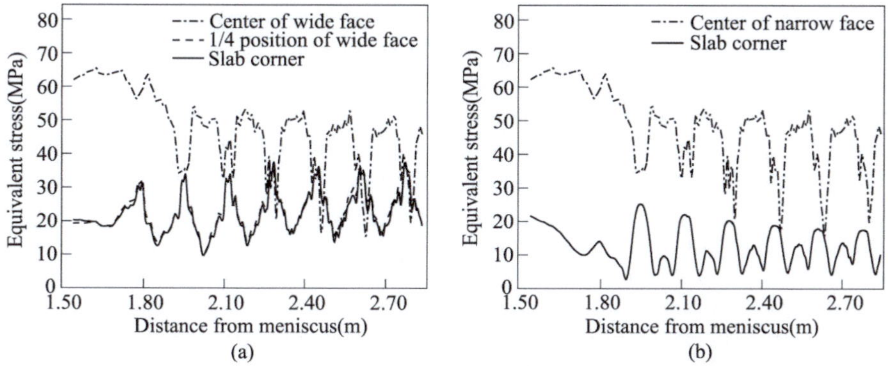

Fig. 3.40 Evolution of slab stress under the mold with narrow face of Gaussian concave surface and the rapid cooling process of narrow face foot roller. **a** Wide face; **b** narrow face

Fig. 3.41 Slab morphology produced by mold with narrow face of Gaussian concave surface

under the new process, the ductility of the slab corner structure is greatly increased, the stress under the liquid core reduction greatly reduced, which is beneficial to the prevention and control of slab corner cracks.

3.7 Implementation of Thin Slab Corner Crack Control Technology

Based on the above strategy for controlling corner crack in thin slab and combined with the actual production of SCP thin slab continuous casting in a domestic steel plant, the narrow copper plate of the mold was transformed into a Gaussian concave curved surface structure as shown in Fig. 3.39. Considering the insufficient water supply capacity of the cooling loop in the narrow foot roller zone of the original mold, a new large flow, dynamically controllable water supply loop with steel grade and casting speed was installed. The offline assembly of the narrow face Gaussian concave curved surface mold and the large flow supply loop and its control system are shown in Figs. 3.42 and 3.43, respectively. The physical structure of the strong cooling spray in the narrow face foot roller zone is shown in Fig. 3.35.

Figures 3.44 and 3.45 are respectively the precipitation and distribution morphology of carbonitrides at different depths of the corner of the Qste380TM low carbon Nb-Ti-bearing micro-alloyed steel slab produced by a steel plant based on the slab corner crack control technology and traditional continuous casting process. As can be seen from Fig. 3.44, due to the rapid cooling effect of the mold and the narrow face foot roller zone, the cooling rate of the microalloy carbonitride precipitation in the corner of the slab is always higher than 10 °C/s, realizing the precipitation of the carbonitride in the corner of the slab within 10 mm from the narrow face with a size less than 10 nm uniformly dispersed in the matrix, thereby improving the ductility of the corner structure of the slab. However, under the traditional continuous casting process, although the size of the microalloy carbonitride precipitation at the same position in the corner of the slab is significantly reduced compared to the carbonitride precipitation size of the corner structure of the slab under the conventional slab continuous casting process conditions, it still precipitates in a chain concentrated form. Therefore, this new process completely solves the problem of the corner crack of the slab caused by the chain precipitation of the precipitates at the grain boundary of the corner structure of the slab, which aggravates the hot ductility.

Figures 3.46 and 3.47 are respectively the morphology of the structure at different positions in the corner of the Qste380TM low carbon Ni–Ti-bearing micro-alloyed steel slab under the new process and the traditional slab continuous casting process. It can be seen that the room temperature structure of the corner of the slab under different continuous casting processes is composed of block ferrite and pearlite, and due to the low carbon content, most of the structure is ferrite. However, after

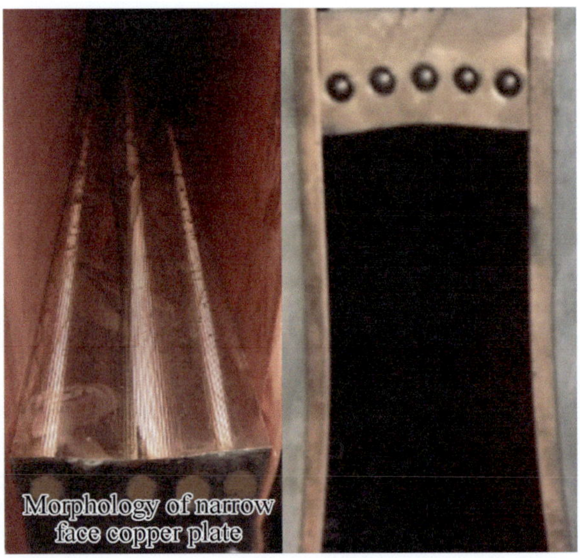

Fig. 3.42 Morphology of the mold with a narrow face of Gaussian concave surface

Fig. 3.43 Water supply pipeline and cooling control system of the narrow face foot roller. **a** Water supply pipeline; **b** cooling control system

(a)

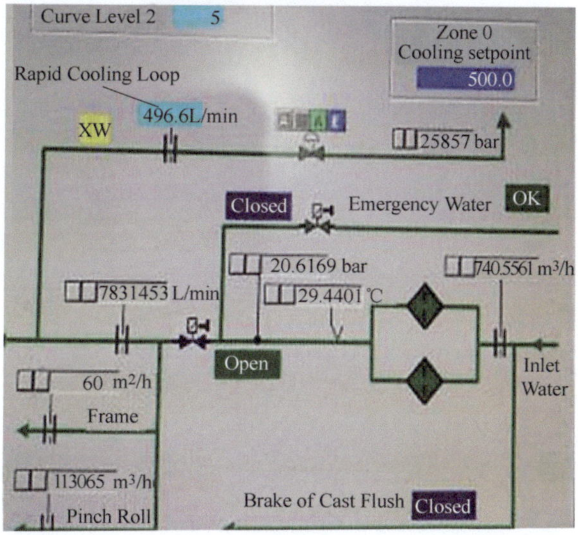

(b)

adopting the new process, the cooling intensity of the corner of the slab during the solidification process in the mold is large, the austenite grains formed by the initial solidification are fine. At the same time, due to the rapid cooling and quick reheating of the corner structure of the slab after exiting the mold, the grain of the corner of the slab is finer than that of the traditional thin slab continuous casting process under the action of phase transition, which further improving the hot ductility of the corner structure of the slab.

Since its first production in the second half of 2017, the thin slab continuous casting technology of a certain steel plant has achieved stable promotion and application realizing high quality and stable production of a full range of microalloy including

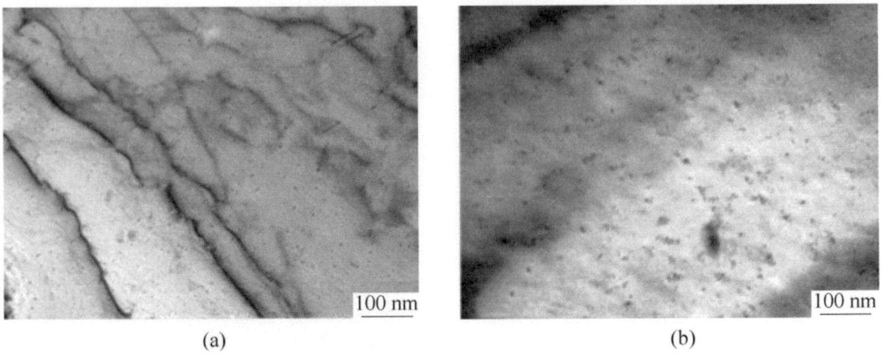

Fig. 3.44 Morphology of carbonitride precipitation at different depth below the slab corner of Qste380TM steel under new process **a** 5 mm; **b** 10 mm

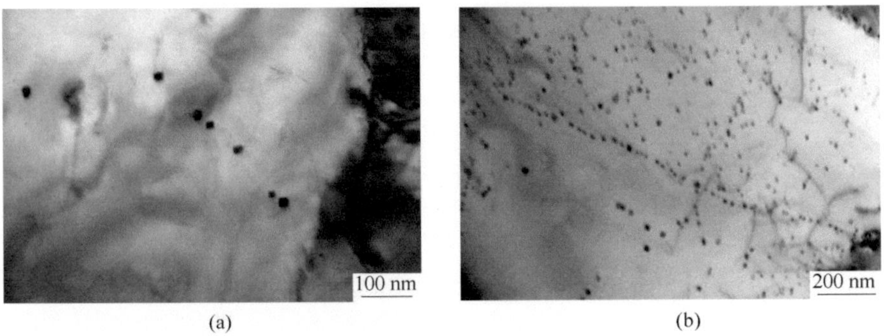

Fig. 3.45 Morphology of carbonitride precipitation at different depth below the slab corner of Qste380TM steel under traditional process. **a** 5 mm; **b** 10 mm

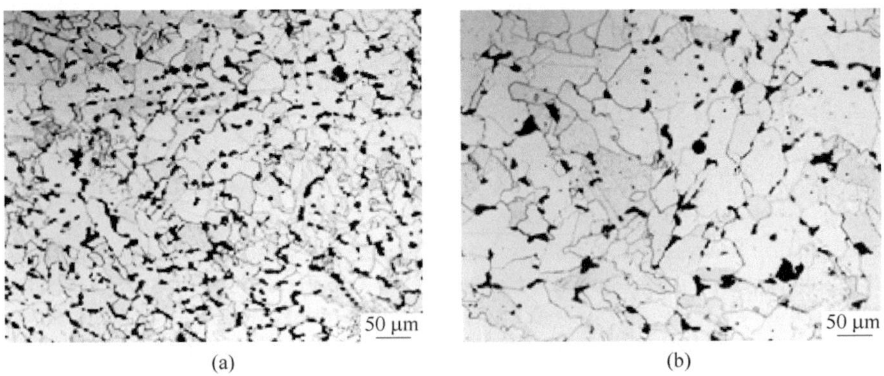

Fig. 3.46 Morphology of micro-structure at different depth below slab corner of Qste380TM steel under new process **a** 5 mm; **b** 10 mm

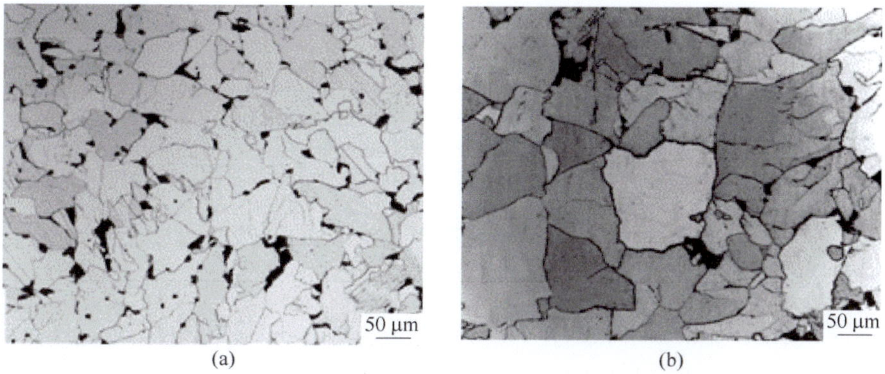

Fig. 3.47 Morphology of micro-structure at different depth below slab corner of Qste380TM steel under traditional process. **a** 5 mm; **b** 10 mm

Qste500TM (as shown in Fig. 3.48, the comparison of edge crack control effects of Qste500TM hot rolled coil before and after the implementation of the technology is shown in Fig. 3.48). The edge cracking occurrence rate of hot rolled coil has decreased from an average of 6.75% to about 0.025%. Table 3.8 shows a comparison of crack occurrence rate before and after the implementation of the corner crack control technology for the CSP continuous caster in a certain steel plant for one year.

Fig. 3.48 Edge morphology of the Qste500TM hot rolled coil before and after use the new process. **a** Before use; **b** after use

Table 3.8 Comparison of edge crack ratio of micro-alloyed steel plate before and after use the new process

Year	August 2017–June 2018 (after use)			June 2016–July 2017 (before use)			Reduction rate of crack (%)
Steel grade	Production amount (t)	Coil with edge crack (t)	Crack rate (%)	Production amount (t)	Coil with edge crack (t)	Crack rate (%)	
QSTE340TM	37,494.6	242.84	0.65	37,554	2397	6.38	89.81
QSTE380TM	59,566.2	222.75	0.37	62,001	3251	5.24	92.94
QSTE420TM	56,079.1	168.35	0.30	87,750	5447	6.21	95.17
QSTE460TM	6038.13	68.94	1.14	2062	487	23.62	95.17
QSTE500TM	4616.43	60	1.30	14,017	4232	30.19	95.69
SAPH400	17,736.1	193.9	1.09	29,384	910	3.10	64.84
SAPH440	73,330.8	425.03	0.58	78,876	4299	5.45	89.36
Total	254,861	1381.81	0.54	311,644	21,023	6.75	92.00

3.8 Summary

This chapter analyzes the micro-structure and precipitation morphology of carbonitride at the corner of typical Nb-Ti-bearing micro-alloyed steel thin slab and calculates the precipitation thermodynamic behavior of carbonitride during the solidification process of typical micro-alloyed steel, the solidification thermo-mechanical evolution of the slab in the funnel mold and secondary cooling casting flow, and clarifies the cause of corner crack in micro-alloyed steel thin slab. On this basis, a thin slab corner crack control technology based on the dispersion of microalloy carbonitride and the reduction of corner stress during the slab liquid core reduction process is proposed, and the Gaussian concave surface mold and narrow face foot roller rapid cooling process and device technology are developed. The following main conclusions were obtained:

(1) The main reason for the corner cracks in the micro-alloyed steel thin slab is that the corner cooling rate in the upper part of the mold and the secondary cooling zone from the exit of the mold to the exit of the liquid core reduction segments is low, and the microalloy carbonitride precipitate concentratedly at the grain boundary of the slab corner structure to form a low ductility structure, which will cause corner crack defects in the slab during the liquid core reduction segments.
(2) Implementing rapid cooling control in the precipitation temperature zone of carbonitride for typical composition micro-alloyed steel can significantly improve the hot ductility at different deformation rates.
(3) A "fast compensation in the upper part, slow compensation in the middle and lower part" Gaussian concave narrow face curved mold and narrow face foot roller rapid cooling process and device technology are developed, which enables

the corner of the micro-alloyed steel slab to be cooled from the molten steel casting temperature to the temperature range of 730°C at a cooling rate higher than 10°C/s. It realizes the dispersion of the carbonitride in the corner structure of the thin slab, improves its hot ductility, promotes the metal flow in the width direction during the liquid core reduction of the slab narrow face, reduces the stress during the liquid core reduction of the slab corner, and effectively prevents the occurrence of corner crack in the micro-alloyed steel thin slab.

References

1. Yin RY, Zhang H (2011) Progress and development direction on thin slab continuous casting and rolling technology under new situation. Steel 46(4):1–9
2. Yang GH, Zhang J, Li HB et al (2016) The key technology of thin slab continuous casting and rolling and strip continuous casting. Metallurgical Industry Press, Beijing
3. Muntin AV (2019) Advanced technology of combined thin slab continuous casting and steel strip hot rolling. Metallurgist 62(9):900–910
4. He AR, Jing FW, Liu C et al (2020) Development, application and prospect of process control technologies of thin slab continuous casting and rolling. Rolling Steel 37(3):1–7
5. Yang T, Yang LD et al (2017) Development of thin slab continuous casting-continuous rolling technology. World Metal Guide 24 Jan 2017
6. Liu J (2003) Engineering and production practice of angang's 1700 mm medium thickness slab continuous casting-rolling line (ASP-Angang Strip Production). Steel 7:8–11
7. Yu Y, Zheng XT (2016) Endless strip production technology in Rizhao steel company. Contin Cast 41(5):1–4
8. Zhang M (2018) Introduction of Shougang Jingtang MCCR finishing mill inlet temperature control system. China Metal Bull 3:170–171
9. Arvedi G, Mazzolari F, Siegl J et al (2010) Arvedi ESP first thin slab endless casting and rolling results. Ironmaking Steelmaking 37(4):271–275
10. Park CJ, Kang SY, Chang HL (2010) Advanced temperature control of high carbon steel for hot strip mills. J Mech Sci Technol 24(5):1011–1017
11. Pan F, Zhou S, Liang X et al (1997) Thin strip casting of high speed steels. J Mater Process Technol 63(1):792–796
12. Wang SZ, Cai Z, Sun YQ et al (2020) Development status and product development practice of thin slab continuous casting and rolling technology. In: Guide to the 10th annual meeting of the Chinese society for metals and the 4th Liaoning young scientist forum, p 79
13. Liu ZP, Min HG (2019) Development of 700 MPa high strength steel plate by thin slab continuous casting and rolling. Metal World 3:73–76
14. Zhou T, Zhang P, Kuuskman K et al (2018) Development of medium-to-high carbon hot-rolled steel strip on a thin slab casting direct strip production complex. Ironmaking Steelmaking 45(7):603–610
15. Lee J, Cooman BD (2015) Development of a press-hardened steel suitable for thin slab direct rolling processing. Metall Mater Trans A 46(1):456–466
16. Tian NY (2009) Continuous casting and direct rolling of thin slab. Metallurgical Industry Press, Beijing
17. Wang XY, Liu XH (2005) Current state and advantages on developing electrical steels in CSP process. China Metall 12:39–43
18. Wang XY, Liu XH, Yao J (2006) Present situation and advantages analysis of electrical steel manufacture by thin-slab casting and rolling. Steel Res 3:58–62

19. Li CS, Yu YM, Wang SZ et al (2007) Technological advantage on electrical steel production by continuous casting and rolling. Modern Manuf Eng 9:9–12
20. Tang D, Mi ZL, Cai QW (1999) The Metallurgical problems of thin slab continuous casting and rolling and its advantages. Rolling Steel 4:56–58
21. Bayoumi LS, Megahed GM (2003) Determination of rolling load in liquid core reduction of continuously cast thin steel slab. Ironmaking Steelmaking 30(5):348–352
22. Hans S (2001) Thin-slab casting with liquid core reduction. Iron Steel Rev 45(3):71–74
23. Sun HD (2008) Numerical study of the liquid core reduction for thin slab continuous casting. Yanshan University, Qinhuangdao
24. Cen YQ (1997) Rolling process for CC slab with liquid core reduction. Shanghai Metal 5:42–47
25. Wu ZG (2001) Liquid core reduction for thin slab casting. Contin Cast 6:4–6
26. Zhang JJ, Zhang H, Xi CS et al (2017) Analysis and control of transverse corner cracking formation on middle carbon steel in thin slab casting. Iron Steel 52(11):32–42
27. Kislitsa VV, Vorozheva EL, Lavrov VN et al (2021) Conditions for increasing the productivity of a thin slab continuous casting caster without quality degradation. Steel Transl 51(5):320–323
28. Wang JC, Zheng W, Wang CF et al (2021) Cause analysis and control of edge cracking of Q345 hot rolled plate produced by CSP. Steelmaking 37(5):60–66
29. Zhang JJ, Zhang H, Xi CS et al (2017) Analysis and control of transverse corner cracking formation on medium carbon steel in thin slab casting. Steel 52(11):32–36
30. Yang J, Yao HM, Li MY et al (2017) Research of surface quality control technology of middle-thin slab caster. Contin Cast 42(5):66–70
31. Han XY (2006) Functions of Nb, V and Ti in micro-alloyed steel. Wide Thick Plate 12(1):39–41
32. Yong QL (2006) Second phase in steel materials. Metallurgical Industry Press, Beijing
33. Maehara Y, Ohmori Y (1984) The precipitation of AlN and NbC and the hot ductility of low carbon steel. Mater Sci Eng 62:109–119
34. Maehara Y, Nakai K, Yasumoto K et al (1987) Hot caracking of low-alloy steels in simulated continuous casting direct rolling process. Tetsu-to-Hagane. 73:876–883
35. Wang XH, Liu XY, Lv WJ et al (1998) Carbon/nitrides precipitation and hot ductility of continuous casting steel slabs containing Nb, V, Ti. J Iron Steel Res 10(6):32–36
36. Barozzi S, Fontana P, Pragliola P (1986) Computer control and optimization of secondary casting during continuous casting. Iron Steel Eng 63(11):21–26
37. Kato T, Ito Y, Kawamoto M et al (2003) Prevention of slab surface transverse cracking by microstructure control. ISIJ Int 43(11):1742–1750
38. Anand L (1982) Constitutive equations for the rate-dependent deformation of metals at elevated temperatures. Trans ASME 104(1):12–17
39. Brown SB, Kim KH, Anand L (1989) An internal variable constitutive model for hot working metals. Int J Ductil 5:95–130
40. Anderson JD, Wendt J (1995) Computational fluid dynamics. McGraw-Hill, New York
41. Mintz B, Crowther DN (2010) Hot ductility of steels and its relationship to the problem of transverse cracking in continuous casting. Int Mater Rev 55(3):168–196
42. Oconnor TG, Dantzig JA (1994) Modeling the thin-slab continuous-casting mold. Metall and Mater Trans B 25(3):443–457
43. Li B, Fumitaka T (2006) Effects of electromagnetic brake on vortex flows in thin slab continuous casting mold. Trans Iron Steel Inst Jpn 46(12):1833–1838
44. Wu CH (2014) Study on thermal-mechanical behavior in secondary cooling process of slab continuous casting. Northeastern University, Shenyang
45. Cai KK (2008) Continuous casting mold. Metallurgical Industry Press, Beijing

Chapter 4
Off-Corner Depression of Micro-alloyed Steel Thick Slab and Its Control

In recent years, with the large scale development of infrastructure and device such as energy petrochemical, transportation, marine engineering, the demand for high performance thick/ultra-thick plates is increasing [1–6]. Driven by this, steel companies at home and abroad have widely developed thick plate products. As the rolling base material of high performance wide-thick slab, the thickness of the slab is required to be larger and larger [7–11]. For this reason, Yingkou wide-thick slab, Xingcheng Special Steel, Xiangtan Steel, Wuyang Steel, Hanye Special Steel, Xinyu Steel, Shougang Steel, and others have newly constructed several continuous casters for ultra-thick plate with the thickness exceeding 400 mm. However, with the significant increase in the thickness of the continuous casting slab, the actual continuous casting process is prone to defects of narrow face bulge and longitudinal depression in wide face off-corner of the slab, which leads to transverse cracks and subsurface longitudinal cracks in the off-corner depression, which is an important common technical problem in the field of thick slab continuous casting [12–15]. Especially in the continuous casting production of micro-alloyed steel thick slab, the traditional continuous casting process often adopts a weak cooling secondary cooling system to prevent surface cracks, and the problem of slab narrow face bulge and longitudinal depression at wide face off-corner is particularly prominent. This chapter will carry out research on the formation mechanism and control of the longitudinal depression in the off-corner of the thick slab based on the most common wide-thick slab continuous casting production with 250 mm thickness in steel companies at home and abroad.

© Metallurgical Industry Press 2025

Z. Cai and M. Zhu, *The Control Technology for Slab Surface Crack During Continuous Casting of Micro-alloyed Steel*, Key Steel Manufacturing Technologies by Collaborative Innovation, https://doi.org/10.1007/978-981-96-7171-7_4

4.1 Characteristics of Thick Slab Off-Corner Depression

Figure 4.1 shows the longitudinal depression in the off-corner and its surface and subsurface crack defects of a wide-thick slab continuous casting slab with 250 mm thickness in a certain steel plant. It can be seen that the depression is mainly distributed in the width range of 50–150 mm from the corner of the slab and runs through the entire continuous casting slab along the longitudinal direction of the slab (the slab casting direction). The width and depth of the depression are not fixed, and its severity is greatly related to the process parameters such as cross-sectional size of the slab, steel composition, mold taper, taper of narrow face foot roller, casting speed, spray of secondary cooling zone, and reduction of solidification end. In general, the depth and width of the longitudinal depression at the off-corner on the outer arc of the slab are greater than those of the inner arc. Generally, the width of the longitudinal depression in the wide face off-corner is generally 50–100 mm and the depth can reach 1.0–5.0 mm for wide-thick slab with a cross-sectional thickness of 230–360 mm in continuous casting production. For ultra-thick plate with a cross-sectional thickness of more than 400 mm, the depth and width of longitudinal depression zone are further significantly increased. When the depression is serious, transverse cracks and longitudinal crack defects will occur on the surface and subsurface of the off-corner depression, as shown in Fig. 4.1, and even breakout in the slab off-corner may occur [16–20].

In the actual production process, the longitudinal depression defect in the wide face off-corner of the thick slab is not a very serious quality defect, and it generally does not cause performance and surface quality defects in subsequent rolling. However, once the slab has a serious longitudinal depression in off-corner during production, the depression zone is often accompanied by surface transverse cracks and subsurface longitudinal crack defects [21–23]. In actual continuous casting production, steel companies often need to flame clean the continuous casting slab with longitudinal depression at wide face off-corner, which greatly reduces production efficiency. The subsurface longitudinal crack defects in the depression zone also affect the mechanical properties of the rolled product. The frequent longitudinal depression defect in the wide face off-corner has become an important common technical problem in the continuous casting production of thick slab, which needs to be solved urgently.

4.2 Causes of the Off-Corner Depression

As mentioned in Chap. 1, in view of the causes of the formation of longitudinal depression defects in the wide face off-corner during the continuous casting process of thick slab, although scholars at home and abroad have proposed the single-stage formation theory formed in the mold [21, 24–26] and the two-stage formation theory formed in the mold and then in the foot roller segments of the continuous caster [27, 28], the control processes and device technology based on taper structure of the

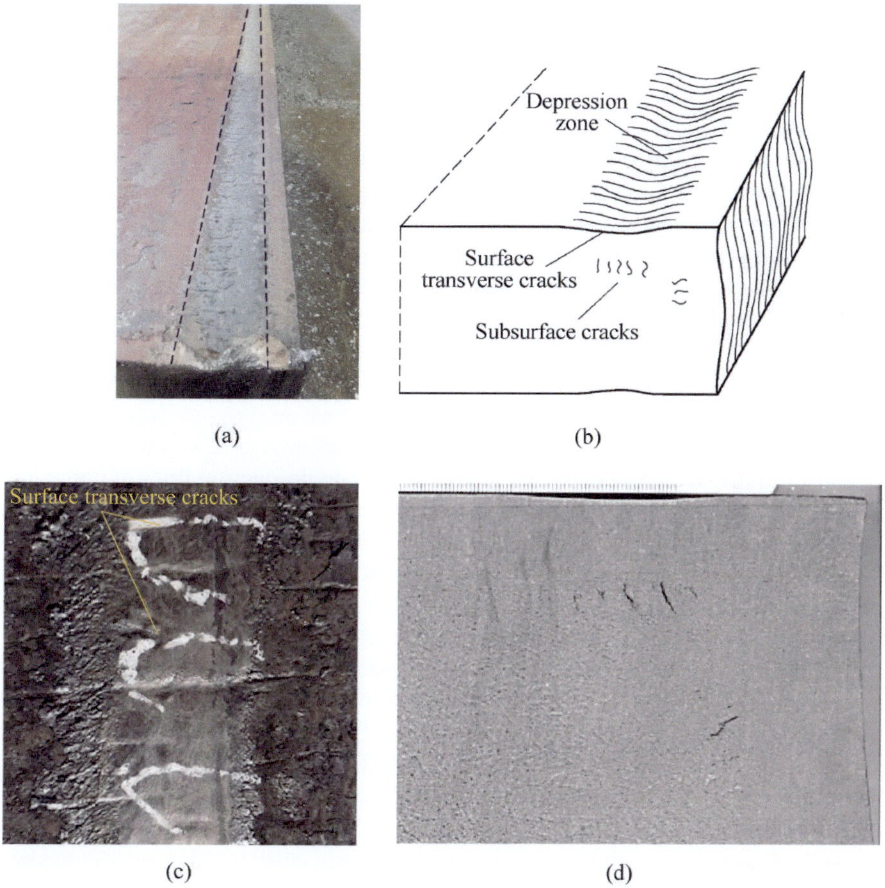

Fig. 4.1 Morphology of off-corner longitudinal depression and cracks caused by the depression. **a** Morphology of off-corner depression; **b** schematic diagram of off-corner depression and its cracks; **c** surface transverse cracks in the depression zone; d subsurface longitudinal cracks below the depression zone

mold, the structure of submerged nozzle and submerge process, physical properties of mold flux, the optimization of secondary cooling spray structure have been carried out [22, 29–33], there are many discrepancies between the formation theory of the wide face off-corner depression of the thick slab and the actual continuous casting production, and so far it is still unable to stably eliminate defect.

4.2.1 Thermo-mechanical Model of Thick Slab Solidification

According to the previous research on the formation mechanism of the off-corner depression of the thick slab and the related metallurgical behavior of the actual thick slab continuous casting production process, it can be seen that the wide face off-corner depression of the thick slab could be formed in the high temperature zone of the continuous caster, and there are complex evolution behaviors in the solidification of the casting flow and the reduction process at the end of the solidification of the slab. This book will develop a multi-physical field coupling model [34, 35] of shell solidification in full continuous casting process based on thick slab mold and secondary cooling casting flow (including solidification end reduction), and explore the behavior of heat transfer and deformation in the thick slab continuous casting process.

The continuous casting slab with the section thickness of 250 mm is currently the most important slab material for wide-thick plate production lines at home and abroad. The corresponding continuous caster is also the main type of continuous caster for major steel companies. Considering the consumption of multi-physics field coupling calculation resources and the typicality of slab size, this book takes the solidification process of the wide-thick slab with a thickness of 250 mm in a certain steel plant as the research object and develops a multi-physics field coupling calculation model that includes the mold and secondary cooling full casting roller array and its interaction with the slab solidification heat transfer and stress. Among them, the solidification behavior of the slab in the mold takes into account the actual copper plate water slot structure and its internal cooling water flow, the dynamic distribution of the flux film and air gap in the interface between shell and copper plate, the flow of molten steel, and the dynamic deformation of the solidified shell. The solidification of the slab in the secondary cooling casting roller array considers the contact heat transfer and contact mechanics of the slab and the bow segment casting roller, soft reduction at the end of the slab solidification and other behaviors. Considering the symmetry of the thermo-mechanical behavior along the width and thickness direction in the actual thick slab continuous casting solidification process (especially in the mold and high temperature zone of secondary cooling), to accelerate the model calculation, only 1/4 of the system of the slab and casting roller array and the mold is taken as the model calculation domain, and considering that the temperature gradient inside the casting flow in the actual continuous casting process is small, the mesh of the slab core is coarsened. The model entity structure and mesh of the mold and the secondary cooling high temperature zone of the continuous caster are shown in Fig. 4.2. In addition, considering that the slab is reduced at the end by multiple rollers of the bow segments at the same time, the length of the bow segments of a certain steel plant is about 2.1 m, so the length of the slab in the model is set to 2 m.

The cooling structure of the mold copper plate is an important factor affecting the uniformity of the solidification heat transfer inside the shell. The model development process fully considers the actual structure of the mold copper plate of a certain

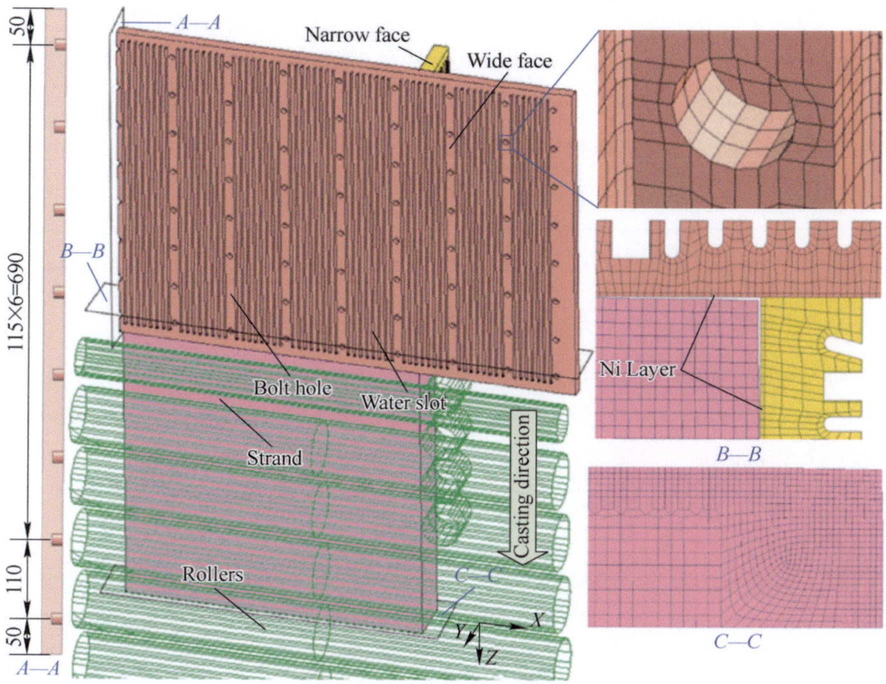

Fig. 4.2 Meshed assembly and geometric contact bodies [35] (unit: mm)

steel plant, and the section structure and size of the wide and narrow face copper plates are shown in Fig. 4.3. The nominal width of the slab selected by the model is 1600 mm. Considering the solidification shrinkage of the slab and the deformation in the secondary cooling zone, the width of the mold top is set to 1617.6 mm and the thickness is 265.5 mm. The taper of the narrow and wide faces of the mold is set to 1.1% and 2 mm respectively, and other detailed geometric parameters of the mold are shown in Table 4.1 [35].

In this calculation model, considering that the tangential friction between the continuous casting slab and the casting roller has a relatively limited effect on the formation and evolution of the longitudinal depression in the wide face off-corner of the slab, the casting roller is set as a meniscus. The spacing and diameter of each roller in the casting flow roller array are given according to the actual parameters of the wide-thick slab continuous caster of a certain steel plant, as shown in Fig. 4.4. It can be seen that the roller gap in the actual production of a certain steel plant gradually decreases along the direction of slab casting, and the roller diameter gradually increases step by step along the casting flow direction. The soft reduction segments at the slab solidification end include three horizontal segments of No.9–No.11 segments, the entrance of the reduction segments corresponds to the position of the casting flow at 21.2 m below the meniscus, and the exit of the No.11 segment is at 28.0 m below the meniscus, with a total reduction amount of 6 mm, and the

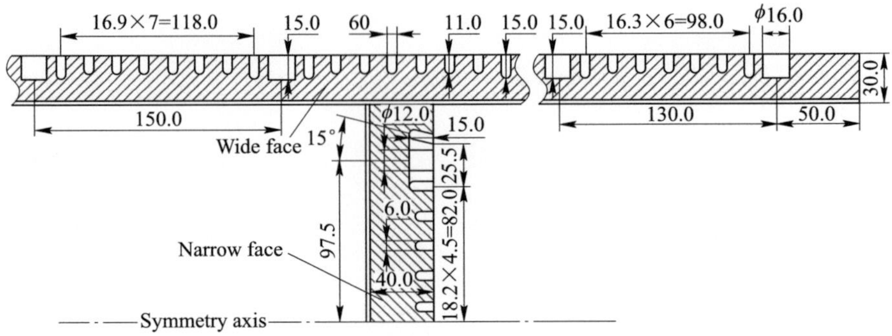

Fig. 4.3 Layout and dimensions of the water slots [35] (unit: mm)

Table 4.1 Geometric structure parameters of the mold (unit: mm)

Item	Value
Mold height	900
Effective height of the mold	800
Mold top size	1617.6 × 265.5
Mold exit size	1600 × 263.5
1/2 wide face copper plate	1230
Thickness of wide face copper plate	30
Thickness of narrow face copper plate	40
Diameter of wide face bolt hole	16
Diameter of narrow face bolt hole	12
Depth of bolt hole	15
Depth of deep water slot	15
Depth of shallow water slot	11
Height of water slot	850
Width of water slot	6

specific distribution of the reduction amount is 3 mm for the No.9 segment, 2 mm for the No.10 segment, and 1 mm for the No.11 segment.

The steel type simulated for casting is Nb-Ti-bearing micro-alloyed steel, and the main components are detailed in Table 4.2. The key continuous casting process parameters are: casting speed 1.05 m/min, molten steel superheat 25 K, mold cooling water volume, mold flux consumption, secondary cooling zone division, and its water distribution process are as shown in Tables 4.3 and 4.4 [35], respectively.

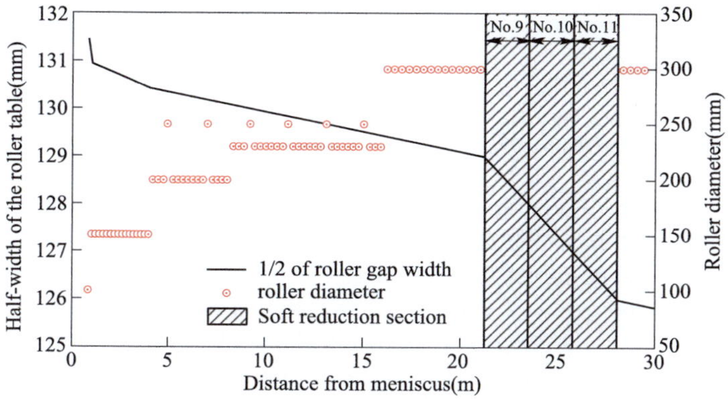

Fig. 4.4 Evolutions of roller gap and diameter [35]

Table 4.2 Composition of peritectic micro-alloyed steel (mass fraction, %)

Element	C	Si	Mn	P	S	Nb	Ti	Al
Content	0.08	0.3	1.5	0.013	0.003	0.01	0.015	0.03

Table 4.3 Mold process of casting micro-alloyed slab in a steel plant

Item	Value
Effective height (mm)	800
Narrow face taper (%)	1.1
Wide face water volume (L/min)	4760
Narrow face water volume (L/min)	530
Inlet water temperature (K)	305
Mold flux consumption (kg/t)	0.45

Table 4.4 Division of secondary cooling zones and water volume (1.05 m/min)

Cooling zone	Segments	Length/m	Water volume (L/min)
Zone 1 L/R	Narrow face foot roller	0.59	110
Zone 1 I/O	Wide face foot roller	0.24	330
Zone 2 I/O	Bender	0.56	420
Zone 3 I/O	Bender	1.11	405
Zone 4 I/O	Bender	1.53	340
Zone 5 I/O	No.1 segment	2.05	290
Zone 6 I	No.2 and No.3 segments	4.01	115
Zone 7 I	No.4 and No.5 segments	3.9	55
Zone 8 I	No.6 to No.8 segments	6.76	45
Zone 9 I	No.9 to No.11 horizontal segments	7.36	25
Zone 10 I	No.12 to No.14 horizontal segments	7.14	0

4.2.2 Thermo-mechanical Behavior of Slab Solidification in Mold

4.2.2.1 Shell Deformation

Figure 4.5 shows the morphology of solidification growth and deformation the shell at different heights under the above mold structure and continuous casting process conditions. Among them, the interfacial gap is magnified 3 times. When the slab solidifies to 100 mm below the meniscus, the shell already has a certain thickness. In this process, the solidifying shell gradually shrinks toward the center of the wide face and the narrow face, causing its corners to gradually leave the copper plate. But in this process, the area where the corner of the shell leaves the copper plate is still small, and it has not yet extended to the off-corner of the corresponding face. However, as the solidified shell continues to cool, its thickness gradually increases, and the thickness and width of the interfacial gap in the shell corner continue to expand. When the shell solidifies to 300 mm below the meniscus, the interfacial gap in the corner of the wide face expands to about 80 mm toward the center of the wide face from the corner, and the narrow face also expands from the corner to a range of 50 mm, and the width range has covered the off-corner of the corresponding face. After that, as the shell solidifies, the shrinkage rate begins to gradually slow down. The interfacial gap in the corner of the narrow face begins to gradually decrease under the continuous compensation effect of the taper of the narrow face copper plate. However, the corner of the wide face continues to increase in thickness and width due to the small amount of taper compensation (the compensation for half the thickness of the slab is only 1 mm).

Figure 4.6 extracts the normal displacement of the wide and narrow faces near the corner of the slab solidification process from Fig. 4.5. From the figure, it can be seen that the solidification shrinkage of the wide and narrow faces of the shell in the mold is mainly concentrated in the range of 0–80 mm and 0–50 mm from the corner of the slab. At the same time, as the slab solidifies, the shrinkage deformation of the shell along the thickness and width directions is slightly severe in the upper part of the mold, while the middle and lower parts are generally gentle. In addition, it can be seen from the figure that there is no depression defects in the wide face off-corner of the solidified shell as shown in Fig. 4.1. This phenomenon indicates that the depression defects in the corner of the slab wide face do not form inside the mold.

4.2.2.2 Heat Transfer Medium Distribution in Shell-Mold Interface

The solidification shrinkage and deformation of the shell will inevitably affect the filling behavior of the flux film in the interfacial gap formed by it and the dynamic formation of the air gap, thereby affecting the heat transfer behavior of the entire system of mold and shell, and once again affecting the solidification shrinkage and

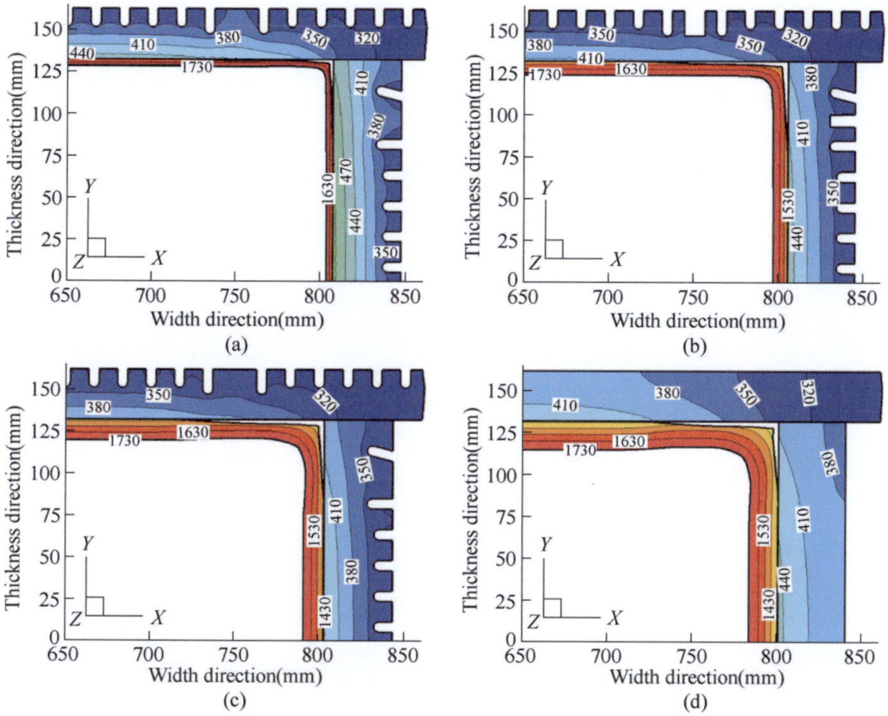

Fig. 4.5 Deformation of the solidified shell at different positions below the meniscus [35]. **a** 100 mm; **b** 300 mm; **c** 500 mm; **d** mold exit

deformation of the shell. Understanding the distribution characteristics of the heat transfer medium inside the interface between shell and mold is the basis for analyzing the non-uniform heat transfer behavior inside the slab mold.

A. Liquid Flux Film

Figure 4.7 shows the distribution of the liquid flux film within the interface of the vicinity of the wide and narrow face corners of the shell in the mentioned mold under deformation and solidification conditions. It can be seen that the liquid flux film in the vicinity of the wide and narrow face corners of the mold mainly exists in the height range of 550 mm below the meniscus, and the thickness decreases rapidly with the increase in the height of the mold. For the distribution of the liquid flux film along the circumference of the mold, it can be seen from the figure that the flux film thickness in the vicinity of the wide and narrow face corners fluctuates greatly, presenting a thicker distribution in the off-corner of the wide and narrow faces of the slab, and a thinner distribution in the area far from the off-corner and in the direction of the center of the wide and narrow faces. Specifically, for the off-corner of the wide and narrow faces of the slab, the thickness of the liquid flux film increases first and then decreases with the increase in the height of the mold, and the change in the off-corner

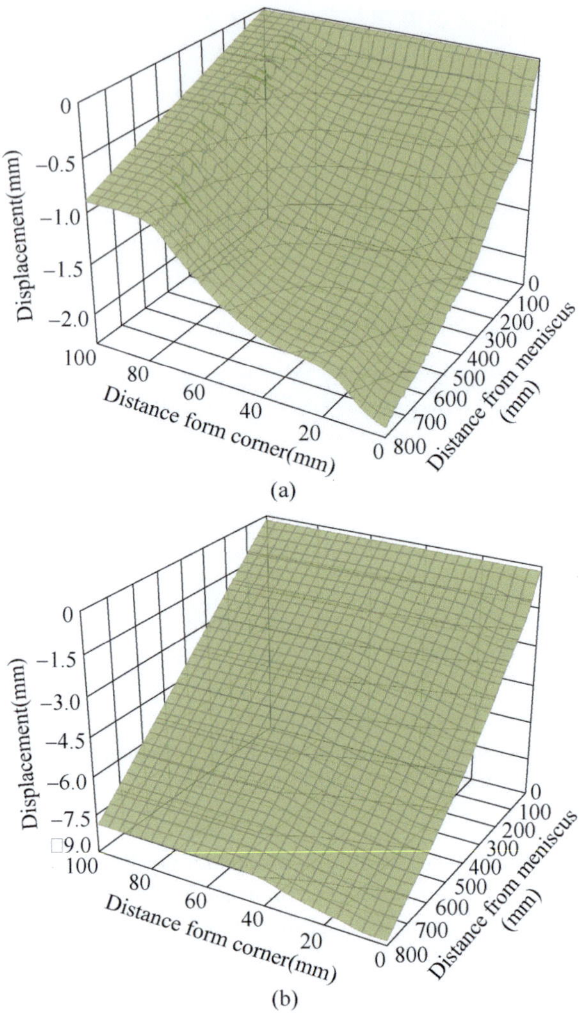

Fig. 4.6 Normal displacements of the corner and off-corner of the shell [35]. **a** Wide face; **b** narrow face

of the narrow face in the upper part of the mold is severer. The main reason is that in the height range of about 100 mm below the meniscus, the corners of the wide and narrow faces of the shell and the vicinity quickly detach from the copper plate due to severe shrinkage (with the shrinkage of the narrow face being more significant), and at this time, the surface temperature of the shell is still higher than the melting temperature of the flux, the flux film flowing into the interface between shell and mold quickly fills the changing interfacial gap, thereby causing the thickness of the liquid flux film in the off-corner to increase rapidly. As a result, the thickness of the liquid flux film in the off-corner of the wide and narrow faces of the solidified shell in the mold is significantly greater than the thickness in the corresponding area far from the shell corner and in the center of the wide and narrow faces, thereby significantly increasing the total thickness of the flux film in the off-corner of the wide and narrow faces of the slab.

B. Solid Flux Film

Figure 4.8 shows the distribution of the solid flux film thickness in the vicinity of the wide and narrow faces corner of the slab in the mold. As can be seen from the figure, due to the shrinkage deformation of the solidified shell in the mold, the filling and heat transfer effects of the liquid flux film, the solid flux film in the vicinity of the wide and narrow faces corner of the slab are mainly concentrated within 0–80 mm

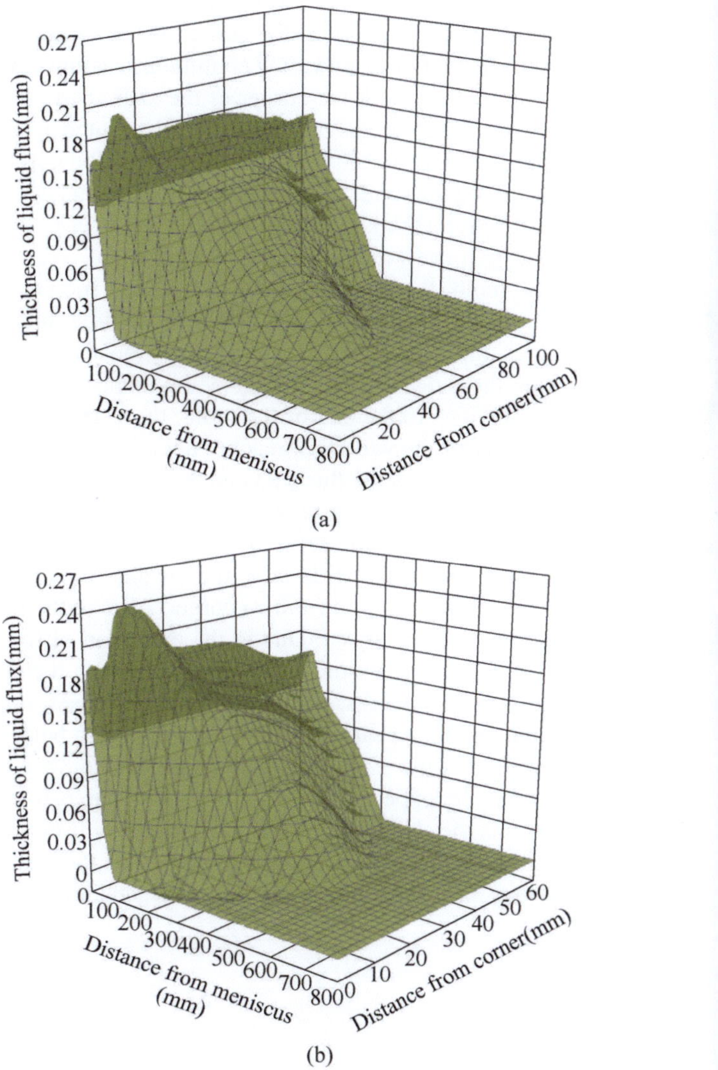

Fig. 4.7 Distribution of liquid flux thickness of slab corner [34]. **a** Wide face; **b** narrow face

and 0–50 mm from the corner. The thickness gradually increases from the meniscus to the mold exit and then tends to stabilize. When the slab solidifies to the lower part of the mold, the thickness difference of the solid flux layer in the off-corner of the wide face of the slab is about 0.9 mm, and the thickness difference of the solid flux layer in the off-corner of the narrow face is about 1.07 mm, greatly increasing the thermal resistance of the heat transfer in the off-corner of the wide and narrow faces.

C. Air Gap

The thermal resistance of the air gap is large, which is the most important factor affecting the heat transfer efficiency of the solidified shell in the mold. Figure 4.9 shows the distribution of air gap under the deformation and dynamic filling of the flux film in the vicinity of the corner of the solidified shell in the mold. As can be seen from the figure, under the continuous casting conditions of wide-thick slab, the air gap first forms at the corner of the slab about 90 mm below the meniscus. For the wide face, the air gap is mainly distributed within 0–90 mm from the corner of the slab. Due to the early detachment of the corner of the slab from the copper plate during the solidification process, and the insufficient taper compensation for the solidification shrinkage of the wide face copper plate along the thickness direction, the air gap shows a continuous growth trend. When the slab solidifies to the exit of the mold, the thickness of the air gap at the corner of the wide face reaches a maximum of about 0.62 mm. At the same time, it can also be seen from the figure that a significant thickness and width of the air gap is also formed in the off-corner within the range of 320 mm below the meniscus to the exit of the mold. When the slab solidifies to the exit of the mold, the maximum thickness of the air gap in the off-corner of the wide face reaches about 0.13 mm. Due to the combined effects of the distribution of the flux film and the formation of the air gap in the off-corner, the heat transfer efficiency of the off-corner of the wide face will be greatly reduced, causing a "hot spot" to form in the off-corner of the wide face.

Compared to the wide face, the air gap of the narrow face of the slab mainly forms within 0–10 mm from its corner. Among them, the thickness of the air gap at the corner is the largest. Due to the evolution of the interfacial gap on the narrow face of the slab mentioned above, the air gap at the corner of the narrow face of the shell generally increases first and then decreases along the casting direction, with the maximum appearing at a height of 375 mm below the meniscus, which is about 0.51 mm. At the same time, due to the overcompensation of the taper of the lower narrow face copper plate, no significant air gap is formed in the off-corner. Since the air gap in the slab corner is only distributed within a narrow range of 0–10 mm from its corner, it only has a significant impact on the heat transfer of the slab corner and does not have a significant impact on the heat transfer in the off-corner of the narrow face.

Fig. 4.8 Distribution of solid flux thickness of slab corner. **a** Wide face; **b** narrow face

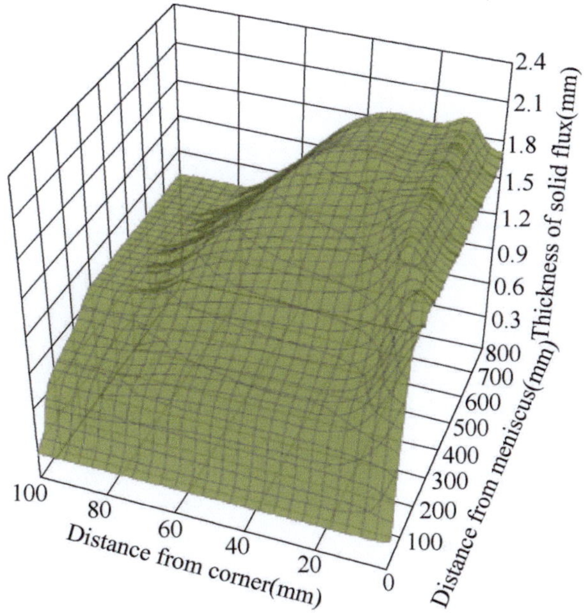

(a)

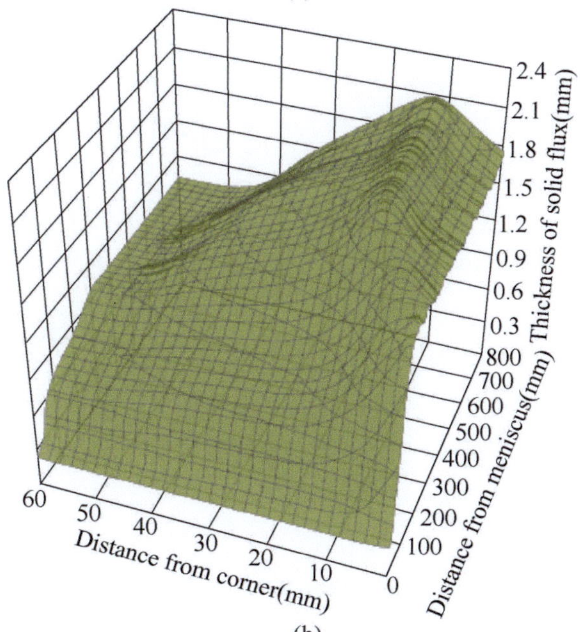

(b)

Fig. 4.9 Distribution of air gap thickness of slab corner [34]. **a** Wide face; **b** narrow face

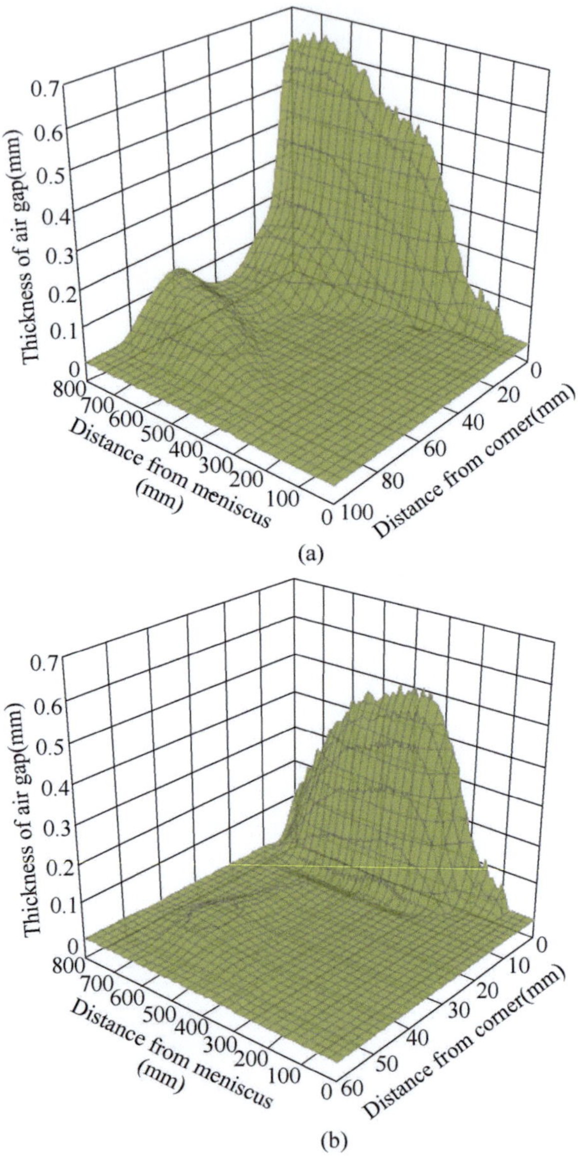

4.2.2.3 Shell Heat Transfer and Growth

Figure 4.10 shows the evolution of heat flux along the height direction of the mold under the filling behavior of the interface heat transfer medium near the wide corner of the slab. Overall, the closer the interfacial gap is to the shell corner, the thicker the flux film and air gap, and the smaller the heat flux of the copper plate, that is, the heat flux at point A is minimum. However, the area at point A is the corner of the slab,

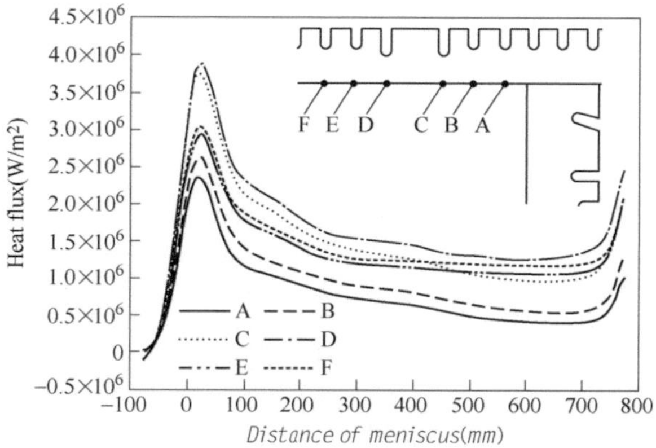

Fig. 4.10 Evolution of heat flux of mold around off-corner

which is subject to 2D heat transfer during solidification process, and the temperature at the same solidification height is the lowest. In the wide face off-corner far from the slab corner, due to the rapid thickening of the flux film and the formation of air gap, the heat flux at point C decreases rapidly when the height is below 400 mm of the mold meniscus. Although the heat flux at other marked positions also decreases, the amplitude decreases.

Figure 4.11 shows the distribution of slab surface temperature at different heights of the mold under the influence of thc interface heat transfer medium filling and heat flux changes. As can be seen that the temperatures of the wide and narrow face off-corner of the solidifying shell in the mold are significantly higher than the temperatures at the corresponding face center and other positions. For the wide face shell, the temperature difference between the off-corner and the centerline of the wide face gradually increases with the solidification of the slab and rapidly increases in the range of 500 mm below the meniscus to the mold exit height. When the slab solidifies to the mold exit, this difference reaches nearly 129 K. For the narrow face of the slab, due to the absence of significant air gap in its off-corner, although the thickness of the flux film in the off-corner is significantly higher than that in the center of the narrow face, the temperature difference between the off-corner and the centerline of the narrow face is significantly reduced compared to the wide face, and it shows a trend of first increasing and then decreasing with the downward movement of the slab. When the slab solidifies to the mold exit, this difference is only about 60 K. At the same time, it can also be seen from the figure that due to the overall larger width of the thick flux film and air gap distribution in the wide face corner than in the narrow face off-corner, the width of the "hot spot" distribution in the wide face off-corner is also significantly larger than that in the narrow face off-corner.

Due to the non-uniform heat transfer in the slab corner and its vicinity, the solidification rate of the wide face off-corner of the slab significantly decreases, leading to the formation of a thin shell in the wide face off-corner of the slab. When the slab solidifies to a height of 500 mm below the meniscus, the solidification thickness of the shell at the centerline of the wide face reaches about 14.0 mm, while the thickness

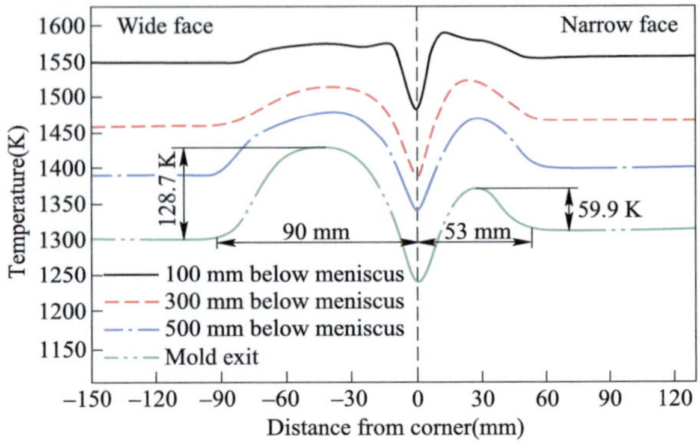

Fig. 4.11 Distribution of surface temperature around shell corner [35]

of the shell in the off-corner about 45 mm from its corner point is only about 13.1 mm, the shell thickness decreases by nearly 1 mm, as shown in Fig. 4.12. When the slab continues to solidify to the mold exit, due to the continuous increase in the thickness of the flux film and air gap in the wide face off-corner, the difference in the solidification thickness of the shell compared to the center of the wide face continues to increase. When the shell solidifies to the mold exit, the thickness difference reaches the maximum, about 2.0 mm, becoming a significant "hot spot" thin area. For the narrow face shell, the difference in shell thickness between its off-corner and the centerline is only about 0.5 mm.

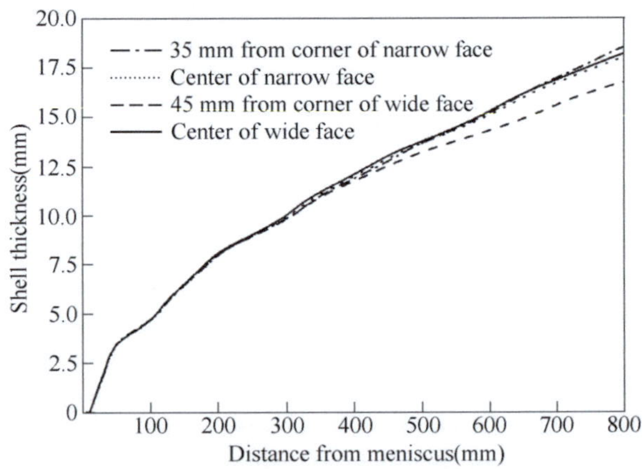

Fig. 4.12 Shell growth of slab wide face

4.2.3 Slab Heat Transfer and Deformation in Secondary Cooling

4.2.3.1 Slab Surface Temperature

The evolution of the slab temperature field in the secondary cooling zone is an important factor, which affects its deformation behavior. As known from Sect. 4.2.2, a significant "hot spot" appeared in the off-corner of the wide face of the slab after exiting the mold. Based on the heat transfer characteristics of the solidified shell in the mold, this chapter will further study the evolution of the temperature field of slab in the secondary cooling casting flow. Figure 4.13 shows the temperature evolution of the wide face center, off-corner, and narrow face center of the slab in the secondary cooling casting flow. As can be seen from the figure, during the solidification process of the slab, intermittently contact and heat transfer occur between the solidified shell in the wide face center and casting roller in the bow segments, which causes the slab surface temperature decreases almost at equal intervals. However, for the corner and off-corner of the slab, their surface temperature evolution is significantly different from the wide face center. For the corner of the slab, within the range of 5.5–21.4 m below the meniscus, due to significant solidification shrinkage, it does not contact the casting roller, and the slab surface continuously contact the casting roller in this casting flow, so the slab surface temperature does not decrease at equal intervals. For the wide face off-corner of the slab, intermittent cooling occurred throughout the solidification process without contact with the bow segment casting rollers in the secondary cooling casting flow, indicating that there is a significant depression in the wide off-corner of the slab throughout the secondary cooling casting flow.

To further analyze the evolution characteristics of the lateral distribution of surface temperature of casting in the secondary cooling flow, the lateral distribution data of surface temperature of casting in different cooling zones are extracted, as shown in Fig. 4.14a. It can be seen that at the mold exit, the maximum temperature difference

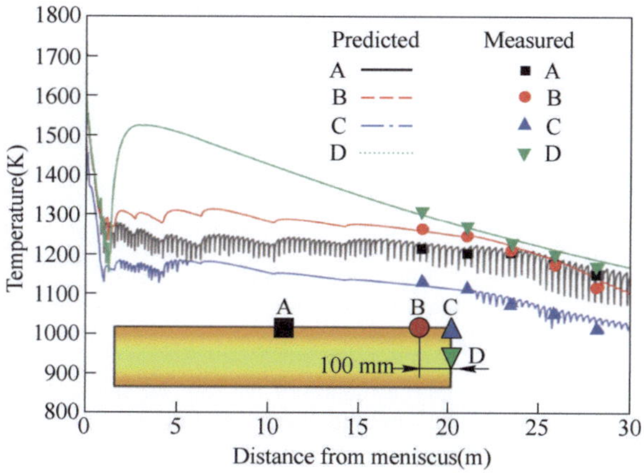

Fig. 4.13 Evolution of slab surface temperature in secondary cooling [35]

between the off-corner and the centerline of the wide face of the slab reaches 123 K. As the slab moves downward, the surface temperature of the slab rapidly decreases overall due to the strong spray cooling effect in the wide face foot roller zone (No.1 wide face foot roller zone), and the temperature drops the fastest in the off-corner. As a result, the temperature difference between the off-corner and the centerline of the wide face of the slab decreases to about 54 K. After the slab exits the wide face foot roller zone, the cooling intensity of the slab surface decreases. At this time, due to the relatively thin shell of the wide face off-corner, the rate at which heat is transferred from the liquid core to the surface of the casting slab is faster. Under the action of uniform cooling water spray, the maximum temperature difference with the centerline of the wide face quickly rises to about 84 K, and then gradually decreases again until the slab exits the bender and drops to about 67 K. After the slab enters the bow segments, the secondary cooling water intensity decreases again, and the temperature in the wide face off-corner increases rapidly again, causing the maximum temperature difference with the centerline of the wide face to increase again to about 91 K. After that, as the thickness of the slab gradually increases and the surface temperature of the slab continues to drop, the maximum temperature difference between the wide face off-corner and the centerline decreases again. When the slab solidifies to the exit of the secondary cooling No.6 zone, this maximum temperature difference drops to about 74 K.

In addition, as shown in Fig. 4.14b, due to the influence of spray cooling in each cooling zone and the temperature evolution in the wide face off-corner, the transverse width of the hot spot zone in the wide face off-corner in the secondary cooling zone shows an overall trend of expansion. When the slab solidifies to the end of the secondary cooling No.4 zone, the width of the hot spot zone in the wide face off-corner rapidly increases from about 93 mm at the exit of the mold to about 190 mm, and then this width gradually stabilizes. When the slab solidifies to No.6 zone, the width of the hot spot zone expands to about 200 mm.

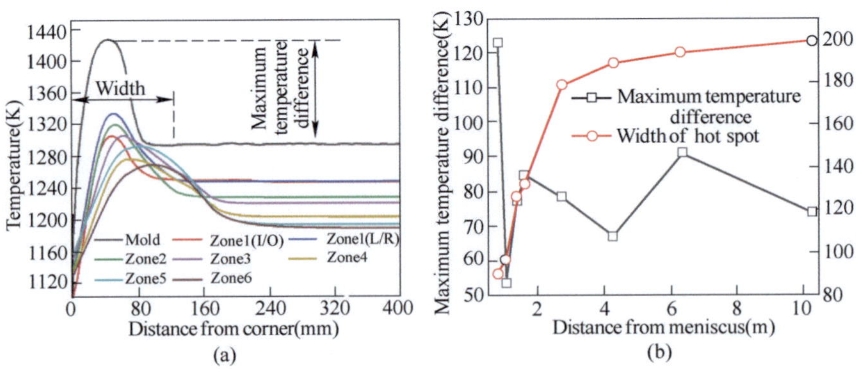

Fig. 4.14 Temperature evolution of the slab surface at the end of different cooling zone [35]. **a** Transverse distributions at the end of cooling zones; **b** temperature difference and width of the hot spot

4.2.3.2 Slab Deformation and Stress in High Temperature Cooling Zone

Figure 4.15 shows the evolution of the deformation of the slab at each exit of the secondary cooling No.1–6 zone. To intuitively analyze of the evolution of the slab contour, a 2 mm offset distance has been added between the contour lines of the slab surface in adjacent cooling zones along the width and thickness directions. As can be seen from the figure, in the wide face foot roller zone, although the cooling water volume in wide and narrow faces of the slab is large, the temperature of the shell is high, and the shrinkage of the solidified shell in the direction of the wide face center is not very significant. Due to the support of the narrow face foot roller, the narrow face shell does not have obvious bulge, so it does not have a significant effect on the wide face corner deformation of the shell. The wide face corner of the slab only shrinks in the direction of the slab thickness reduction under the effect of its solidification shrinkage in the direction of the narrow face center, until the slab shell solidifies to the exit of the narrow face foot roller zone. When the slab solidifies to the exit of the secondary cooling No.2 zone, the narrow face shell, which is still at high temperature, loses the support of the foot roller and begins to gradually form a narrow face bulge. At this time, under the combined effect of the solidification shrinkage of the shell in the direction of the wide face center and the bulge of the narrow face, the off-corner of the wide face slab begins to gradually twist and form a depression with the corner as the fulcrum (the corner of the slab is relatively hard and not easy to deform). Afterward, as the slab continues to solidify and move downward, the bulge of the narrow face of the shell aggravates with the increase of the ferrostatic pressure, causing the depth and width of the off-corner depression of the wide face to increase. When the slab solidifies to the exit of the secondary cooling No.4 zone since the strength of the narrow face shell can fully support the ferrostatic pressure, the bulge of the slab narrow face is basically stable, and the off-corner depression of the wide face also gradually stabilizes.

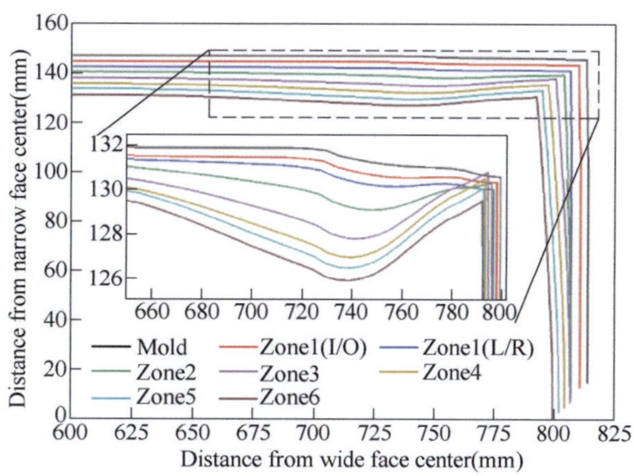

Fig. 4.15 Profile of the slab corner and off-corner at the end of different cooling zone [35]

Figure 4.16 shows the relationship between the depth of the off-corner depression and the bulge of the slab under the deformation conditions in secondary cooling zones. As can be seen from the figure, there is a good corresponding relationship between the depth of the off-corner depression of the solidified shell wide face in the high temperature secondary cooling zone and the bulge of the narrow face, the depth of the depression increases with the increase of the narrow face bulge of the slab and gradually stabilizes as the bulge tends to be stable. The narrow face bulge of the slab mainly forms from the exit of narrow face foot roller zone to the exit of secondary cooling No.4 zone, and the closer to the mold exit, the faster the bulge changes. Within the height range of the narrow face foot roller zone, the narrow face of the slab is supported by the foot roller extrusion, the bulge amount is slightly reduced, and no obvious depression forms in the wide face off-corner. However, after the slab exits the narrow face foot roller zone, the narrow face bulge of the slab quickly increases from about 0.4 mm at the No.2 zone exit to about 6.4 mm at the No.3 zone exit. As can be seen from Fig. 4.15, after the narrow face shell bulges, the off-corner hot spot zone of the shell wide face will twist to form a depression with the slab corner as the fulcrum. Under this effect, the depth of the off-corner depression of the wide face at the secondary cooling No.3 zone exit quickly increases to about 3 mm. When the slab enters secondary cooling No.4 and No.5 zones, the shell gradually grows to a certain thickness and can gradually resist the deformation caused by the ferrostatic pressure, and the increasing rate of the narrow face shell bulge gradually decreases and tends to be smooth. The depth of the off-corner depression of the wide face of the slab also shows a slight increase and then tends to be stable. When the slab solidifies to the secondary cooling No.6 zone exit, the bulge of the narrow face shell and the depth of the off-corner depression of the wide face reach 7.24 and 3.63 mm.

Figure 4.17 is a 3D morphology diagram of the heat transfer and deformation behavior of the slab within the height range from the mold exit to the secondary cooling No.2 zone exit. As can be seen from Fig. 4.17a, due to the "hot spot" inheritance effect of the off-corner of the wide face and narrow face on the solidified shell

Fig. 4.16 Evolution of bulge amount and depth of longitudinal depression [35]

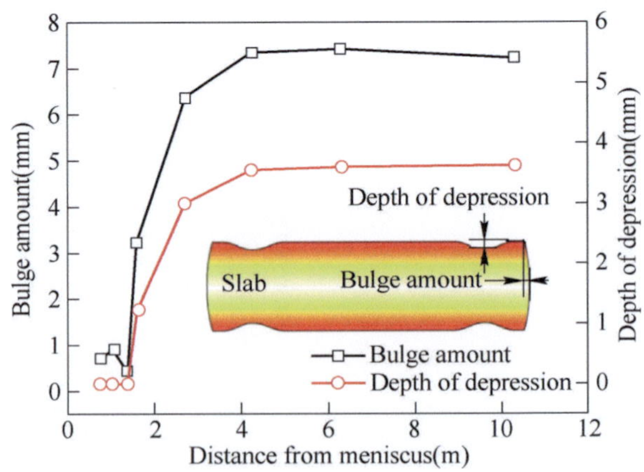

in the mold, the significant hot spot zone is preserved in the casting flow of the high temperature secondary cooling zone, laying an important temperature condition for the formation of the off-corner depression of the wide face. The gradual formation of the off-corner depression of the wide face causes the slab at this position to detach from the wide face casting roller grasping of the bow segments. Therefore, as can be seen from Fig. 4.17c, the formation of the off-corner depression of the wide face of the slab should be located in the casting flow of the high temperature zone from below the narrow face foot roller zone exit to the bender exit.

Figure 4.18 shows the evolution of ductile strain in the cross section of the slab at the exit of different zones of the secondary cooling zones. As can be seen from the figure, at the exit of the wide face foot roller zone, only slight deformation occurs in the off-corner of the wide face, and no obvious ductile deformation occurs within the cross section of the slab. However, when the slab solidifies to the exit of the secondary cooling No.2 zone, the off-corner of the narrow face and wide face of the shell gradually form a bulge and a depression. Since the process of bulge forming on the narrow face of the shell is outwardly curved, a slight tensile strain is formed on the surface of the slab, and a more obvious compressive strain is formed at the corresponding solidification front. At the same time, because the process of forming a depression in the off-corner of the wide face is twisted with its corner as the fulcrum, a significant compressive strain is formed on the surface of the slab in the depression area, and a significant tensile strain is formed at the solidification front. Subsequently, as the bulge on the narrow face of the slab increases, the depth and width of the depression in the off-corner of the wide face increase, the surface compressive strain in the depression gradually aggravates, the subsurface tensile strain also gradually increases, and the area of ductile deformation gradually widens, and the position gradually deepens. When the slab solidifies to the exit of the secondary cooling No.6 zone, the surface compressive strain and subsurface tensile strain in the depression reach about -0.037 and 0.048, respectively. The above strain evolution behavior of the slab also explains the reason why longitudinal cracks are prone to occur on the subsurface of the depression area of the off-corner of the wide face in the actual wide-thick slab production process.

4.2.3.3 Off-Corner Depression Evolution During Soft Reduction

As an effective method to improve the center segregation and porosity of the slab, the solidification end soft reduction process will cause a certain degree of deformation along the thickness and width directions. Under this effect, the depression shape of the off-corner on the wide face of the slab will undergo a certain degree of change. Figure 4.19 shows the morphology evolution of the slab at the exit of each bow segment from No. 7 to No. 12. As can be seen from the figure, before the slab enters the reduction segments, the width and depth of the off-corner depression of the wide face are basically stable, and the depression starts from the corner point, with the width and depth of about 165 and 3.9 mm, respectively. When the slab enters the reduction process, significant deformation occurs on the inner arc side of the slab

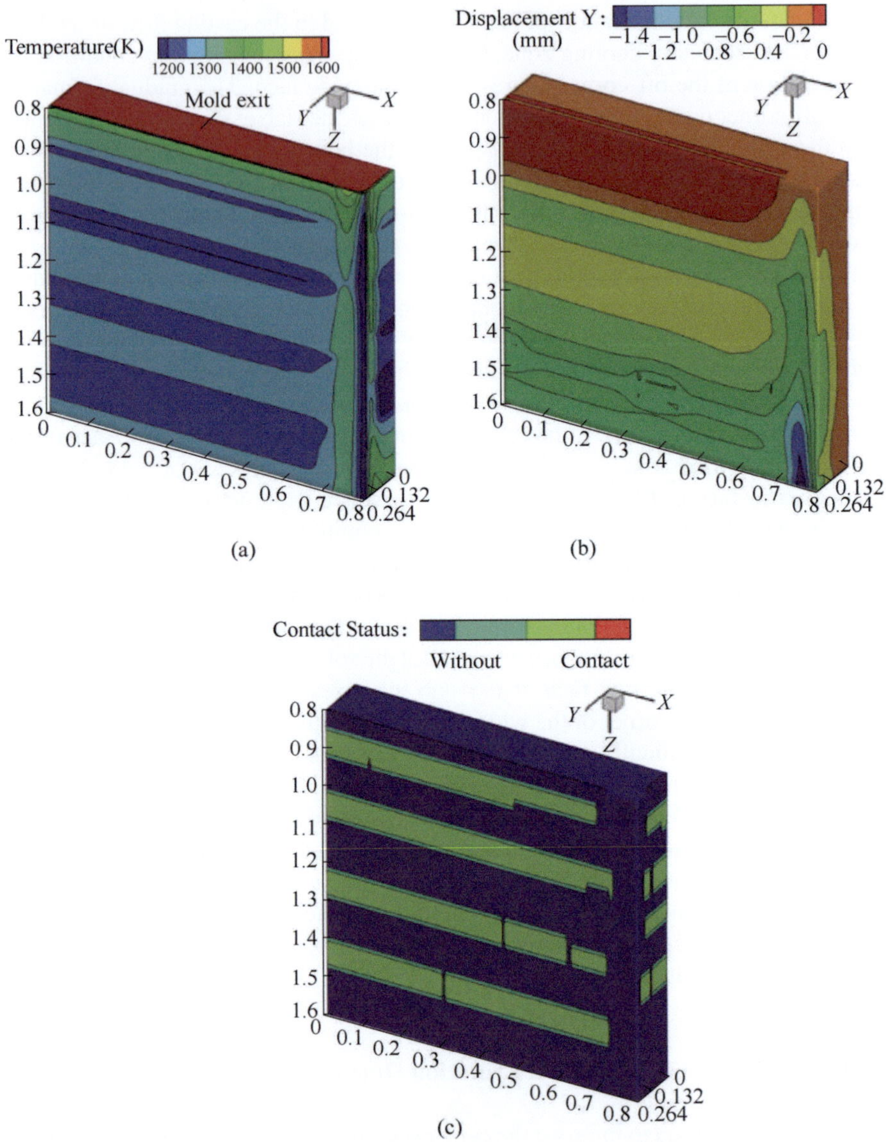

Fig. 4.17 Thermo-mechanical behaviors of the slab below the mold [35]. **a** Temperature; **b** displacements in the thickness directions; **c** contact state

under the reduction of 3, 2, and 1 mm applied to the No.9, No.10, and No.11 segment of the bow segments, respectively. Especially in the No.1 reduction segment, due to its large reduction, the wide face corner of the slab is flattened, and the depth and width of the off-corner depression rapidly decrease. When the slab solidifies to the exit of No.9 segment, the width and depth of the off-corner depression of the wide

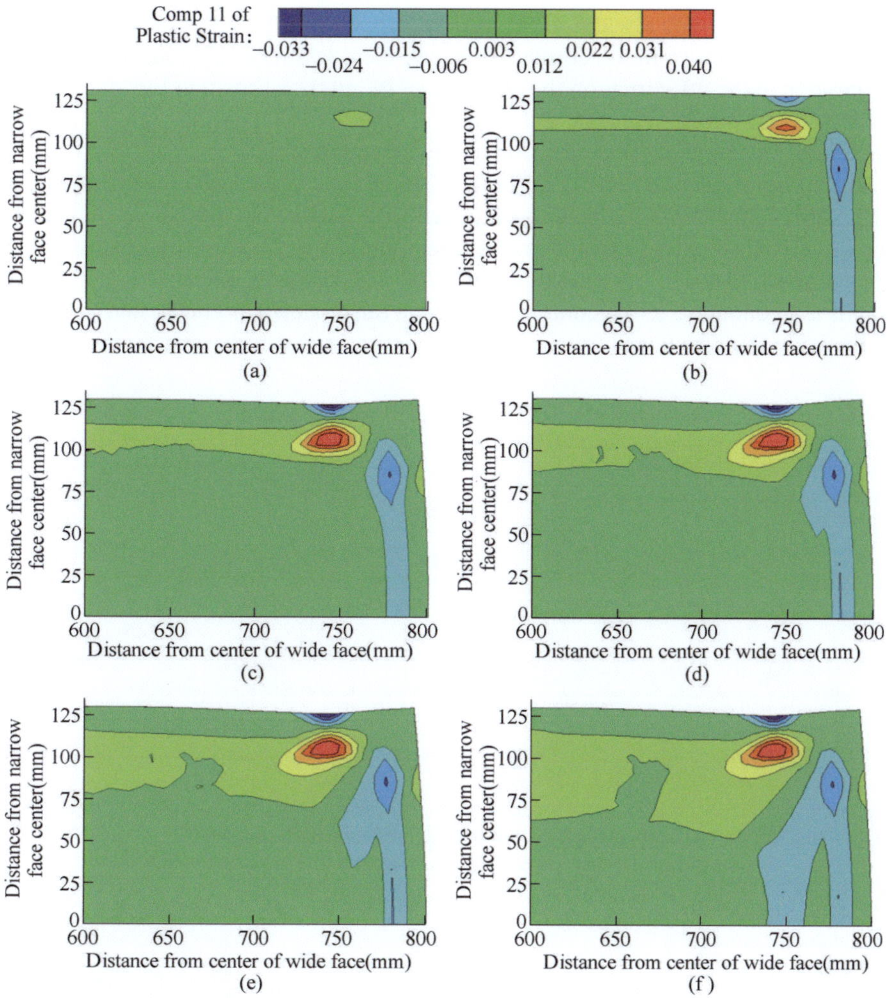

Fig. 4.18 Plastic strain of slab at the end of the secondary cooling zones [35]. **a** Zone 1; **b** Zone 2; **c** Zone 3; **d** Zone 4; **e** Zone 5; **f** Zone 6

face are reduced to about 153 and 3.6 mm, respectively. Then, under the reduction in bow segments No. 10 and No. 11, the slab in the direction of the corner and the center of the wide face is further reduced and thinned, reducing the width and depth of the off-corner depression of the wide face to about 129 and 2.8 mm in No.10 segment and about 84 and 1.8 mm in No.11 segment, as shown in Fig. 4.19b. The reduction process resulted in a reduction of approximately 49.1 and 53.8% in the width and depth of the off-corner depression on the wide face, respectively. Thus, in the actual production process, it is often found that the oscillation marks at the off-corner of the wide face are relatively coarse and no obvious depression is found, or only a slight depression with a width of 50–100 mm appears in the off-corner of the wide face of 30–50 mm from the corner. However, there is a more obvious longitudinal depression in the off-corner on the outer arc of the slab, as shown in Fig. 4.20. After the slab enters No. 12 segment, under the clamping of the horizontal segment rollers, the width and depth of the depression slightly decrease.

In addition, as can be seen from Fig. 4.19, under the reduction of the bow segments (especially in the last two segments of the reduction segments), the corner of the slab shows an obvious width expansion deformation, making the narrow face of the slab appear as an "S shape" as shown in Fig. 4.20. The shape of the narrow face of the slab is not conducive to the control of the edge (also known as "black line") defects of the wide-thick plate.

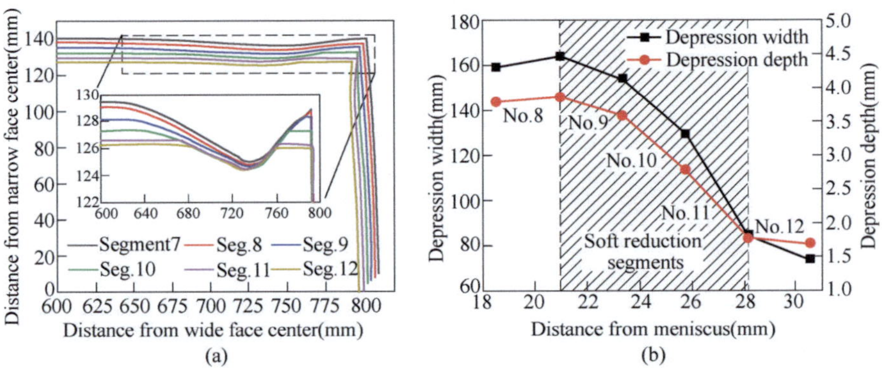

Fig. 4.19 Evolution of slab profile during soft reduction [35]. **a** Slab profile of wide face longitudinal off-corner depression; **b** evolution of depression width and depth

Fig. 4.20 Morphology of slab with off-corner depression

4.3 Formation Mechanism of the Off-Corner Depression

Based on the above solidification heat transfer and mechanical behavior evolution of the thick slab in the full casting flow, the formation mechanism of the consecutive longitudinal depression in the off-corner of the wide face of the thick slab can be summarized as follows: during the solidification process of the thick slab in the mold, the shell near the corner of the wide face and its vicinity lacks efficient compensation from the mold copper plates and the deformation of the corner is not constrained. It causes the corner of the wide face on the shell to separate from the copper plate, forming a large interfacial gap, and thick flux film and air gap are concentrated within it. This hinders efficient heat transfer in the off-corner of the slab, thus forming a significant hot spot zone, as shown in Fig. 4.21a. In the narrow face foot roller zone after slab exits the mold, the shrinkage of the shell in the wide face direction is supported by the narrow face foot roller, the corner of the shell does not form a torsional deformation. The shell in the corner of the wide face only shrinks to the center of its thickness as a whole, as shown in Fig. 4.21b. However, when the high temperature solidified shell exits the narrow face foot roller zone of the mold, under the ferrostatic pressure, the narrow face of the slab quickly forms bulge. Under the action of its deformation, the solidified shell is twisted with the corner as the fulcrum, and the wide face deviates from the high temperature area of the corner to bend inward to form a depression, which is intensified in the secondary cooling No.2 and No.3 zones, until the depression is gradually stabilized after the bender, as shown in Fig. 4.21c.When the slab enters the solidification end reduction segments, the shell on both sides of the off-corner of the wide face is gradually flattened by the casting roller, causing the width and depth of the depression gradually decrease, forming an off-corner depression of the wide face morphology of the thick slab with a width generally of 50–100 mm, depth of 1.0–5.0 mm, and the defect morphology of off-corner depression of thick slab with flat shell on both sides of depression zone, as shown in Fig. 4.21d.

It is worth noting that the formation mechanism of the longitudinal depression in the off-corner of the wide face of the slab proposed in this chapter has certain similarities with the two-stage theory proposed by Thomas et al. [26, 30], but there is a significant difference in the position of the initial formation of the depression. Thomas et al. believe that the longitudinal depression in the off-corner of the wide face of the slab is formed in the foot roller zone within the mold exit, while this book believes that the depression is formed in the secondary cooling No.2–No.4 zones after the narrow face foot roller of the mold. In addition, the formation process of the depression and its evolution in the casting flow (especially in the reduction process at the end of the solidification of the slab) are given in detail, which explains the formation of the longitudinal depression in the wide face off-corner of the slab, and the phenomena such as the large oscillation mark in the wide face off-corner of the slab and the larger depression in the outer arc off-corner than in the inner arc off-corner in the actual continuous casting process.

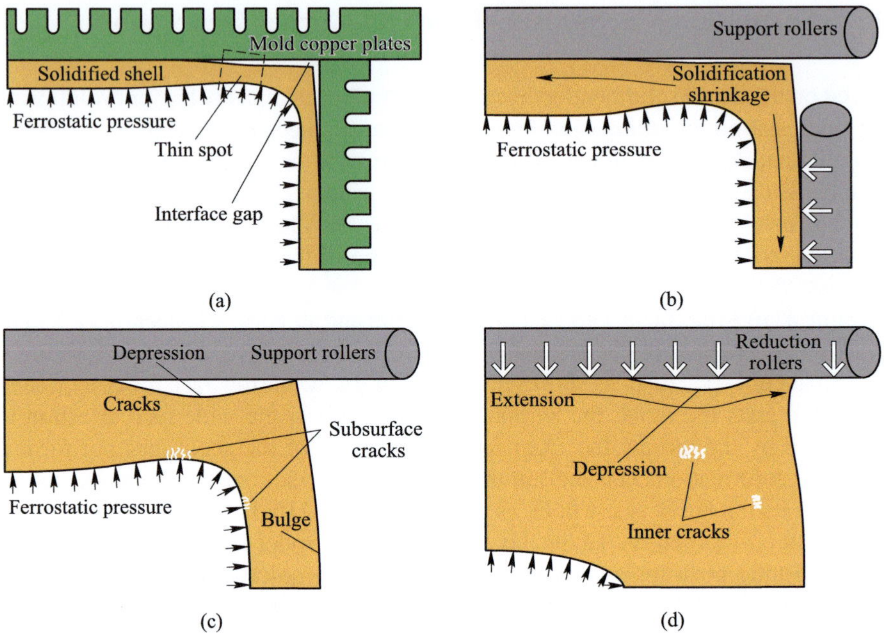

Fig. 4.21 Formation and evolution of the longitudinal depression in different position of casting flow [35]. **a** Mold; **b** foot roller, **c** bender; **d** reduction segment

4.4 Control Strategies for the Off-Corner Depression

From the formation mechanism of longitudinal depression in the off-corner of the wide face of the mentioned thick slab continuous casting process, it is known that preventing the formation of significant "hot spots" in the off-corner of the wide face of the solidified shell in the mold is the key to controlling the formation of this depression. On this basis, it should be ensured that the high temperature solidified shell after the mold gets enough support in the narrow face foot roller zone of the continuous caster and solidifies to form a high strength shell, preventing the slab from forming a large bulge after exiting the narrow face foot roller zone. From the deformation shrinkage of the shell in the mold and the dynamic filling behavior of heat transfer medium such as flux film and air gap at the shrinkage interface in Sect. 4.2.2, it is known that the key to homogenizing the solidification heat transfer in the off-corner of the wide face of the shell in the mold is to reduce the shrinkage gap in the off-corner of the wide face of the mold and prevent the thick flux film and air gap (especially the air gap in the middle and lower part of the mold) from filling in it. The most effective method is to increase the taper of the mold in the direction of the slab thickness, enlarging the width difference between the top and the exit of the narrow face copper plate. However, in the actual continuous casting production process, each mold needs to meet the requirements of universality, and needs to meet the continuous casting production of all steel grades including ultra-low carbon, low carbon, peritectic, medium carbon and even high carbon steel. At present, the taper compensation amounts of the mold in the thickness direction of the main thick slab continuous casting caster at home and abroad generally adopt 1 and 2, mainly 2 mm. For the continuous casting of ultra-thick slab, although it is possible to consider appropriately increasing the taper compensation amount of the mold in the thickness direction (at present, the mold of 300–420 mm thick slab at home and abroad still adopts 2 mm compensation amount in the thickness direction), but to rely on this method to eliminate the shrinkage gap in the off-corner of the wide face, taking the 250 mm thick slab continuous casting mold as an example, as can be seen from Fig. 4.6, the taper of the mold in the thickness direction needs to reach more than 5 mm, which will greatly increase the wear of the wide face copper plate of the mold and is not feasible.

Similarly from Sect. 4.2.2, it is known that in the solidification process of the shell in the thick slab mold, if the narrow face copper plate can efficiently compensate and support the solidification shrinkage of the narrow face shell in the direction of the wide face center, it can "hold up" the narrow face shell, thereby limiting the free deformation of the shell corner (restricting the twisting of the shell corner in the mold), can reduce the interfacial gap thickness and width in the off-corner of the wide face of the slab, and prevent the formation of "hot spot" area on the off-corner of the wide face with excessively large width and high temperature.

However, the current domestic and foreign thick slab continuous casting casters mostly use traditional flat-plate mold. In actual continuous casting production, the solidification process of the slab in the mold is affected by high temperature phase transition characteristics, etc., often shrinking quickly in the upper part of the mold and relatively slowly in the middle and lower parts. Using this flat-plate mold, it often fails to compensate for the shrinkage of the shell in the upper part of the mold, thus causing the narrow face corner of the solidified shell to detach the copper plate early, and then "drag" the shell in the wide face corner to detach the copper plate and form a larger interfacial gap. The shrinkage of the shell in the middle and lower parts is often overcompensated, making its service process often offline repair or scrap due to excessive wear of the mold exit. If based on the flat-plate mold, further increasing its narrow face taper will help to compensate for the shrinkage of the shell in the upper part of the mold, but it will further aggravate the wear of the copper plate at the mold exit. At the same time, it will also increase the friction between the copper plate and the shell in the lower part of the mold, which is not conducive to the control of transverse corner cracks. Therefore, it is necessary to develop a curved mold with a fast compensation structure in the upper part and a slow compensation structure in the middle and lower parts according to the solidification characteristics of the steel grade, so that the slab can efficiently attach to the copper plate for heat transfer during the entire solidification process in the mold, prevent the slab corner in the upper part of the mold from twisting and reducing the interfacial gap in the corner of wide face, and at the same time increase the solidification thickness of the shell of the narrow face.

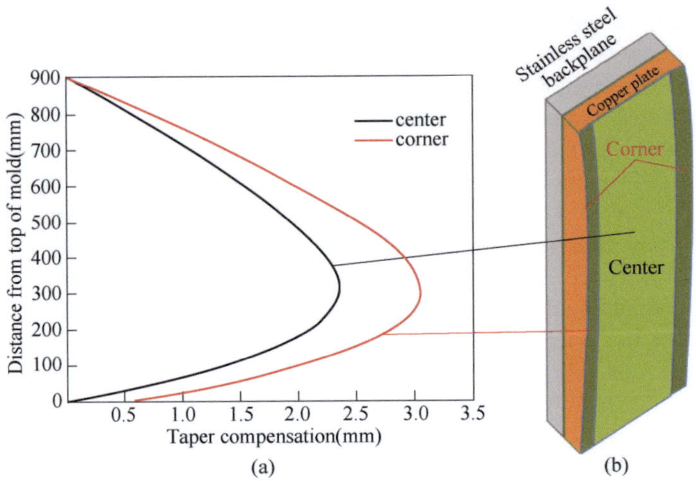

Fig. 4.22 The curve of its working face and schematic diagram of the convex-structure mold. **a** Working face curve; **b** structural schematic

Therefore, this book has designed a curved mold as shown in Fig. 4.22b, based on the mentioned idea of uniform control of the wide face corner of the shell in the mold. The compensation area of the mold along the height direction adopts the taper compensation curve of the novel convex-structure mold developed in Chap. 2. At the same time, to limit the deformation of the narrow face corner of the slab in the upper part of the mold, a structure of the mold inner cavity that is concave toward the deformation constraint of the slab corner has been added. Considering the severe wear in the vicinity of the mold exit during the actual continuous casting production process, the taper of the copper plate corner in the middle and lower parts has been gradually reduced to 0 or a negative value, and the compensation curve of the middle and corner of the copper plate is shown in Fig. 4.22a.

Based on the novel convex-structure mold, it is another key to control the longitudinal depression in the off-corner of the wide face of the thick slab to ensure the cooling of the shell in the narrow face foot roller zone of the continuous caster and the support of the casting roller and to prevent the slab from bulge in the narrow face foot roller zone and the area below it.

At present, for micro-alloyed steel, especially the peritectic micro-alloyed steel with high crack sensitivity in the continuous casting process often adopts secondary cooling with weak cooling control technology. However, as shown in Chaps. 2 and 3 of this book, the mechanism of corner crack formation in micro-alloyed steel slab and thin slab, and the production practice in more than ten domestic and foreign companies, based on ensuring that the device accuracy meets the requirements, appropriately increasing the cooling strength of the wide face and narrow face of the slab after exiting the mold will not cause surface quality defects such as transverse corner cracks. Therefore, for the continuous casting of thick slab of main compositions such as low carbon, peritectic, and medium carbon steel, the conditions are met for implementing rapid cooling in the wide face high temperature zone (secondary cooling No.1–4 zones) and the narrow face foot roller zone.

At the same time, thanks to the convex-structure mold shown in Fig. 4.22, the actual thick slab continuous casting process can implement the slab narrow face large taper compensation control process, increasing the taper compensation of the mold and its narrow face foot roller zone, controlling free deformation of the shell in the mold, and further restraining the deformation and bulge of the slab in the narrow face foot roller zone. Similarly, taking the 250 mm thick slab continuous casting mentioned in this book as an example, after a steel plant implemented the slab narrow face large taper compensation control process for a certain width slab production, the single-sided shrinkage compensation amount of the slab in the effective solidification height within the mold and the narrow face foot roller zone along the width direction will increase by 1.03 mm and 1.47 mm, respectively, as shown in Fig. 4.23, thus effectively controlling the formation of narrow face bulge. For other cross-sectional thick slab continuous casting, the production practice of multiple companies shows that the basic principle of setting the taper of the narrow face foot roller is: to ensure that the narrow face of cold slab retains about 1–2 mm depth of foot roller mark.

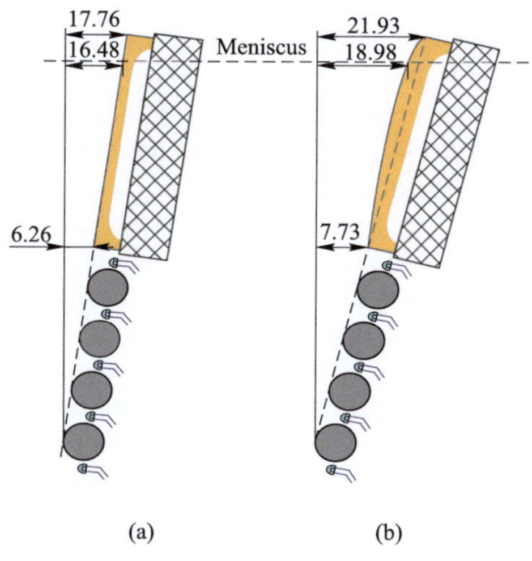

Fig. 4.23 Schematic diagram of compensation amount by using the different mold (unit: mm). **a** Traditional mold; **b** novel convex-structure mold

4.5 Slab Solidification Behavior Under New Process and Its Application Effect

4.5.1 Shell Heat Transfer and Deformation in Novel Mold

Figure 4.24 shows the morphology of solidification shrinkage and deformation of the shell at different heights in the mold under the narrow face convex-structure copper plate shown in Fig. 4.22. It can be seen from the figure, during the entire solidification process of the slab in the mold, the narrow face convex-structure copper plate efficiently compensates for its shrinkage along the center direction of the wide face, and the shell of the narrow face is in close contact with the copper plate at all heights in the mold. At the same time, due to the restriction of zone near the corner of the narrow face copper plate to the inner cavity depression, the wide face corner of the slab only forms shrinkage deformation along the thickness direction, and there is no twisting and "dragging" deformation at the slab corner similar to that the traditional planar narrow face copper plate, the thickness and width of interfacial gap near the slab wide face corner are significantly reduced.

Figure 4.25 shows the distribution of the flux film near the slab corners of the wide and narrow face at different heights of the mold under the novel convex-structure narrow face copper plate. It can be seen from the figure, under the conditions of the novel convex-structure mold, the thickness of mold flux and distribution width near the wide corner of the slab significantly decreases. The maximum thickness of the mold flux film in the zone near the wide corner and its nearby under lower of the mold is only about 0.7 mm, and the distribution width is only in the range of 0–60 mm from the slab corner. At the same time, it can be seen that the maximum thickness of the mold flux film at the corner of the slab wide face appears at 10 mm from the corner. From Sect. 4.2.2, this maximum thickness of mold flux film is relatively close to the corner of the slab, which will only reduce the heat transfer efficiency of the shell in the wide face off-corner of 40–60 mm from the slab corner to a certain extent.

For the narrow face of the slab, due to the efficient compensation of the shell shrinkage along the wide face center direction by the novel convex-structure copper plate, there is no obvious shrinkage gap near the corner of the shell, so there is no phenomenon of concentrated filling of the mold flux film. The mold flux film is evenly distributed at all heights of the mold.

Figure 4.26 shows the distribution of the air gap in the area near the wide and narrow face corners of the slab in the mold under the novel convex-structure narrow face copper plate. The solidification process of the slab in the novel convex-structure mold only forms a large thickness air gap in the wide face corner with a width range of 0–15 mm from the corner, which eliminates the phenomenon of a large thickness air gap formed in the wide face off-corner of the shell under the traditional flat-plate mold. Similarly, although the thickness of the air gap is increased compared with the traditional narrow face flat-plate mold, the air gap is relatively close to the corner of the slab, which will only significantly reduce the heat transfer efficiency of the wide face corner of the slab, but will not have a significant impact on the heat transfer

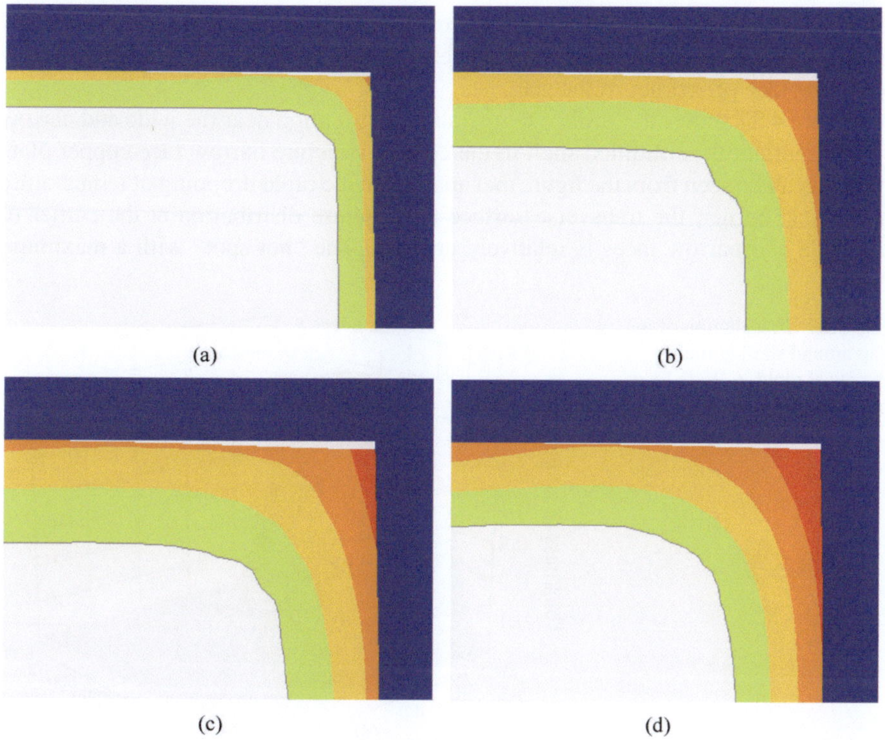

Fig. 4.24 Shell deformation at different positions below the meniscus. **a** 100 mm; **b** 300 mm; **c** 500 mm; **d** mold exit

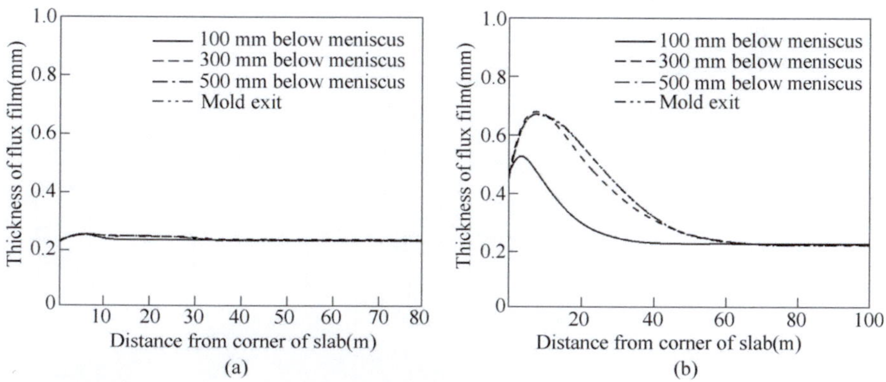

Fig. 4.25 Distribution of mold flux film around shell corner in the novel mold. **a** Narrow face; **b** wide face

of the shell in the off-corner. For the narrow face of the slab, its air gap is basically eliminated, which greatly improves the heat transfer efficiency of the entire narrow and wide face off-corner of the slab.

Figure 4.27 shows the surface temperature distribution near the wide and narrow face corners of the solidified shell in the convex-structure narrow face copper plate mold. It can be seen from the figure that apart from the rapid dropping of temperature at the slab corner, the transverse surface temperature distribution at the corner of the wide and narrow faces is relatively uniform. The "hot spot" with a maximum

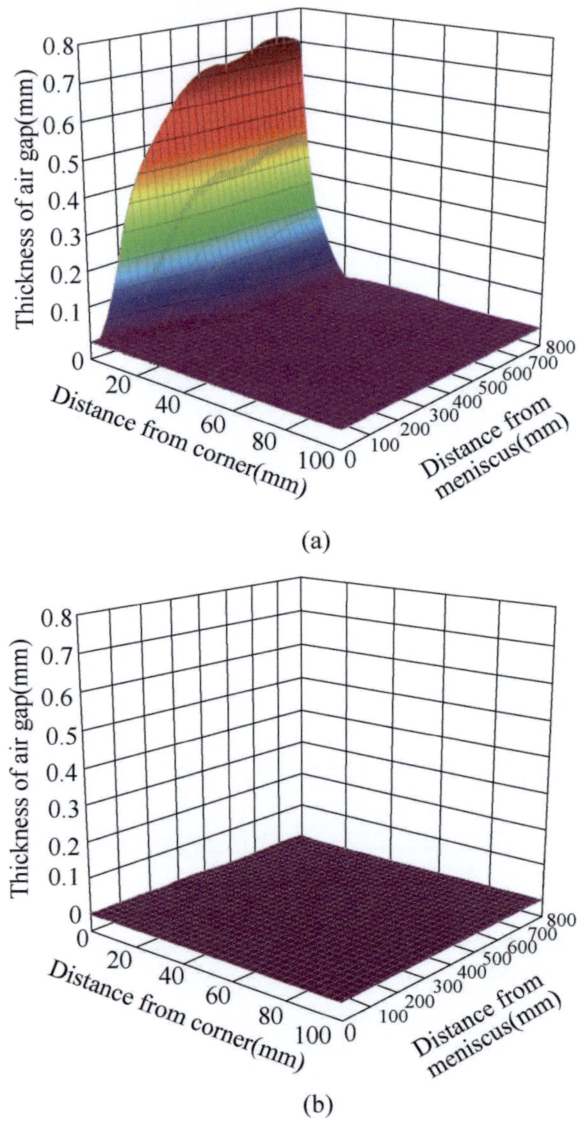

Fig. 4.26 Distribution of air gap around shell corner in the novel mold. **a** Wide face; **b** narrow face

temperature difference of about 50 °C from the center of the slab wide face is only formed in 20–60 mm from the corner. Compared with the significant "hot spot" area with a width of about 90 mm and a maximum temperature difference of about 129 °C from the center of the wide face under the traditional flat-plate mold, the amplitude and width of the "hot spot" zone of the wide face off-corner of slab under the novel mold are significantly reduced, and the temperature that causes the depression of the wide face off-corner of the slab is basically eliminated.

4.5.2 Slab Heat Transfer and Deformation in Secondary Cooling Under New Process

Figure 4.28 shows the surface temperature distribution of the slab at the exit of each cooling zone from the mold exit to the bender exit under the condition of appropriately strengthening the cooling intensity of the slab in the wide face and narrow face foot roller zone and the bender in the secondary cooling zones (No.2–No.4 zones) of the novel convex-structure mold. As can be seen, due to the small "hot spot" in the off-corner of the wide face of the slab exits the mold, and the effect of strong spray cooling on the wide face, the "hot spot" in the off-corner of the wide face almost disappears at the exit of the wide face foot roller zone, and then with the 2D heat transfer at the slab corner and the continuous rapid cooling of the wide face, the shell surface temperature in the off-corner of the wide face gradually becomes lower than the center area of the wide face, which is beneficial to strengthen the deformation resistance of the off-corner of the wide face.

Figure 4.29 shows the deformation contour curve of the slab at the exit of each segment of the continuous caster under the novel convex-structure mold and secondary cooling water distribution process (without offset drawing processing). As can be seen, except for a slight depression in the off-corner of the wide face within the casting flow range before the exit of the bender, during the whole solidification process of the slab in the following secondary cooling casting flow, the depth and width of the depression in the wide off-corner do not expand and deepen significantly (as shown in Fig. 4.29a), and the slight depression will be flattened and disappeared under the action of soft reduction at the solidification end of the No.9-No.11 bow segments of slab as shown in Fig. 4.29b.

At the same time, as can be seen from Fig. 4.29 that there is a certain degree of bulge deformation in the narrow face shell before the slab enters the reduction segments of solidification end. However, under the solidification end soft reduction process, the slab has a more obvious width expansion deformation during the thinning process, and the deformation rate of its surface is faster than the middle of the narrow face, which will cause the slab narrow face to form a microdepression structure. The narrow face of the slab after the implementation of the new process is shown in

Fig. 4.27 Distribution of the shell surface temperature around shell corner in the novel mold. Wide face; **b** narrow face

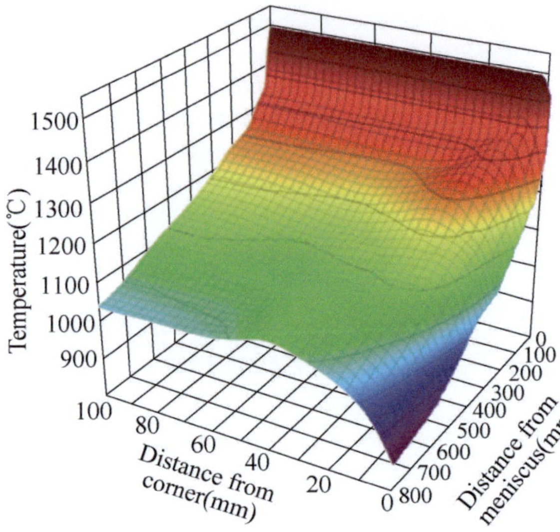

(a)

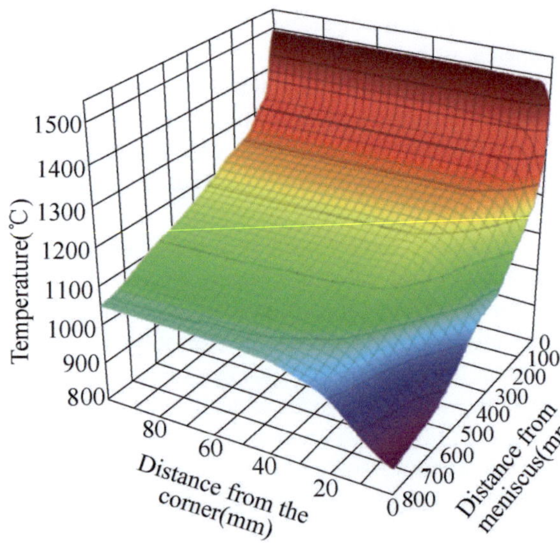

(b)

Fig. 4.30, which presents a microdepression shape. The cross-sectional morphology and solidification evolution of the slab at the outlet of each segment are shown in Fig. 4.31.

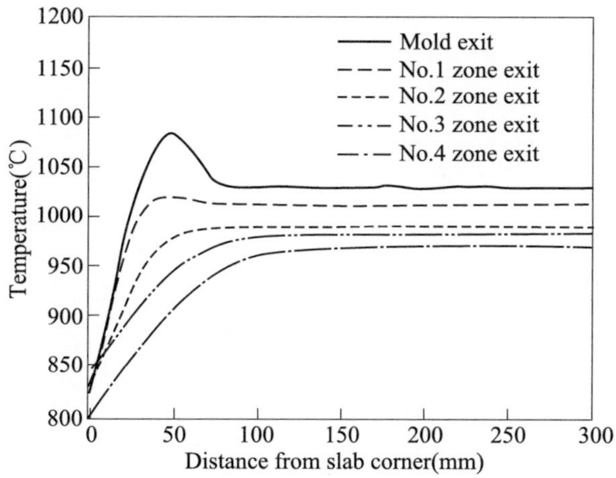

Fig. 4.28 Evolution of slab surface temperature at different positions of secondary cooling

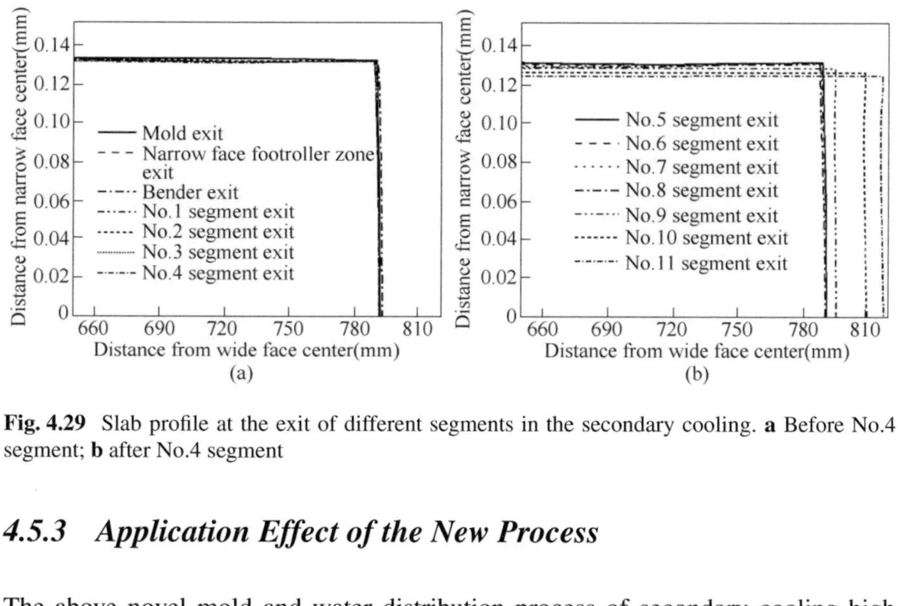

Fig. 4.29 Slab profile at the exit of different segments in the secondary cooling. **a** Before No.4 segment; **b** after No.4 segment

4.5.3 Application Effect of the New Process

The above novel mold and water distribution process of secondary cooling high temperature zone is applied to the thick slab continuous caster. The edge morphology of the 250 mm thick slab produced before and after the application of the new technology is shown in Fig. 4.32. It can be seen that before the new technology is put into operation, the peritectic wide-thick slab produced by the steel plant mostly forms depression of the wide face off-corner as shown in Fig. 4.32a, and the narrow face has an obvious bulge. After the implementation of the new technology, the narrow face of the slab production process is effectively compensated by the shrinkage and supported of the novel convex-structure mold, and rapid cooling process of the wide

(a) (b)

Fig. 4.30 Morphology of slab narrow face under the new process. **a** Left side; **b** right side

and narrow faces is implemented in the secondary cooling high temperature zone. After exiting the narrow face foot roller zone, the shell has a strong ability to resist the ferrostatic pressure, and there is no obvious bulge phenomenon on the narrow face of the slab and no depression in the wide face off-corner. At present, this technology has been stably used in a domestic steel plant for more than three years, which effectively eliminates the formation of the longitudinal depression at the wide face off-corner and its surface and subsurface crack.

Figure 4.33 is a comparison of the morphology of the slab edge before and after the implementation of the technology in a 300 mm slab continuous caster of a certain steel plant. It can be seen that before the implementation of the technology, the narrow face of thick slab production process obviously bulges, and wide face off-corner also has a certain depth of depression defect, as shown in Fig. 4.33a. Affected by this, transverse cracks often formed on the surface of the wide face off-corner depression zone of the slab in the process of Nb-bearing micro-alloyed steel, and the thick slab often needs to be offline flame cleaned on the off-corner, as shown in Fig. 4.34. At the same time, longitudinal crack defects often occur at a depth of about 20–30 mm under the subsurface of the off-corner, which restricts the high quality and efficient production of high-end wide-thick slab. After the implementation of the technology, the bulge of the thick slab narrow face has improved to a certain extent, and the depression in the wide face off-corner is basically been eliminated, as shown in Fig. 4.33b. At present, this technology has also been steadily implemented in a certain steel plant for nearly three years.

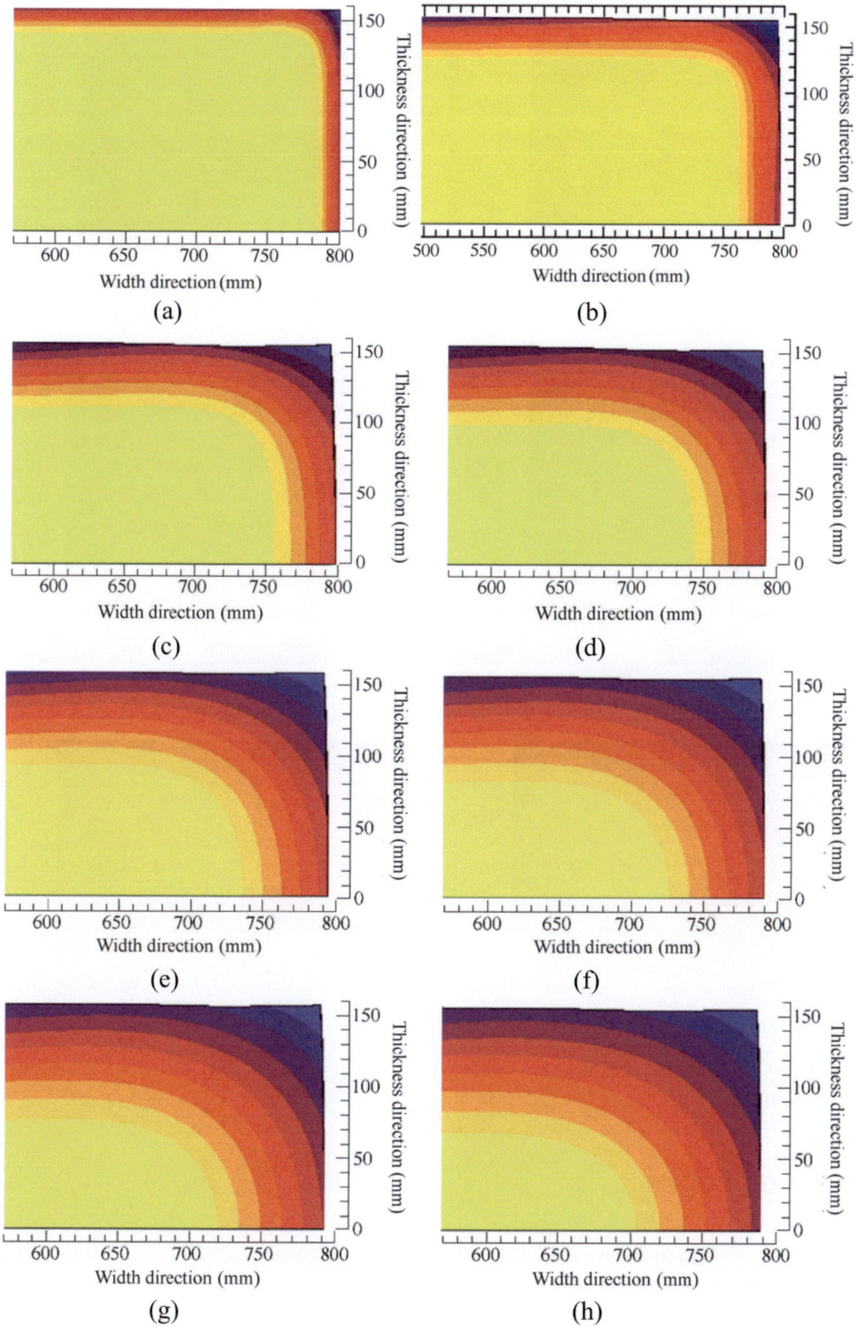

Fig. 4.31 Evolution of slab temperature contour and profile at the exit of different segments in secondary cooling. **a** Mold exit; **b** wide face foot roller exit; **c** Narrow face foot roller exit; **d** No.1 segment exit; **e** No.2 segment exit; **f** No.3 segment exit; **g** No.4 segment exit; **h** No.5 segment exit; **i** No.6 segment exit; **j** No.7 segment exit; **k** No.8 segment exit; **l** No.9 segment exit; **m** No.10 segment exit; **n** No.11 segment exit

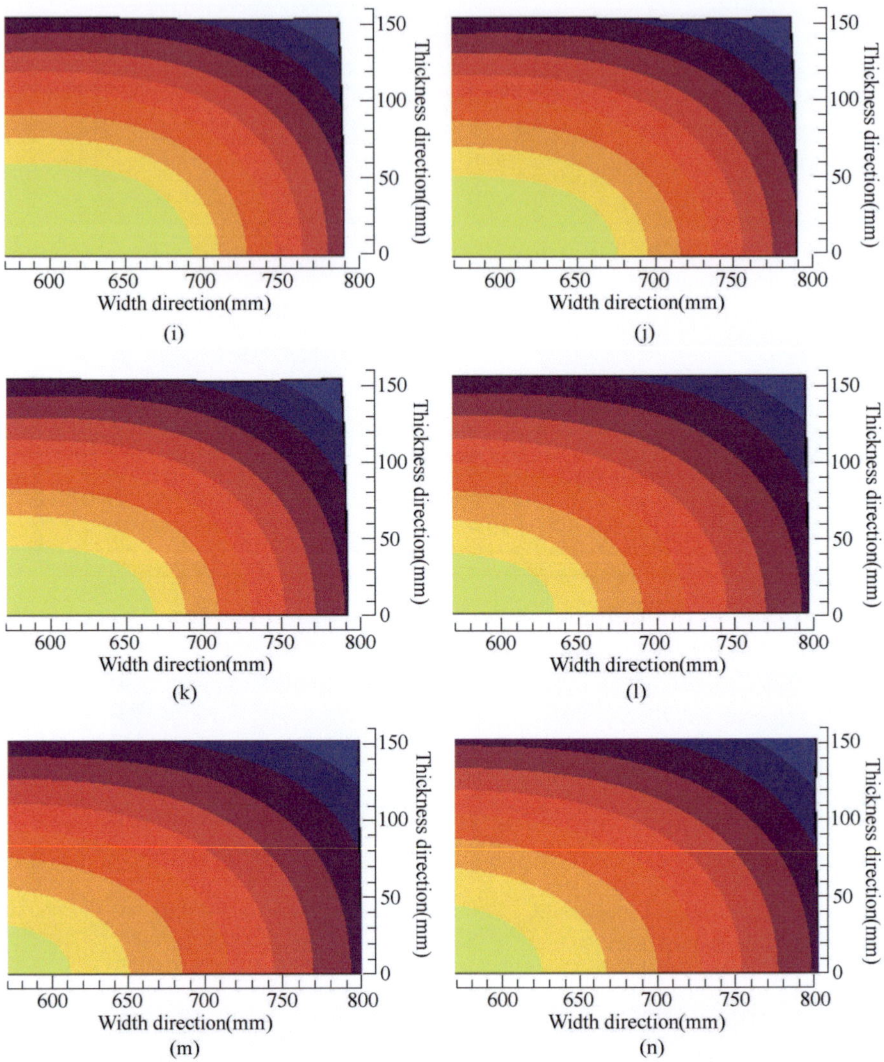

Fig. 4.31 (continued)

4.6 Summary

In this chapter, the thermo-mechanical behavior evolution analysis of the full casting flow solidification process of thick slab is carried out to solve the problem of longitudinal depression defects in the wide face off-corner during the continuous casting process of thick slab. The formation mechanism of longitudinal depression in the wide face off-corner is revealed, and its control method is proposed. The main conclusions are as follows:

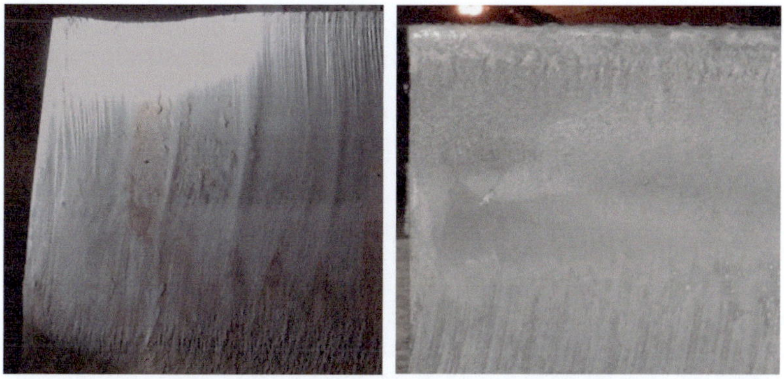

(a) (b)

Fig. 4.32 Comparison of the profile of 250 mm thickness slab under different casting process in a plant. **a** Traditional process; **b** new process

(a) (b)

Fig. 4.33 Comparison of the profile of 300 mm thickness slab under different casting process in a plant. **a** Traditional process; **b** new process

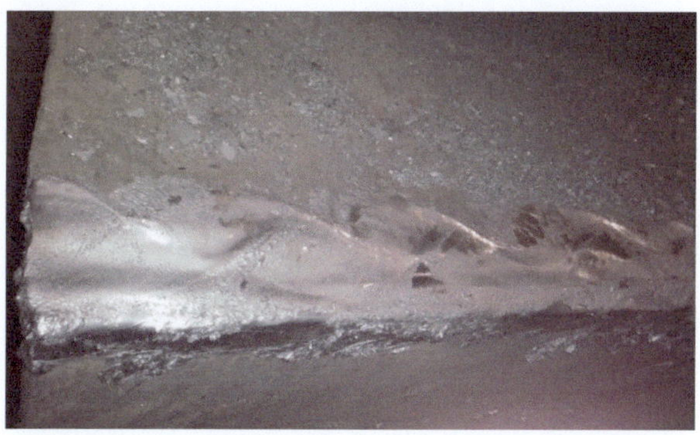

Fig. 4.34 Flame cleaning of the off-corner of the 300 mm thickness slab before the application of technology in a plant

(1) In the traditional flat-plate continues casting mold of narrow face thick slab, the solidification process of the slab in the upper part of the slab is not effectively compensated and constrained due to the shrinkage and deformation of the narrow face slab shell, and a large interfacial gap is formed in the corner of the wide face of the slab and its vicinity. Due to the concentrated filling of the flux film and the air gap, the solidification of off-corner of the slab wide face forms a significant "hot spot," which lays a temperature condition for the formation of the longitudinal depression in the off-corner.

(2) The bulge of the narrow face of the slab after the narrow face is the "dynamic" factor for the formation of the longitudinal depression in the wide face off-corner. After the high temperature narrow face slab shell exits the narrow face foot roller segment of the mold, it loses the support of the casting roller, and gradually bends outward under the ferrostatic pressure to form the bulge. Driven by this deformation, the off-corner of the wide face of the slab twists around the corner as a fulcrum, bending toward the slab thickness direction, forming a depression that starts from the slab corner and develops toward the wide face center direction. Then, under the solidification end soft reduction, both sides of the depression area are flattened to finally form longitudinal depression in the off-corner of the wide face shown in the slab.

(3) The key to the control of the longitudinal depression of the slab wide face off-corner is to eliminate the significant "hot spot" area formed by the solidification process of the slab in the mold, and to prevent the formation of obvious narrow face bulge after it exits the narrow face foot roller zone. When the narrow face curved-structure mold is adopted, this can compensate and support the shrinkage of the narrow face of the slab during the solidification process in the mold. It can prevent the slab wide face corner from being "dragged" to form excessive interfacial gap, thereby eliminating the condition of forming excessive "hot spot" area in the off-corner of the wide face, and homogenizing the heat transfer of the shell in the mold. The implementation of proper rapid cooling in the high temperature zone of continuous casting secondary cooling will not increase the crack sensitivity of the continuous casting process of micro-alloyed steel. It can strengthen the ability of the narrow face of the slab to resist bulging to a certain extent and further eliminate the "hot spot" of the wide face off-corner of the slab and eliminate the longitudinal depression of the wide face off-corner of the thick slab.

References

1. Chang YF (2002) Development of HSLA wide and heavy plate. Wide Thick Plates 8(5):1–5
2. Qian ZL (2008) The development and review of wide and heavy plate production technology and equipment in China. Metall Manage 3:57–60
3. Shen WR, Qiu SN, Qian HJ (2009) The process and technology equipment for wide and heavy plate production in Shagang. Wide Thick Plates 2:38–44

4. Qian ZL (2010) Wide and heavy plate skin continuous casting technology and wide and heavy plate production. China Steel Ind 7:30–32
5. Deng W, Zhao D, Qin X et al (2009) Simulation of central crack closing behavior during ultra-heavy plate rolling. Comput Mater Sci 47(2):439–447
6. Wang DW (2007) Analysis of the world's 5-meter extra-wide and heavy plate binding machine situation. Metall Manage 8:53–55
7. Wu DQ, Hu CZ, Huang B (2006) Features of production process for wide and heavy plate in Xinyu Iron and Steel Co·Ltd. Wide and Thick Plates 12(1):45–47
8. Chen Y (2002) Development of levelling technology for medium and heavy plate. Wide Thick Plate 8(6):1–5
9. Hu J, Du LX, Xie H et al (2014) Microstructure and mechanical properties of TMCP heavy plate micro-alloyed steel. Mater Sci Eng, A 607:122–131
10. Fu TL, Wang ZD, Li Y et al (2014) The influential factor studies on the cooling rate of roller quenching for ultra-thick plate. Appl Therm Eng 70(1):800–807
11. Zheng-Wei S (2006) Development tendency of process and equipment of heat treatment for medium and heavy plate. Steel Rolling 4
12. Matsumiya T (2006) Recent topics of research and development in continuous casting. ISIJ Int 46(12):1800–1804
13. Lu YJ, Wang Q, Li YG et al (2011) Prevention of longitudinal off-corner depression in continuously cast steel slabs. Contin Cast 27(6):33–37
14. Li Y, Zhang L (2017) Causes and measures of surface depressions in continuous slab. Baosteel Technol 35(6):53–59
15. Hu H (2020) Analysis and control of surface depression of continuous casting slab. Strait Sci Technol Ind 31(6):40–44
16. Dippenaar R, Moon S, Szekeres ES (2007) Strand surface cracks-the role of abnormally large prior-austenite grains. Iron Steel Technol 4(7):105–115
17. Lee GG, Shin HJ, Kim SH et al (2009) Prediction and control of subsurface hooks in continuous cast ultra-low-carbon steel slabs. Ironmaking Steelmaking 36(1):39–49
18. Cheng J, Wu Y, Wang Y et al (2018) Thermomechanical analysis of triangular zone cracks in vertical continuous casting slabs based on viscoelastic—plastic model. J Iron Steel Res Int 25(8):813–820
19. Tang W, Meng LT, Sun LB et al (2017) Cause analysis and improving measures of central crack in continuous casting slab of Q235B steel. Contin Cast 42(5):59–65
20. Wang SL, Wang HF (2019) Formation mechanism of internal cracks in continuous casting slab and the controlling practice. Contin Cast 44(2):53–57
21. Brimacombe JK, Weinberg F, Hawbolt EB (1979) Formation of longitudinal, midface cracks in continuously-cast slabs. Metall Trans B 10(2):279–292
22. Zhu ZY, Zhang YL, Wang KZ (2019) Causes analysis and control of slab bulging and depression. In: National seminar on application technology of efficient continuous casting and slab quality control. Hebei Metal Society, Shijiazhuang, pp 241–244
23. Wang E, He J (2001) Finite element numerical simulation on thermo-mechanical behavior of steel billet in continuous casting mold. Sci Technol Adv Mater 2(1):257–263
24. Storkman WR, Thomas BG (1988) Mathematical models of continuous slab casting to optimize mold taper. In: Modeling of casting and welding processes. Minerals, Metals & Materials Society, Pittsburgh, pp 287–297
25. Mahapatra RB, Brimacombe JK, Samarasekera IV et al (1991) Mold behavior and its influence on quality in the continuous casting of steel slabs: Part I. Industrial trials, mold temperature measurements, and mathematical modeling. Metall Mater Trans B 22(6):861–874
26. Zappulla MLS, Thomas BG (2017) Thermal-mechanical model of depression formation in steel continuous casting. In: TMS 2017 146th annual meeting & exhibition supplemental proceedings. Springer Nature, Cham, pp 501–510
27. Thomas BG, Moitra A, Zhu H (1995) Coupled thermo-mechanical model of solidifying steel shell applied to depression defects in continuous-cast slabs. In: Proceedings of the 1995 7th conference on modeling of casting, welding and advanced solidification processes. The Minerals, Metals & Materials Society, Pittsburgh, pp 241–248

28. Meng Y, Thomas BG (2003) Heat-transfer and solidification model of continuous slab casting: CON1D. Metall Mater Trans B 34(5):685–705
29. Ji C, Cui Y, Zeng Z et al (2015) Continuous casting of high-Al steel in shougang jingtang steel works. J Iron Steel Res Int 22(S1):53–56
30. Thomas BG, Moitra A, McDavid R (1996) Simulation of longitudinal off-corner depressions in continuously cast steel slabs. ISS Trans 23(4):57–70
31. Zappulla MLS, Thomas BG (2020) Simulation of longitudinal surface defect in steel continuous casting. In: Modelling of casting, welding and advanced solidification processes. IOP Publishing, Bristol, pp 861–868
32. Fix C, Elixmann S, Senk D (2020) Design of As-cast structures of continuously cast steel grades: modeling and prediction. Steel Res Int 91(11):1–7
33. Zhao JP, Wang S, Feng S et al (2020) Reason and control of surface depressions defects in slab of low-carbon low-silicon aluminum-killed steel. Contin Cast 45(02):31–35
34. Niu Z, Cai Z, Zhu M (2021) Effect of mold cavity design on the thermomechanical behavior of solidifying shell during micro-alloyed steel slab continuous casting. Metall and Mater Trans B 52(3):1556–1573
35. Niu Z, Cai Z, Zhu M (2021) Formation mechanism of a wide face longitudinal off-corner depression during thick slab continuous casting. Metall Mater Trans B, pp 1–16
36. Cai ZZ, Zhu MY (2011) Simulation of thermal behavior during steel solidification in slab continuous casting mold I. Mathematical model. Acta Metall Sin 47(6):671–677
37. Cai ZZ, Zhu MY (2011) Simulation of thermal behavior during steel solidification in slab continuous casting mold II. Model verification and result analysis. Acta Metallurgica Sinica 47(6): 678–687
38. Cai ZZ, Zhu MY (2014) Simulation of air gap formation in slab continuous casting mold. Ironmaking Steelmaking 41(6):435–446
39. Yamasaki N, Shima S, Tsunenari K et al (2016) Numerical simulation of the continuous casting process and the optimization of the mold and the strand. Nippon Steel Sumitomo Metal Tech Rep 112(4):64–70

Chapter 5
Hot Charging Crack of Micro-alloyed Steel Slab and Its Control

In recent years, with the intensification of market competition and the implementation of national carbon emission policies in the steel industry, the efficient, low cost, low emission steel production processes and their supporting technologies have attracted much attention in the steel industry. Hot charge of continuous casting slab, the high temperature slab produced by the continuous caster is directly charged to the heating furnace for heating and rolling through roller table and other means without offline cooling, thereby improving the efficiency of the steel production process and realizing the full utilization of the heat of the high temperature slab, is an important process for efficient and green steel production. As an important component of special steel products in domestic and foreign steel companies, the proportion of micro-alloyed steel containing Al, Nb, V, and other alloying elements is increasing year by year. In the actual production process, the micro-alloyed steel continuous casting slab sent to the heating furnace adopting the hot charging process is prone to form surface reticular and other cracks ("hot charging" cracks), which is a common problem restricting the stabilization of the implementation of this process. This chapter will focus on the temperature field evolution and high temperature casting structure characteristics of the micro-alloyed steel slab production process, and introduce hot charging technology based on continuous quenching of the slab surface at the horizontal segments or exit of the continuous caster.

© Metallurgical Industry Press 2025

Z. Cai and M. Zhu, *The Control Technology for Slab Surface Crack During Continuous Casting of Micro-alloyed Steel*, Key Steel Manufacturing Technologies by Collaborative Innovation, https://doi.org/10.1007/978-981-96-7171-7_5

5.1 Hot Charging of Micro-alloyed Steel Slab and Its Cracks Causes

5.1.1 Evolution of Slab Temperature During the Hot Charging Process

Currently, the steel plants and rolling plants of most steel companies at home and abroad are seamlessly connected by roller table. The high temperature continuous casting slab produced by the steel plant are directly sent to the heating furnace for heating and rolling through the roller table, realizing the efficient utilization of the heat of the high temperature continuous casting slab [1–7]. In this process, the main heat transfer mode of the high temperature slab after the continuous caster is radiation heat transfer. Therefore, the main factors affecting the decrease in the surface temperature of the hot charging slab are the length and speed of the roller table (or the residence of slab on the roller table) and the ambient temperature around the roller table.

Figure 5.1 shows the temperature evolution of the wide face and narrow face of the slab under the continuous casting production and roller hot charging process of a certain steel plant wide-thick slab. Because the continuous caster of this wide-thick slab is short, containing only one bender and 12 bow segments, the surface temperatures of the wide and narrow faces of the slab exiting casting caster are both high, reaching about 940 and 850 °C as shown in Fig. 5.1. Afterward, the continuous casting slab is accelerated to the secondary cutting position by the roller after the first fire cutting, and the secondary cutting to size of the continuous casting slab is carried out. During the secondary cutting pause process, the surface temperatures of the wide and narrow faces of the slab decreased by about 22 and 15 °C, respectively, then the slab was accelerated again and transported to about 25 m in front of the entrance of the heating furnace. After that, it is heated in the furnace according to the heating rhythm. Under the simulated calculation conditions, the surface temperatures of the wide face and narrow face of the wide-thick slab entering the heating furnace in a certain steel plant are about 642 and 578 °C, respectively. From the evolution of the surface temperature during the continuous casting and roller table process, it can be seen that the surface structure of the wide face of the slab entering the heating furnace is in the dual-phase zone structure of austenite and ferrite. In the actual production process, the hot charging crack defects occur frequently on the wide face of the slab after heating.

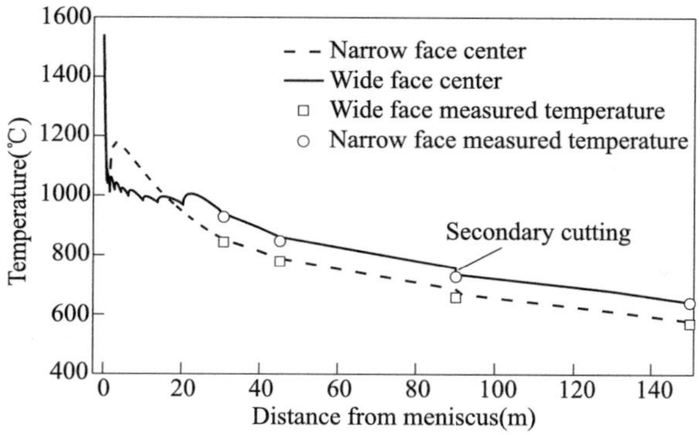

Fig. 5.1 Temperature evolution during the continuous casting and hot charging process of wide-thick slab

5.1.2 *Morphology of Hot Charging Crack of Micro-alloyed Steel Slab*

The reticular hot charging crack on the surface of micro-alloyed steel continuous casting slab is a common crack defect in the production process of steel companies at home and abroad [8–12]. The surface crack of the slab is often presented as the surface quality of the continuous casting slab is intact in the roller table before entering the heating furnace, and no crack defect is found. When the slab is heated after entering the heating furnace, reticular crack defect occurs on the slab surface. Once the continuous casting slab forms the crack, it will appear in batches, and the cracks are dense and deep. After the slab is rolled, the crack is generally left on the surface of the steel plate, forming defects such as steel plate or wire rod surface warping, pockmarked surface, and crack defect.

A certain steel plant Nb-bearing micro-alloyed steel slab, which is heated by the roller table to the heating furnace, its inner arc surface temperature at the entrance of the heating furnace is mostly 600–650 °C. After being heated in the heating furnace, the slab inner arc surface frequently forms crack defects on the surface of the slab and the medium-thick plate as shown in Fig. 5.2. It can be seen that the surface crack of the slab after exiting the furnace are densely distributed, and there is no obvious regularity in the direction and distribution of the crack. At the same time, the opening degree and depth of this crack are different, the crack width with larger opening degree can reach more than 2 mm, and the corresponding crack depth can also exceed 4 mm, as shown in Fig. 5.2a, b. Slab surface cracks with large opening degree and depth generally cannot be eliminated on the surface of the steel plate after rolling. For medium-thick plate, the surface hot charging crack of the finished steel plate often extend and stretch in the rolling direction, which is difficult to be grinded offline, as shown in Fig. 5.2c. For hot rolled coil plate, it is mostly manifested as severe surface warping defects.

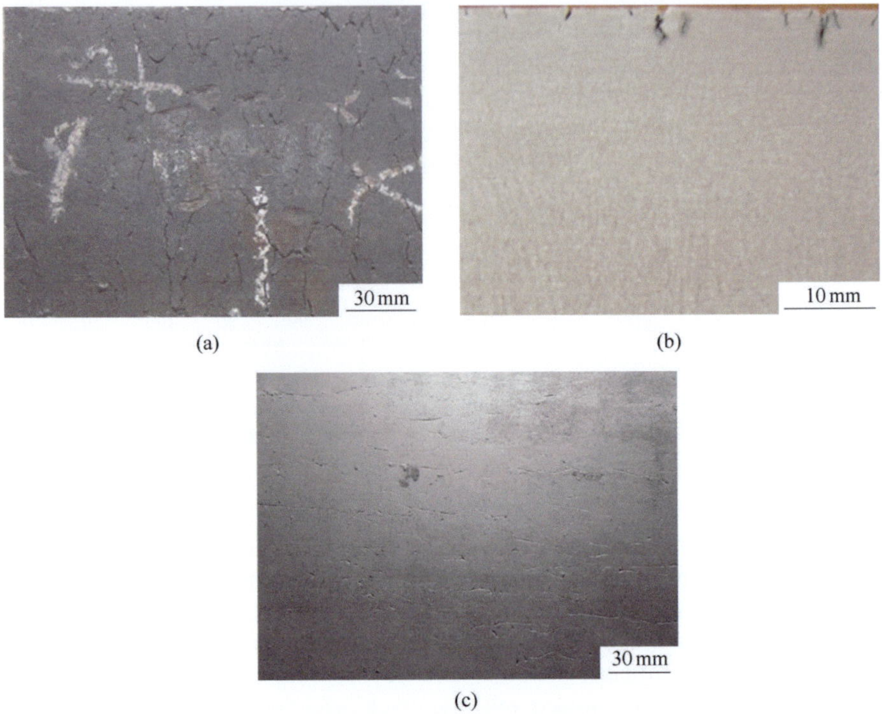

Fig. 5.2 Surface crack morphology of hot charging Nb-bearing micro-alloyed steel slab and rolled steel plate. **a** Surface crack of the cast slab discharged from the heating furnace; **b** Cross-section crack of the cast slab discharged from the heating furnace; **c** Surface crack of the medium-thick plate

5.1.3 Evolution of Micro-alloyed Steel Slab Surface Structure During Hot Charging Process

In order to find out the structural characteristics of the surface structure of the slab at different temperatures, this book sampled the Nb-V-bearing micro-alloyed steel wide-thick slab produced by a certain steel plant, and processed its surface 50 mm thick into test samples, heated it to 1100 °C in a muffle furnace, and then took out the samples and air cooling to 650, 600, 550, 500, 450 and 400 °C, respectively. Quenched in water to retain the slab structure morphology at the corresponding cooling temperature. The composition of the sample steel is shown in Table 5.1.

Table 5.1 Main components of the sample steel (mass fraction, %)

C	Si	Mn	P	S	Cr	Mo	V	Nb	Ti	Al
0.069	0.041	1.51	0.016	0.003	0.21	0.002	0.002	0.046	0.024	0.03

The samples quenched at different temperatures were ground and polished, then corroded with a 3% nitric acid alcohol solution and structure was observed under a microscope. The structure of the slab at different quenching temperatures is shown in Fig. 5.3. It can be seen from the fig. that when the surface temperature of the slab drops to 650 and 600 °C, the structure of the steel is mainly ferrite and lath martensite, and the proportion of lath martensite is higher, as shown in Fig. 5.3a, b. The main reason for this structure is that the austenite in the steel has not been completely transformed into ferrite and pearlite at 650 and 600 °C. When it is quenched and cooled, the austenite structure transforms into lath martensite. Therefore, the surface structure of slab is in the dual-phase structure of austenite and ferrite when the continuous casting slab is hot charged above 600 °C. When the slab temperature drops to 550 and 500 °C, its quenched structure is mainly ferrite, pearlite and a small amount of lath martensite, as shown in Fig. 5.3c, d. Especially, it can be seen from Fig. 5.3d that the number of lath martensite in the structure is very fine. It indicates that at this temperature, the austenite structure in the slab has been almost completely transformed into ferrite and pearlite. When the slab temperature is further reduced to 450 and 400 °C, the lath martensite has completely disappeared, and the structure of the slab is all ferrite and pearlite.

5.1.4 Cause of Hot Charging Crack of Micro-alloyed Steel Slab

Based on the evolution characteristics of steel structure at different temperatures, when the surface of the slab is cooled to 650–550 °C (the subsurface temperature of the slab is slightly higher than the surface), and the surface structure is in the dual-phase zone of $\gamma \rightarrow \alpha$ transition [13–15]. The size between the newly formed austenite and the original austenite during the heating process of the slab in the heating furnace is quite different [16], resulting in a significant mixed grain structure in the surface structure of the heated slab [17–19], thereby greatly reducing the hot ductility of the structure. At the same time, because the grain boundaries position of the original austenite of the slab surface does not move significantly during the heating process, the microalloy carbonitrides precipitated in a chain along the austenite grain boundary during the continuous casting production process are still located at the grain boundary. The strong pinning effect further restricts the ductile deformation of the surface structure of the slab [20–24], so that the surface crack defects of the slab are formed by the heating stress during the heating process. However, when the surface temperature of the slab is put into the furnace at below 500 °C (especially below 450 °C), the surface structure of the slab is in the α single-phase zone. The grains of the austenite structure formed by heating competitive growth are relatively uniform, and the grain boundaries between the newly formed austenite structure and original austenite are no longer coincident. The hot ductility of the surface structure of the slab is high, and the surface crack defect is not easy to form during the

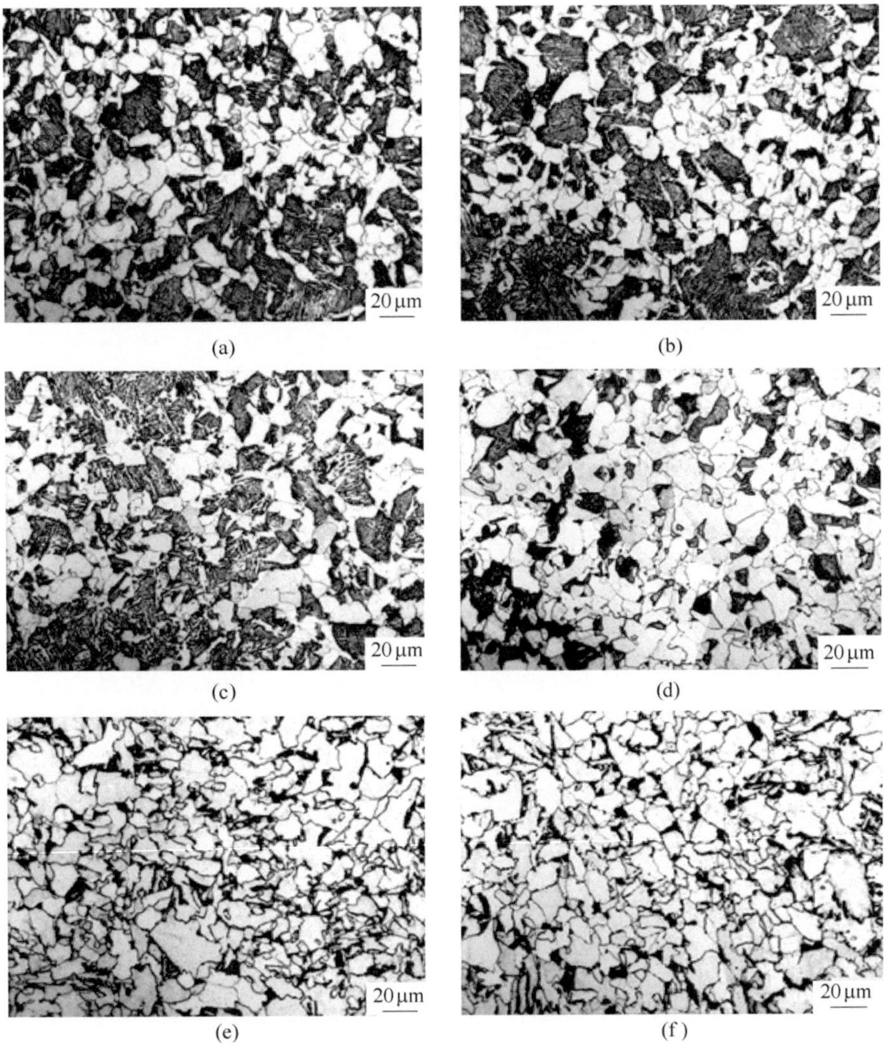

Fig. 5.3 Micro-structure of slabs at different temperatures. **a** 650 °C; **b** 600 °C; **c** 550 °C; **d** 500 °C; **e** 450 °C; **f** 400 °C

heating process. Therefore, significantly reducing the temperature of the slab surface structure or making the austenite structure of the slab surface significantly refined and dispersing the carbonitride precipitation before entering the heating furnace is the key to solving the problem of hot charging cracks of micro-alloyed steel continuous casting slab.

5.2 Control Strategy for Hot Charging Crack of Micro-alloyed Steel Slab

Figure 5.4 is a schematic diagram of the surface micro-structure transformation of micro-alloyed steel slab under different charge processes [25]. When the slab is charged into the furnace above the A_{r3} of the corresponding steel, because its surface structure is all austenite, and there is less precipitation of microalloy carbonitrides at this temperature, the micro-structure ductility of the slab entering the heating process of the heating furnace is high, and its surface generally does not produce hot charging crack defects. However, in the current domestic and international main cross section (especially large cross section) continuous casting slab production process, due to the restriction of low continuous casting speed, it is difficult to ensure that the surface temperature of the slab charged to the heating furnace entrance is higher than A_{r3} temperature, mostly in the temperature range of A_{r1}-A_{r3}. The surface temperature of the slab between A_{r1}–A_{r3} is affected by the surface structure of the slab in the $\gamma \rightarrow \alpha$ phase transition zone and the chain concentration of microalloy carbonitrides at the grain boundaries. The slab is prone to the high incidence of surface crack defects under thermal stress due to low hot ductility during the heating process in the furnace [26–28]. However, by adopting the slab cooling process, the temperature of the slab surface structure has dropped below Ar_1, and its structure is completely transformed into ferrite and pearlite. The newly formed austenite grains in this structure during the heating process are relatively uniform, and the newly formed austenite structure grain boundaries are no longer concentrated in a chain distribution of microalloy carbonitrides, greatly improving the hot ductility of the slab surface structure, making it uneasy to form surface crack defects during the heating process.

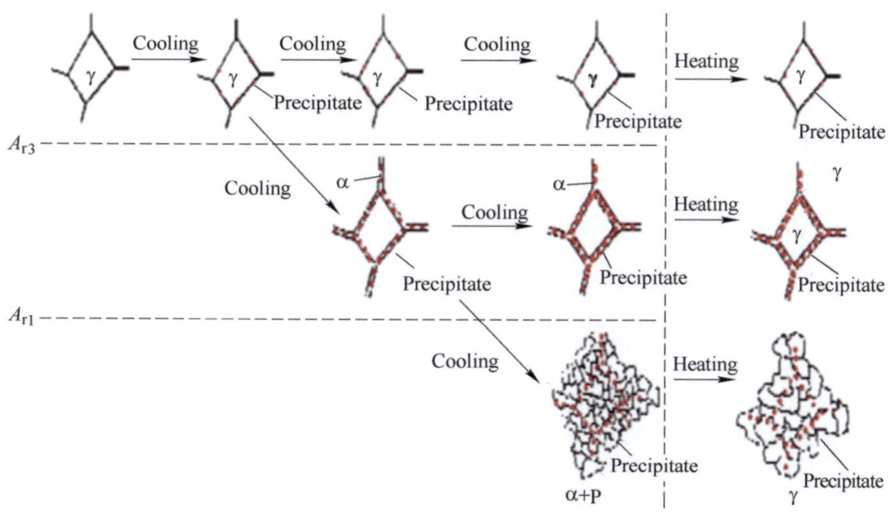

Fig. 5.4 Schematic diagram of steel structure transition under different charging processes [25]

According to the above characteristics of steel micro-structure transition under different continuous casting slab charge processes and the evolution of surface temperature field during continuous casting slab roller table process, it is known that the micro-alloyed steel continuous casting slab produced by most continuous casting production lines at home and abroad are prone to form surface cracks in the heating furnace. However, the production process of cooling the slab offline or reheating after online cooling of the continuous casting slab cannot efficiently utilize the heat in the center of the slab, which is uneconomical and environmentally friendly, and has low production efficiency of offline stacking cooling and the long production cycle.

In recent years, driven by environmental protection and production costs, the production process of efficient connection continuous casting and rolling processes has attracted much attention in the industry. Based on the mechanism of hot charging crack formation of micro-alloyed steel continuous casting slab, if the size of the austenite grains on the surface of the slab before the heating furnace can be significantly refined, and the precipitation of microalloy carbonitrides can be dispersed, the hot charging slab will not form surface crack defects during the heating process. According to this target, rapid quenching is implemented on the surface of the high temperature continuous casting slab, so that the temperature of the surface layer in the depth range of 0–10 mm decreases rapidly to the complete ferrite or bainite temperature of the corresponding steel grade, and then the heat in the core of the slab is used to refine the surface micro-structure of the slab. At the same time, during the rapid quenching process, the carbonitrides of the high temperature slab surface structure will be dispersed and precipitated, thereby achieving a high ductility transition of the structure and eliminating the occurrence of hot charging cracks [29–32]. In recent years, this technology has gradually been promoted and applied to the continuous casting production lines of bloom and wide-thick slab.

5.3 Development of Slab Surface Quenching Technology During Continuous Casting

5.3.1 Introduction of Typical Slab Surface Quenching Technology

The surface quenching technology of continuous casting slab originated from abroad. Because it can effectively control the formation of hot charging crack defects on the surface of micro-alloyed steel continuous casting slab, it promoted the high quality, high efficiency and green production of the steel production process.

The basic principle of the surface quenching process of continuous casting slab is: adopting strong cooling spray on the surface of the high temperature slab when exiting continuous caster, so that the temperature of the slab surface is quickly reduced to below Ar_1 [33, 34], to disperse the precipitation of carbonitrides in its steel structure, the rapid transition of austenite to ferrite is completed at the same time, thereby improving the hot ductility of the surface structure of the slab. After the surface quenched slab is heated by the heating furnace, the uniform and fine austenite will be formed. In this process, the newly formed austenite encapsulates the carbonitrides in its grain, thereby eliminating the problem of cracking caused by the low ductility of the surface structure of the slab due to the formation of mixed grain and the chain precipitation of carbonitrides at the grain boundaries during the heating process of traditional micro-alloyed steel continuous casting slab. Because the heat in the center of the slab after surface quenching is basically retained, it has a good effect of energy saving [35, 36]. At present, the typical surface quenching technologies of continuous casting slab at home and abroad are as follows.

5.3.1.1 Bloom Surface Quenching System of Danieli [37]

Danieli is a company that adopted the surface quenching technology of continuous casting slab in actual continuous casting production earlier. Based on the principle of slab surface quenching technology, the company has developed the first slab surface quenching device for bloom continuous casting as shown in Fig. 5.5, which was applied to the ABS factory in Italy in 1994. This slab surface quenching device is installed after the withdrawal and straightening machine, and by adding a strong spray cooling water tank composed of several rows of nozzles, and the cooling water flow rate of each row of nozzles can be individually controlled, the slab surface is continuously strongly sprayed and quenched. In the actual production process with this technology, it requires sufficient amount of cooling water, and the appropriate quenching device length and quenching spray process are designed according to the actual continuous casting process, to meet the ultra-strength uniform quenching of the slab at the maximum casting speed. On this basis, according to the changes in the steel grade and casting speed in the actual production process, automatically control the quenching cooling water, transform the slab surface with high ductility structure at the same time, as far as possible to reduce the surface crack caused by excessive thermal stress of the slab. At present, Danieli has promoted this technology to a number of bloom continuous casters, including Jiyuan Steel in China and Hyundai Steel in South Korea.

5.3.1.2 Immersion Blank Quenching in Nippon Steel [38]

The blank surface quenching technology developed by Nippon Steel Sumitomo Metal during immersion slab quenching is mainly adopted in the continuous casting of billet. Through the studies of frequent surface crack during the heating process of hot charge alloy steel billet, researchers found that the surface structure of the slab charged to the heating furnace is mainly a low ductility dual-phase structure of "austenite + grain boundary ferrite," and a large amount of carbonitrides is precipitated on the austenite grain boundaries. Due to the grain boundary pinning effect of

Fig. 5.5 Surface quenching system of bloom continuous caster in Danieli

the microalloy carbonitride, the grain boundaries of the blank become brittle during the heating process. Under the effect of thermal stress, reticular crack occurs on the surface of the blank. By installing an immersion blank quenching device as shown in Fig. 5.6 on the side of the roller table after the cutting machine of the continuous caster, the high temperature blank is immersed in the water slot, and the quenching structure transition effect of different steel grades is achieved by controlling the immersion time. This quenching device can significantly refine the surface structure of the blank, making the structure within the 0–20 mm range below the slab surface a fine and uniform ferrite + pearlite structure. After quenching the blank, due to the surface structure of the quenched blank eliminates the austenite + ferrite dual-phase structure of the traditional hot charging blank surface crack during the heating process is avoided. Currently, this technology is mainly used by companies such as Shijiazhuang Steel for the hot charging and surface quality control of billet continuous casting. However, in the actual production process of this blank surface quenching technology, the utilization rate is not high due to factors such as failing to achieve the ideal quenching effect and low production efficiency.

5.3.1.3 Slab Surface Quenching Technology After Secondary Cutting

Steel Plant A, in order to solve the problem of high occurrence rate of surface hot charging crack during the heating process of its hot charging wide-thick slab, installed a slab surface quenching device with a total length of about 10 m after deburring machine of continuous caster, as shown in Fig. 5.7. In the actual quenching process, by controlling the residence of the slab in the quenching device and the water volume of its inner and outer arcs, the surface temperature of the slab is quickly reduced from 710 to 730 °C before entering the quenching device to 350–450 °C after quenching. After exiting the quenching device, the surface of the continuous casting slab is first quickly reheated about 100–120 °C, and then slowly cooled to 330–420 °C charging to the heating furnace, effectively eliminating the surface crack of the Nb-bearing steel wide-thick slab during the heating process.

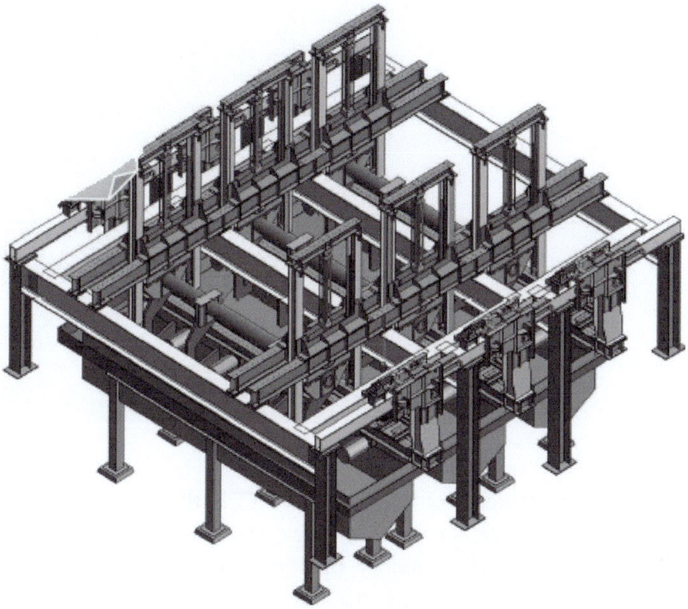

Fig. 5.6 Immersion blank quenching in Nippon steel

(a) (b)

Fig. 5.7 Surface quenching device and slab morphology after quenching. **a** Quenching device, **b** Slab morphology after quenching

The slab surface quenching technology of B Steel plant is also applied to the wide-thick slab continuous caster. Its installation position is the same as that of A Steel plant, which is located after the deburring machine of the continuous caster, as shown in Fig. 5.8. The wide-thick slab with a thickness of 300 and 260 mm produced by this wide-thick slab continuous caster, the temperature of the inner arc of the continuous caster is about 860–900 °C, after the first cut, second cut, deburring and other processes of the flame cutting machine, the temperature of the slab before

Fig. 5.8 Slab surface quenching device

entering the quenching device is mostly 700–730 °C. In the actual quenching production process, the slab surface is cooled to black, the resident time of the slab in the quenching device is controlled manually, the inner and outer arc surfaces of the slab are subjected to strong spray cooling, and then it is charged to the heating furnace by the roller table. It is similar to the slab surface quenching of A Steel Plant, the surface temperature of the hot charging to the heating furnace is about 350–450 °C, thus controlling the hot charging crack of the wide-thick slab.

In addition to the above two companies using the surface quenching technology of the slab to solve the problem of hot charging crack in micro-alloyed steel slab, domestic steel companies such as Puyang Steel and An Steel have installed similar slab surface quenching devices after the deburring machine of their wide-thick slab continuous casters [29]. However, adopting this surface quenching technology of continuous casting slab, the surface temperature of the slab charged to the quenching device has mostly reduced to below 730 °C. At this temperature, the slab surface structure containing Nb, Al and other microalloy carbonitrides has been precipitated, and the surface temperature of the slab after quenching can be reheated to below 600 °C, the slab surface structure cannot be reheated to achieve structure transition again, which only play the role of online cooling. In addition, due to the cooling method of the slab in the quenching device and spraying, the inhomogeneous spraying of the quenching nozzle affects the insufficient cooling of the local surface structure temperature of the slab, and the slab charged to the heating furnace still form hot charging crack [36, 37].

5.3.2 Determination of the Optimum Quenching Parameters of Micro-alloyed Steel Slab

It can be seen from Sect. 5.1, the cause of hot charging crack on the surface of micro-alloyed steel continuous casting slab and the control principle of slab surface quenching, to fundamentally eliminate the occurrence of hot charging crack on the surface of continuous casting slab, the key is to disperse the precipitation of microalloy carbonitrides in the surface structure of the slab before entering the heating furnace and refine its grain structure. At present, the slab surface quenching technology developed in China based on the strong spray of the roller table at the rear of the continuous caster is mostly due to the installation of the quenching device is too late, and the surface temperature of the slab is too low to realize the control of the carbonitride dispersion precipitation and grain refinement. Therefore, it is necessary to clarify the key quenching parameters of typical continuous casting steel grades and processes in each production line in combination with the actual continuous casting production of various steel companies, so as to develop a more reasonable surface quenching control technology for continuous casting slab.

The key parameters of the slab surface quenching process include: the starting quenching temperature, quenching cooling rate, and quenching termination temperature. Among them, the starting quenching temperature determines the online installation position of the quenching device, which is mainly determined by the precipitation of carbonitrides and the transformation characteristics of the micro-structure of the steel grade. The quenching cooling rate determines the volume of cooling water that needs to be provided on site, and the quenching termination temperature is closely related to the length of the quenching zone and the transformation characteristics of the micro-structure of the steel grade. Therefore, various factors need to be considered comprehensively to determine the optimal quenching process parameters.

5.3.2.1 Starting Quenching Temperature

The key to determine the optimal starting quenching temperature of the continuous casting slab surface is to ensure that most of the carbonitrides in a certain depth range of its surface layer are not precipitated, and the micro-structure does not transition from austenite to ferrite. At the same time, it is also necessary to ensure that the surface temperature of the slab after quenching still meets the requirements for its micro-structure to be austenitized again.

Therefore, in this chapter, based on the typical continuous casting slab of Nb-bearing micro-alloyed steel in a domestic steel plant, the thermal simulation detection method was used to prepare the test samples of the micro-structure evolution of quenched steel at different temperatures, and the metallographic detection was used to explore the effect of quenching temperature on the micro-structure transformation.

On this basis, combined with the thermodynamic and kinetic behaviors of the precipitation of carbonitrides in typical Nb-Al-bearing micro-alloyed steel and other components at home and abroad in Chap. 2, the optimal starting quenching temperature of the surface of the micro-alloyed steel slab is comprehensively determined.

The experimental steel is taken from a 2100 mm × 250 mm section Nb-bearing continuous casting slab produced by a certain steel plant, and the detailed composition is shown in Table 5.2.

The thermal simulation experiment scheme is shown in Fig. 5.9. The sample is heated to 1300 °C at a rate of 5 °C/s and preserve 5 min to make the sample fully austenitized and alloyed solid solution, then cooled to 950, 900, 850, 800, 750, and 700 °C at a cooling rate of 0.2 °C/s, and then the test sample was quenched. The cross-sectional structure was detected by metallographic detection to determine the optimal starting quenching temperature.

The metallographic structure of the sample section under the above different quenching temperatures is shown in Fig. 5.10. It can be seen from the fig., when the quenching temperature is not below 900 °C, the micro-structure after quenching is mainly lath martensite, and the original austenite grain boundary profile is roughly visible, and no proeutectoid ferrite is formed at the grain boundary, as shown in Fig. 5.10a, b. This phenomenon indicates that when the test sample is rapidly cooled from 1300 to 900 °C, the steel structure does not undergo $\gamma \rightarrow \alpha$ phase transition, that is, when the temperature is not below 900 °C, the slab structure is all high temperature austenite. When the starting quenching temperature is reduced to 850 °C, the original austenite grain boundary of the sample section is clearly visible, and the micro-structure is also mainly martensite, and no proeutectoid ferrite is

Table 5.2 Composition of the micro-alloyed steel to be tested (mass fraction,%)

Steel grade	C	Si	Mn	P	S	V	Ti	Nb	N
NQ345B-4	0.16	0.30	1.05	0.017	0.003	0.006	0.016	0.027	0.003

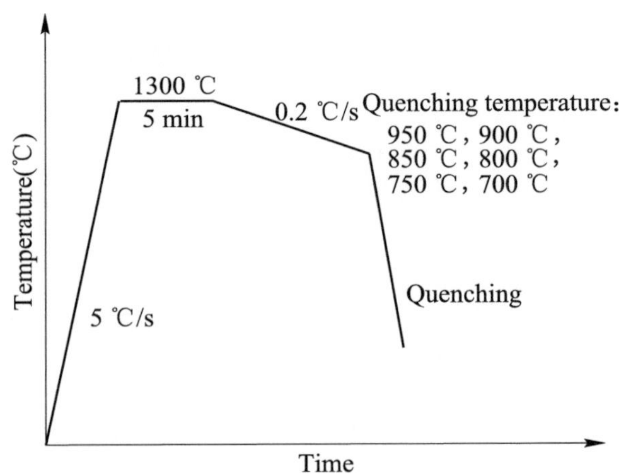

Fig. 5.9 Test scheme of quenching thermal simulation at different temperatures

formed at the original austenite grain boundary, as shown in Fig. 5.10c. When the quenching temperature drops to 800 °C, proeutectoid ferrite is formed at the original austenite grain boundary, but the ferrite layer is thin, and there is no lamellar or feathery Widmanstatten growing from grain boundary nucleation to intragranular growth, as shown in Fig. 5.10d. This phenomenon indicates that a small amount of austenite has been transformed when the test sample is cooled from 1300 to 800 °C, and a proeutectoid ferrite film is formed at the grain boundary. When the quenching temperature is further reduced to 750 °C, the proeutectoid ferrite layer on the original austenite grain boundary is significantly thickened, and a feathery Widmanstatten is formed from the austenite grain boundary to the grain interior, as shown in Fig. 5.10e. And when the starting quenching temperature is further reduced to 700 °C, the grain boundary proeutectoid ferrite layer is further thickened, and ferrite structure is also formed in the grain, as shown in Fig. 5.10f. From the evolution of the steel structure under different quenching temperatures, it can be known that the optimal starting quenching temperature on the surface of the Nb-bearing steel continuous casting slab should be higher than 800 °C.

It can be known from Chap. 2, from the thermodynamic and kinetic of the precipitation of carbonitrides in main Nb-Al-bearing micro-alloyed steel in domestic and foreign steel companies, the temperature at which the grain boundary microalloy carbonitrides begin to precipitate is generally higher than 1050 °C, and the "nose point" temperature is about 900–950 °C. According to the precipitation characteristics of this microalloy carbonitrides, to ensure that the carbonitrides of this Nb-bearing micro-alloyed steel are dispersedly precipitated, the optimal starting quenching temperature of the hot charging slab surface structure should be higher than 950 °C. At the same time, according to the actual heat transfer characteristics of the continuous casting slab, the surface temperature of the slab can be reheated to the level of not below 820 °C under the condition of a quenching layer depth of about 10 mm, which promotes the re-austenitization and refines the grain structure of the slab surface, meeting the requirements of the surface quenching temperature of the slab.

5.3.2.2 Quenching Cooling Rate

The cooling rate is a key factor in determining the structure of steel, the size and distribution of precipitation [39, 40]. In order to explore the impact of quenching cooling rate on the structure and the carbonitride precipitation of steel, the thermal simulation scheme was used for simulation test as shown in Fig. 5.11. In this test scheme, the sample was also heated to 1300 °C at a rate of 5 °C/s for 5 min to make the sample fully austenitized and alloy solid solution, then cooled to 950 °C at a cooling rate of 0.2 °C/s, which is the optimal starting quenching temperature, and then cooled the steel structure at 1, 3, 5, 7 °C/s to determine the optimal quenching cooling rate.

Figure 5.12 shows the metallographic structure of the sample fracture at the initial quenching temperature of 950 °C and cooling rates of 1, 3, 4, 5 and 7 °C/s. It can

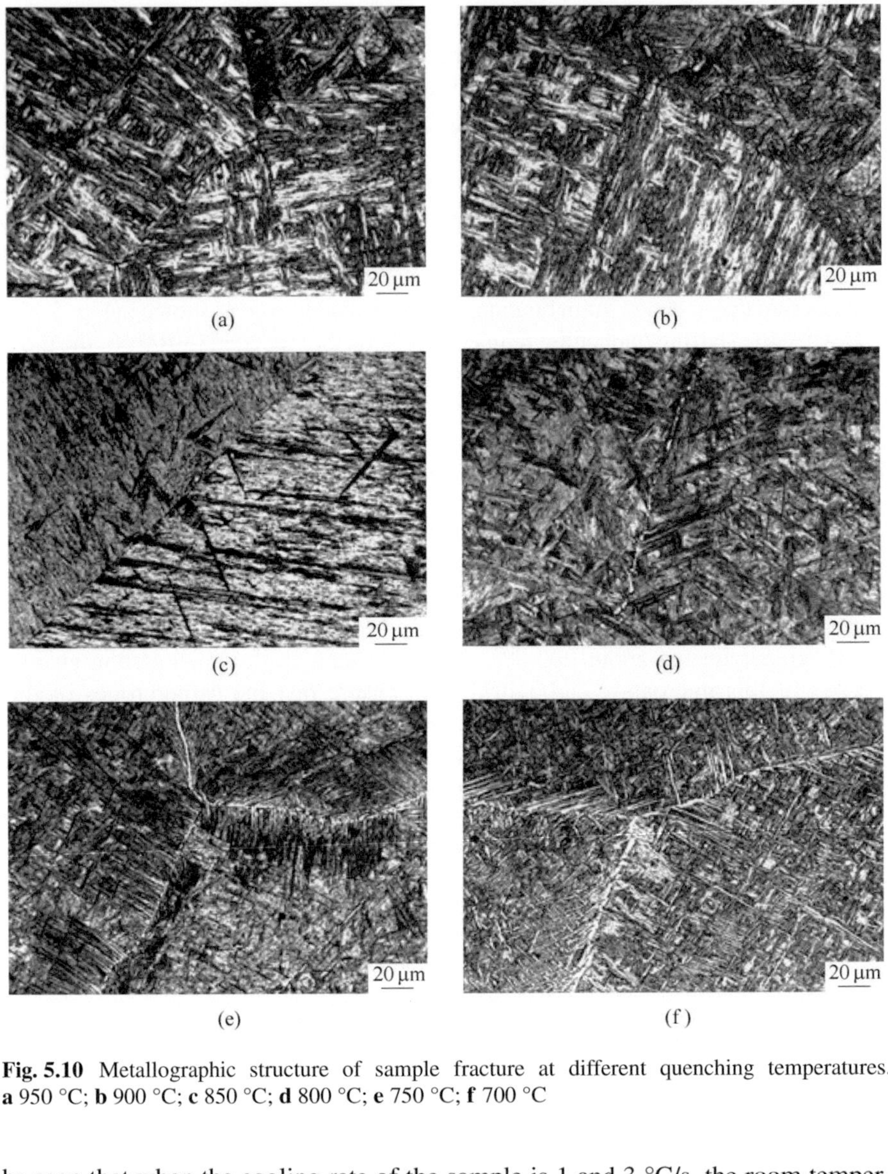

Fig. 5.10 Metallographic structure of sample fracture at different quenching temperatures. **a** 950 °C; **b** 900 °C; **c** 850 °C; **d** 800 °C; **e** 750 °C; **f** 700 °C

be seen that when the cooling rate of the sample is 1 and 3 °C/s, the room temperature structure is mainly ferrite + pearlite, and presents a certain Widmanstatten. However, with the increase of cooling rate, the number of acicular ferrites increases, the block ferrites decrease, the tendency to form Widmanstatten weakens, and the ferrite grains gradually refine. The ferrite grains of the sample are further refined and the Widmanstatten is basically eliminated when the cooling rate reaches 4 °C/s. The bainite structures gradually appear in the test samples when the cooling rate further reaches 5 °C/s. At this time, almost all the ferrite is transformed into a fine

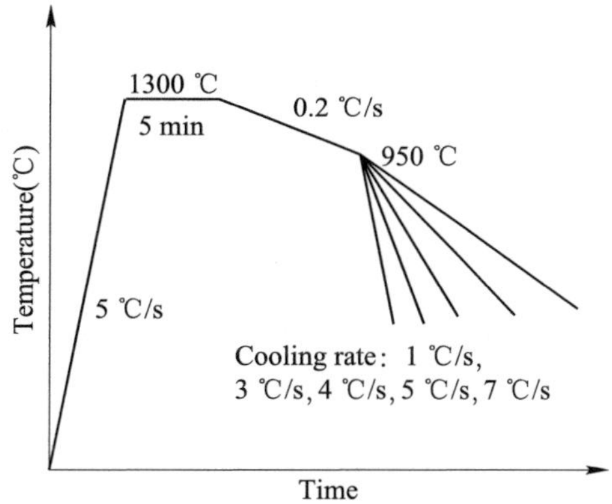

Fig. 5.11 Thermal simulation scheme at different cooling rates

acicular structure and the orientation becomes more complex, with good strength and toughness matching, and the grain refinement effect is also obvious. When the cooling rate reaches 7 °C/s, the number of bainites in the structure increases, but there are also a large number of fine acicular ferrites.

It is also known from Chap. 2 of this book, the implementation of different controlled cooling processes on the structures of high temperature steel not only affects the growth of the structure, but also affects the precipitation size and distribution of microalloy carbonitrides in the steel, thereby affecting the grain size of the reheating austenite. Figure 5.13 shows the precipitation morphology of Nb-bearing carbonitrides in the structure of the test samples at the above different cooling rates. Under the condition that the cooling rate is not greater than 4 °C/s, although the precipitation size of the microalloy carbonitrides in the sample is gradually refined with the increase of the cooling rate, all of them are distributed in a chain form, as shown in Fig. 5.13a–c. However, when the cooling rate of the sample is greater than 5 °C/s, the microalloy carbonitrides in the steel structure are gradually dispersed and distributed finely. Under this effect, the cooling ferrite and reheating austenite process of the steel structure will promote grain nucleation and refine the austenite grains after reheating.

Figure 5.14 shows the metallographic structure morphology of the samples with the above different cooling rates, reheated to 830 °C at a reheating rate of 3.0 °C/s. It can be seen from Fig. 5.14, the size of the austenite grains in the sample is larger and the largest is more than 1 mm under a cooling rate of 1 °C/s, as shown in Fig. 5.14a. When the cooling rate of the sample reaches 3 and 4 °C/s, the austenite grains formed by reheating are significantly refined by rapid cooling to refine the ferrite and the increase in the number of precipitates, and the maximum size is only about 0.5 mm, as shown in Fig. 5.14b, c. When the cooling rate of the sample further increases to 5 and 7 °C/s, the maximum size of the reheated austenite grains is refined to ≤ 0.2 mm, as shown in Fig. 5.14d, e. Combined with the distribution of microalloy carbonitride precipitation and the evolution of structure grain size at different cooling rates, it can be determined that the optimal quenching rate of Nb-bearing micro-alloyed steel should ≥ 5 °C/s.

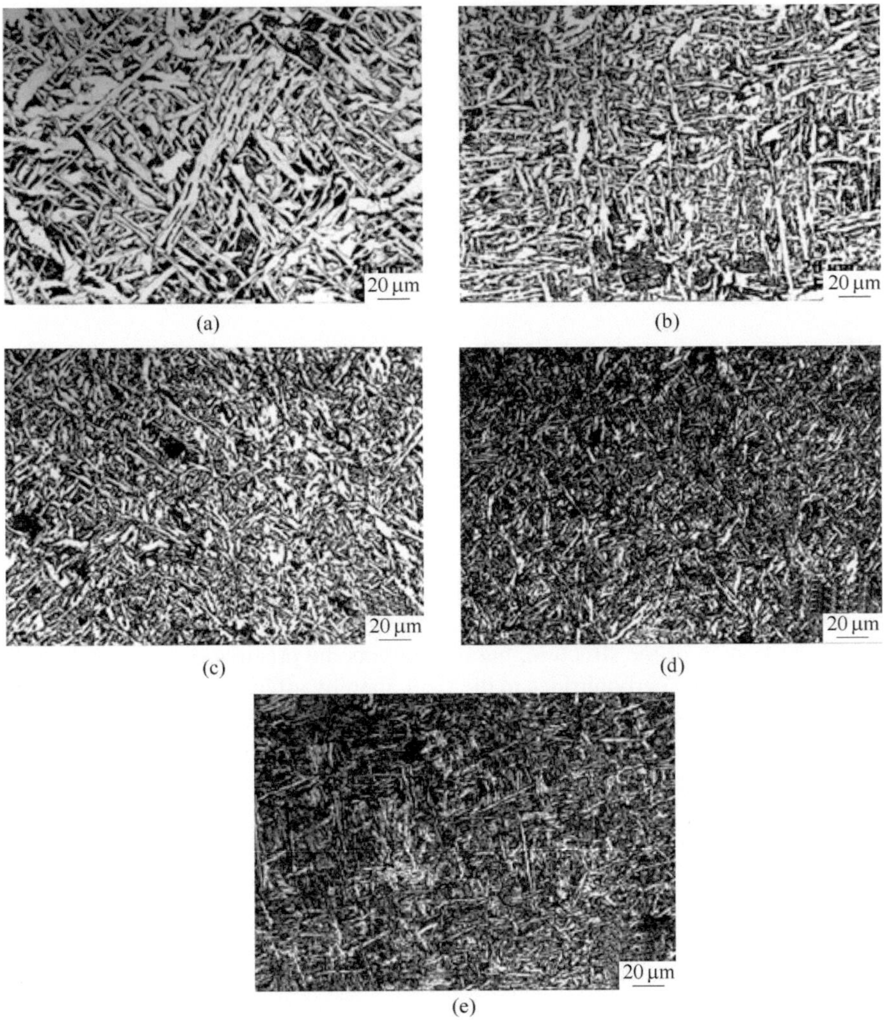

Fig. 5.12 Metallographic structure of Nb-bearing steel samples at different cooling rates. **a** 1 °C/s; **b** 3 °C/s; **c** 4 °C/s; **d** 5 °C/s; **e** 7 °C/s

5.3.2.3 Quenching Termination Temperature

The appropriate slab surface quenching termination temperature is an important process parameter that determines the subsequent surface reheating temperature of the continuous casting slab, the heat of slab left to the heating furnace, and the thickness of the slab surface quenching layer. The key to determine the optimal quenching termination temperature for different steel structures is to determine the temperature of its $\gamma \rightarrow \alpha$ phase transition. On this basis, according to the thickness of the surface layer of the slab to be quenched and combined with the actual production

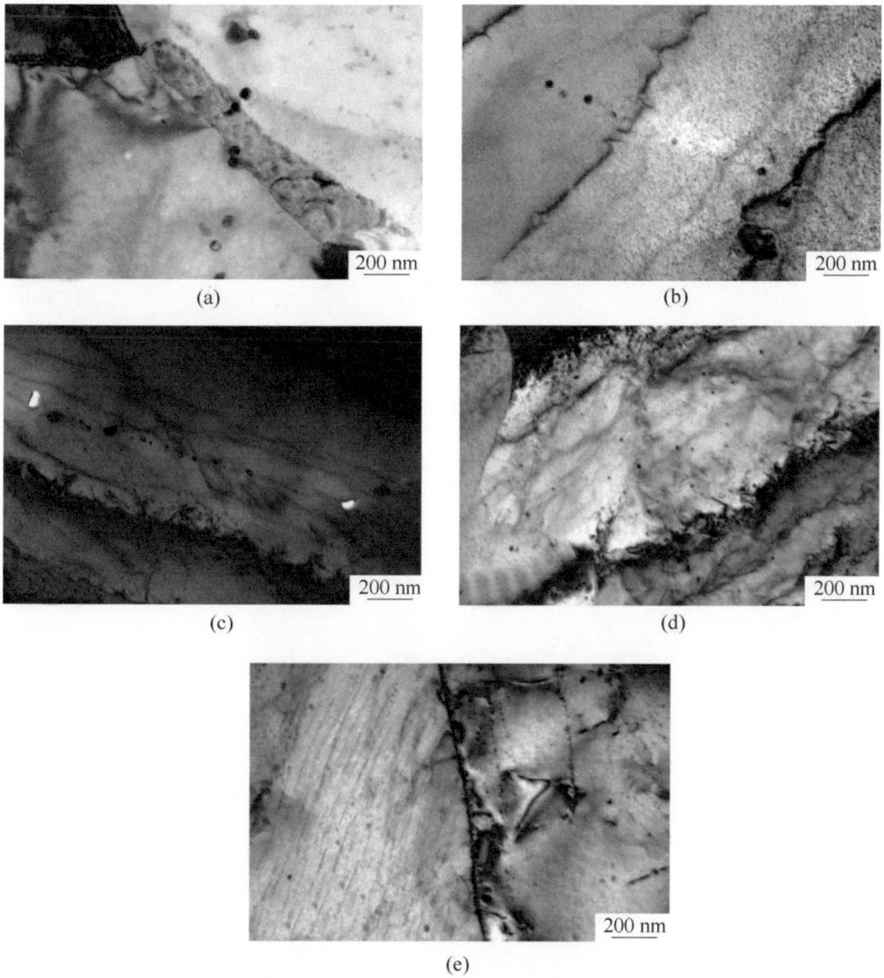

Fig. 5.13 Precipitation morphology of Nb (C, N) in steel samples at different cooling rates. **a** 1 °C/s; **b** 3 °C/s; **c** 4 °C/s; **d** 5 °C/s; **e** 7 °C/s

of continuous casting, the surface temperature of the slab whose surface layer is completely transformed into ferrite, pearlite or bainite structures is determined.

During the cooling process of the steel structure, when a phase transition occurs, there will be a discontinuous volume change, thus forming an inflection point [41]. According to this characteristic of the thermal expansion curve, the phase transition temperature of each grade of steel can be accurately determined [42]. Therefore, this book uses the Ni-bearing micro-alloyed steel shown in Table 5.2 as the research object. The continuous cooling transformation curve (CCT curve) of the overcooling austenite is determined by the thermal expansion method, and combined with the actual measured micro-structure evolution of the steel to determine the end

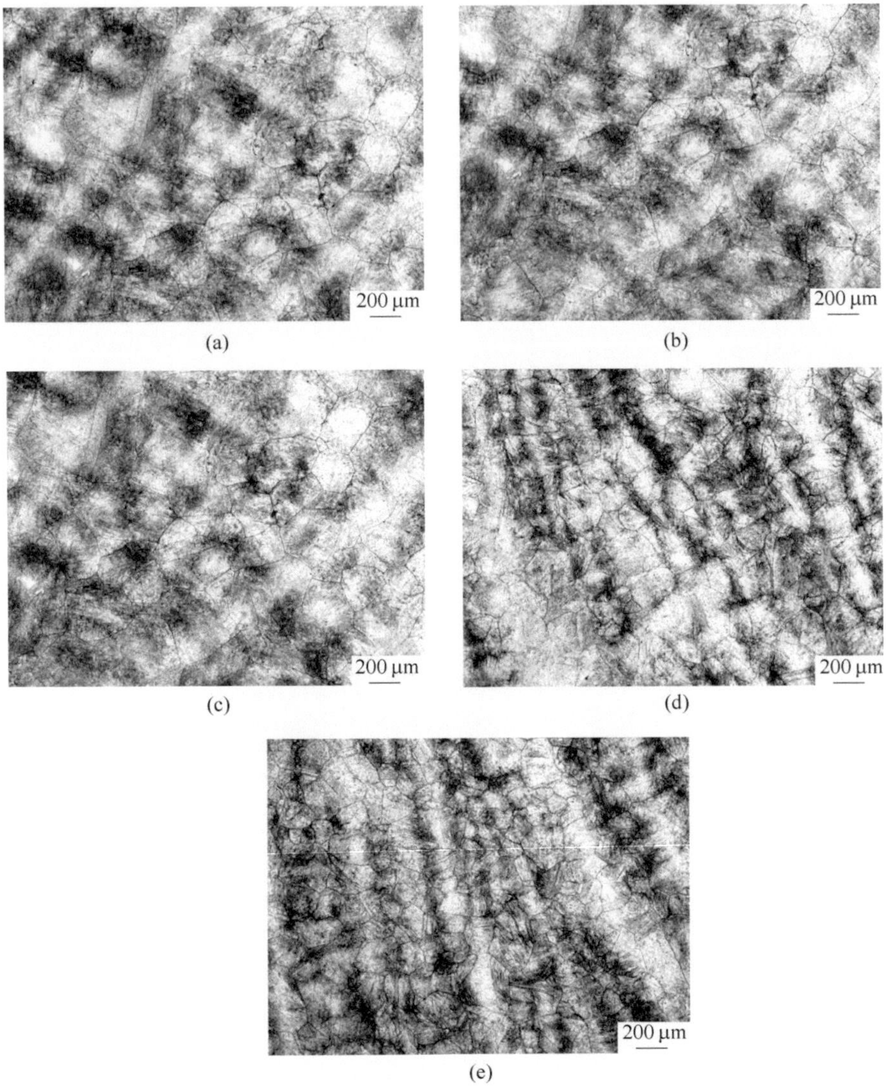

Fig. 5.14 Metallographic structure morphology of reheated austenite at different cooling rates. **a** 1 °C/s; **b** 3 °C/s; **c** 4 °C/s; **d** 5 °C/s; **e** 7 °C/s

temperature of the phase transition of the steel structure under different cooling rate conditions.

During the testing process, the Gleeble thermal simulator was also used to heat the slab samples to 1000 °C at a rate of 5 °C/s. After preserve for 10 min, the samples were cooled to room temperature at different cooling rates such as 0.2, 0.5, 1, 2, 5, 10, 15, 20, and 30 °C/s. The thermal expansion curves of each sample during the experiment were recorded. The test scheme is shown in Fig. 5.15.

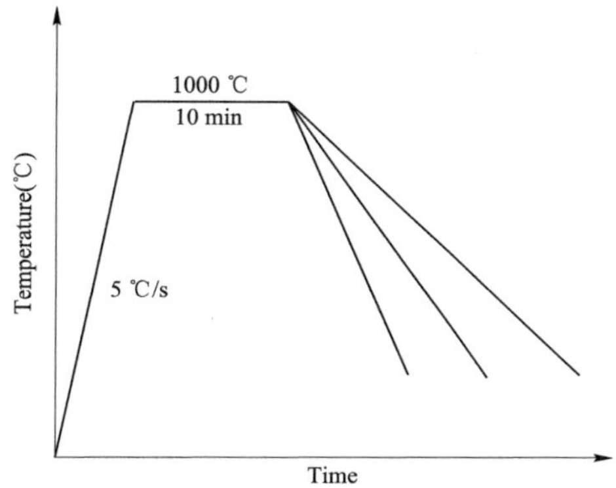

Fig. 5.15 Experimental scheme of static CCT curve

Figure 5.16 shows the metallographic structure of the thermal simulation samples under the above different cooling rates. It can be seen from the fig., when the cooling rate is not greater than 4 °C/s, the micro-structure of sample is mainly ferrite + pearlite, and there is an obvious grain boundary network proeutectoid ferrite structure. However, with the increases of cooling rate, the number and size of pearlite decrease, the intragranular ferrite becomes fine, and the network proeutectoid ferrite becomes thin, as shown in Fig. 5.16a–d.

As the cooling rate continues to increase, the network proeutectoid ferrite at the grain boundaries and so on in the tensile sample continues to thin until it disappears, and gradually appears as bainite and martensite. When the cooling rate reaches 5 °C/s, the tensile sample mainly consists of a large amount of intragranular acicular ferrite and a small amount of pearlite structure. When the cooling rate reaches 10 °C/s, the number of intragranular acicular ferrite significantly decreases, and a large amount of bainite and a small amount of martensite begin to appear. When the cooling rate continues to increase to 15 °C/s, the number of intragranular acicular ferrite almost completely disappears, and the number of martensite significantly increases. When the cooling rate reaches 20 °C/s, the micro-structure of the sample is mainly martensite, with a large amount of black acicular lower bainite; when the cooling rate reaches 30 °C/s, the room temperature structure almost completely transforms into lath martensite, it can be considered that when the cooling rate reaches 30 °C/s, all the austenite of the sample is directly transformed into martensite.

Combining the metallographic structure at different cooling rates with the critical point of the thermal expansion curve, draw the static CCT curve of the experimental steel grade as shown in Fig. 5.17. Among them, A represents austenite; F represents ferrite; P represents pearlite; B represents bainite; M represents martensite. The statistics of the critical temperatures of the phase transition at different cooling rates are shown in Table 5.3. Among them, the subscript s represents the starting temperature of the phase transition, and f represents the end temperature of the phase transition.

As can be seen from Fig. 5.17 and Table 5.3, the phase transition starting and end temperatures of this Ni-bearing micro-alloyed steel under different cooling rates

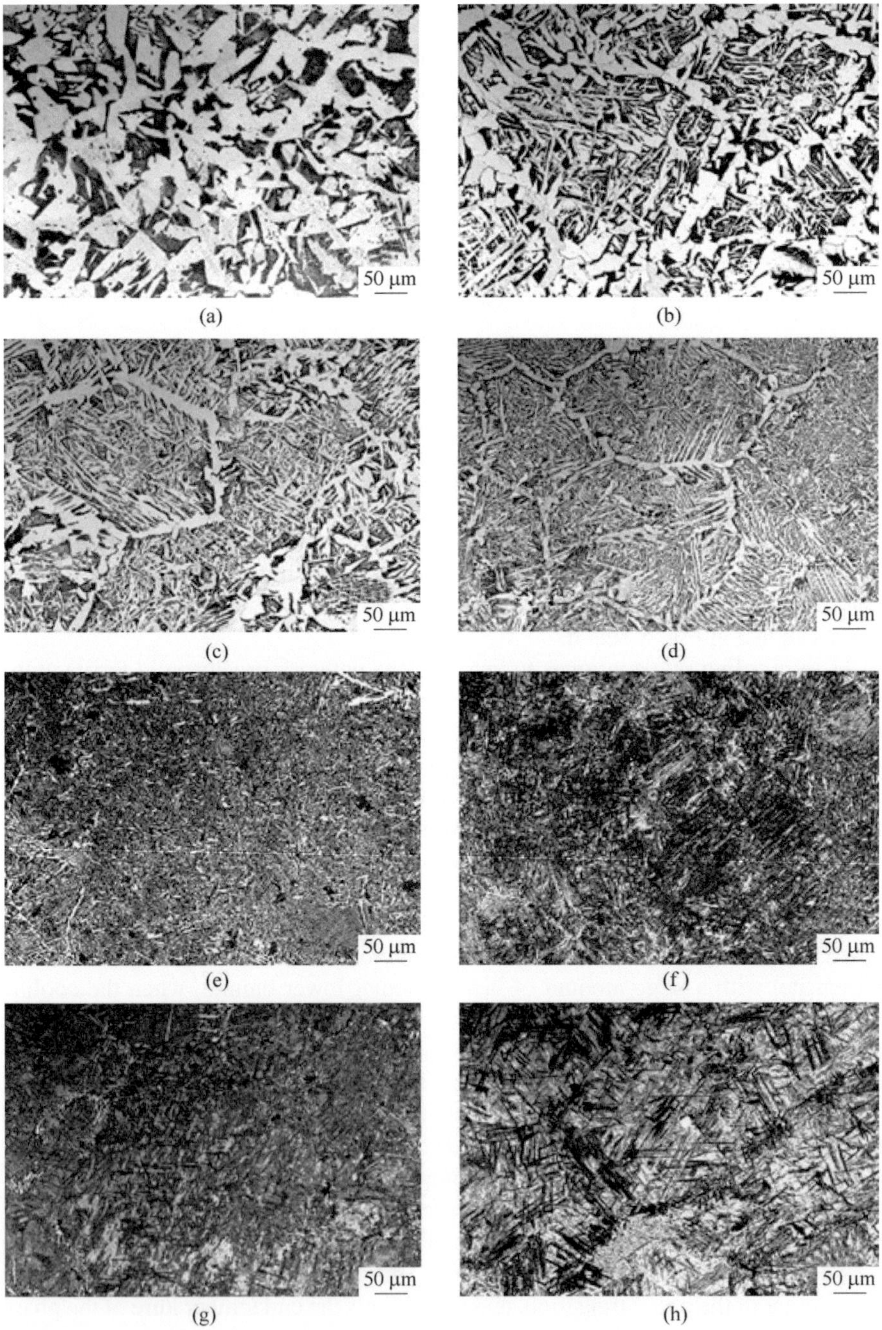

Fig. 5.16 Structure morphology of tensile samples at different cooling rates. **a** 0.2 °C/s; **b** 0.5 °C/s; **c** 1 °C/s; **d** 2 °C/s; **e** 5 °C/s; **f** 10 °C/s; **g** 15 °C/s; **h** 20 °C/s; **i** 30 °C/s

Fig. 5.16 (continued)

(i)

Fig. 5.17 Static CCT curve of a certain Nb-bearing steel

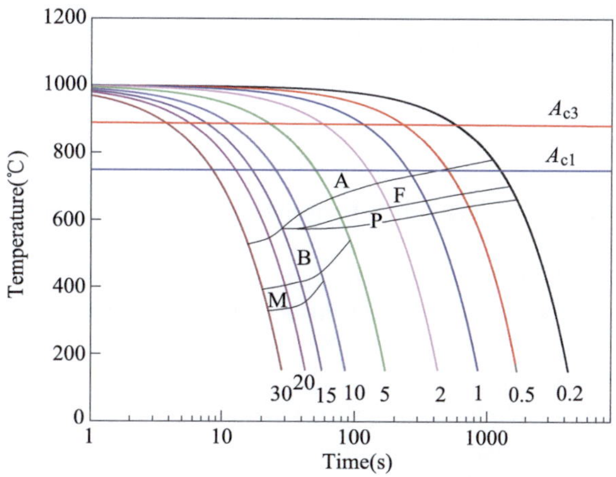

Table 5.3 Critical temperature of micro-structure transition at different cooling rates (°C)

Cooling rate (°C/s)	A-F_s	A-F_f	A-P_s	A-P_f	A-B_s	A-B_f	A-M_s	A-M_f	Metallographic structure
0.2	782	695	695	662	–	–	–	–	F + P
0.5	755	679	679	643	–	–	–	–	F + P
1	732	657	657	620	–	–	–	–	F + P
2	712	637	637	592	–	–	–	–	F + P
5	665	610	610	565	565	539	–	–	F + P + Few GB
10	619	572	572	568	568	428	428	418	F + GB + M
15	–	–	–	–	574	415	415	350	GB + LB + M
20	–	–	–	–	542	403	403	336	LB + M
30	–	–	–	–	528	392	392	328	Few LB + M

conditions decrease with the increase in the cooling rate. The main reason is that the ferrite transition is a diffusion transition [43], and the cooling rate is inversely proportional to the atomic diffusion rate. At the same time, as the cooling rate increases, the difference in free energy between the overcooling new and old phases decreases, resulting in a decrease in the driving force for phase transition, and the phase transition temperature also decreases [41]. According to the requirement that the optimal cooling rate not less than 5 °C/s, the A_{r3} of this steel is about 665 °C, and A_{r1} is about 610 °C, which means its quenching termination temperature should be less than 610 °C. In actual production, due to the high crack sensitivity of martensite, it is generally necessary to prevent the slab surface structure from drop to the temperature zone where martensite structure is formed. Therefore, the quenching termination temperature of the slab surface should be greater than 400 °C. Combined with the intended quenching depth of the continuous casting slab surface, the optimal quenching termination temperature of the slab surface is generally adopted to be 400–450 °C.

5.3.2.4 Mechanical Properties of Quenched Steel

The high ductility of the slab surface structure is the purpose of implementing the continuous casting slab surface quenching process. To explore the hot ductility and other mechanical properties of the slab surface structure after the implementation of the quenching process, the tensile strength and RA of the steel under the above optimum quenching process parameters were tested by thermal simulation tensile test. The test scheme of quenching process and traditional hot charging process to be carried out is shown in Fig. 5.18. In the traditional hot charging process thermal simulation test, the sample is heated to 1300 °C at a rate of 5 °C/s, preserved for 5 min, then cooled to 650 °C at a cooling rate of 0.2 °C/s (simulation of continuous casting and roller table cooling process), preserved for 5 min, and then heated to different test temperatures at a rate of 3 °C/s. The testing temperatures are 1200, 1100, 1000, 900, 800 and 700 °C, respectively, after preserving for 1 min, the sample was tensioned fracture at a deformation rate of 1×10^{-3} s^{-1}, as shown in Fig. 5.18a. In the thermal simulation test of the quenching process, the sample was also heated to 1300 °C at a rate of 5 °C/s, and then cooled to the optimal starting quenching temperature of 950 °C at a cooling rate of 0.2 °C/s after preserving for 5 min. Then, it was cooled to 400 °C at a quenching rate of 5 °C/s, heated to 800 °C at a rate of 3 °C/s, and then cooled to 600 °C at a cooling rate of 0.2 °C/s again (simulation of transportation and cooling process of the slab on the roller after reheating). After preserving for 5 min, it was tensioned fracture when heated to 1200, 1100, 1000, 900, 800 and 700 °C at a rate of 3 °C/s, as shown in Fig. 5.18b.

Tensile strength refers to the maximum stress that steel structure can resist uniform ductile deformation [44]. When the stress of continuous casting slab exceeds its tensile strength, the slab is easy to form crack defects. Therefore, tensile strength is an important parameter to measure whether high temperature slab will crack. Figure 5.19 shows the stress–strain curve of the sample stretching process at 1000 °C.

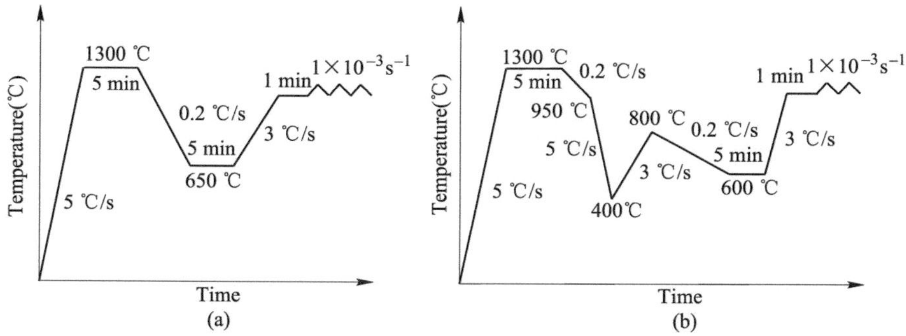

Fig. 5.18 Test scheme of mechanical properties of slab surface structure under different charging processes. **a** Traditional hot charging process. **b** Surface quenching process

As can be seen from Fig. 5.19, when the strain increases to a certain value, the stress reaches a maximum (tensile strength), if the deformation continues to increase, the tensile stress of the sample decreases. At this time, the sample undergoes non-uniform ductile deformation, and gradually begins to crack.

The relationship between the tensile strength of Nb-bearing steel continuous casting slab sample under the above two charging processes and temperature is shown in Fig. 5.20. As can be seen that the tensile strength of the samples under the two different processes decreases with the increase of temperature. However, at each temperature, the tensile strength of the samples under the slab surface quenching process is higher than that of the traditional hot charging process. Therefore, after implementing the slab surface quenching process, the surface structure of the micro-alloyed steel continuous casting slab is less prone to crack due to thermal stress during the heating process, which is beneficial to the control of slab hot charging cracks.

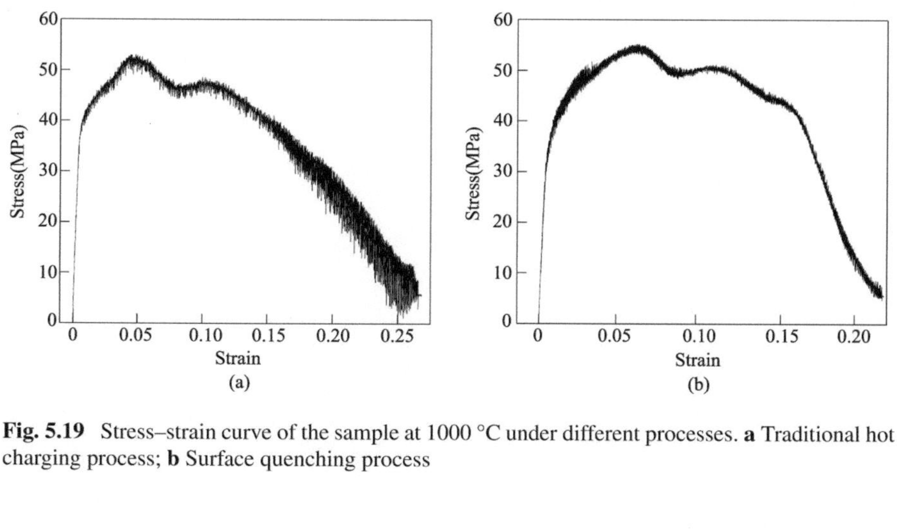

Fig. 5.19 Stress–strain curve of the sample at 1000 °C under different processes. **a** Traditional hot charging process; **b** Surface quenching process

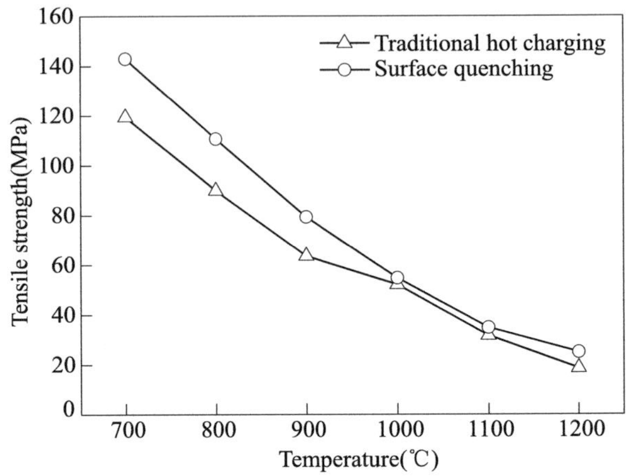

Fig. 5.20 Comparison of tensile strength of samples under different charging processes with temperature

The RA is an important index to measure the hot ductility of steel structure. At each temperature, the larger the RA of the steel structure, the better its hot ductility, the stronger its crack resistance. The relationship between the RA of the tensile sample of the slab under different charging processes and the temperature is shown in Fig. 5.21. Research [45] shows that RA of 60% can be regarded as the boundary between high ductility and low ductility of steel structure (it is common to set RA of 40% as the boundary for whether continuous casting slab cracks). As can be seen that RA at each temperature of the samples under the traditional hot charging process is lower than that of the quenching process. Especially when the slab temperature is 800 °C, the hot ductility of a certain Nb-bearing steel is the worst, and the RA is only about 37.5%, at this time the slab surface is prone to form crack defects under the action of thermal stress. However, when the slab surface quenching process is adopted, the RA of the tensile samples at different test temperatures is significantly higher than that of the traditional hot charging process, and the RA of the samples at each test temperature is higher than 60%. It significantly improves the hot ductility of the slab surface structure. Therefore, the surface of the slab after quenching is generally less prone to form surface crack defects during the heating process in the heating furnace.

5.3.3 *Optimal Quenching Position in Casting Flow*

The optimal quenching casting flow position of the continuous casting slab surface is mainly determined by the solidification characteristics of the micro-alloyed steel produced by the main continuous casting of the steel plant (the optimal starting quenching temperature) and its continuous casting production process (especially the casting speed and secondary cooling process), that is, it should be determined

Fig. 5.21 RA of tensile samples under different charging processes with temperature evolution

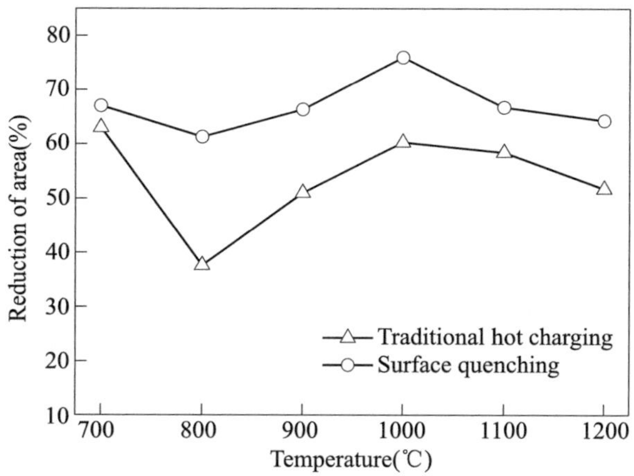

based on the optimal starting quenching temperature of the main steel grade and the surface temperature evolution of the continuous casting slab in the casting flow.

The casting flow roller array of wide-thick slab continuous caster in a steel plant is shown in Fig. 5.22. Using the cooling water volume at the main casting speed of 1.4 m/min shown in Table 5.4, continuous casting is shown in Table 5.2 for the wide-thick slab of Nb-bearing steel. The surface temperature evolution of the process is shown in Fig. 5.23. In the production of a Nb-bearing steel under this condition of continuous casting water distribution, the temperature of the center of the slab inner arc surface is greater than 950 °C in the casting flow at the end of the No.12 bow segment and subsequent segments. Considering the actual continuous casting production process, the liquid core solidification end position of the slab often extends to the No.13 segment at the highest casting speed of 1.6 m/min, and the No.12–14 bow segments are the same cooling zone, so the optimal quenching casting position for the wide-thick slab continuous caster in the steel plant is the No.15 and No.16 bow segments.

5.3.4 Slab Temperature Evolution During New Quenching Process

Based on the optimal quenching casting flow position of the wide-thick slab continuous caster in a certain steel plant, the temperature evolution of the Nb-bearing steel wide-thick slab quenching process shown in Table 5.2 under the main casting speed of 1.4 m/min is shown in Fig. 5.24. During the slab surface quenching process, the slab surface temperature decreases rapidly in the early stage of quenching (0–20 s). In this early stage, the average cooling rate of the slab surface exceeds 15 °C/s, and the temperature quickly decreases to below 600 °C. At this temperature, the structure at the slab surface will undergo a transition from austenite to bainite.

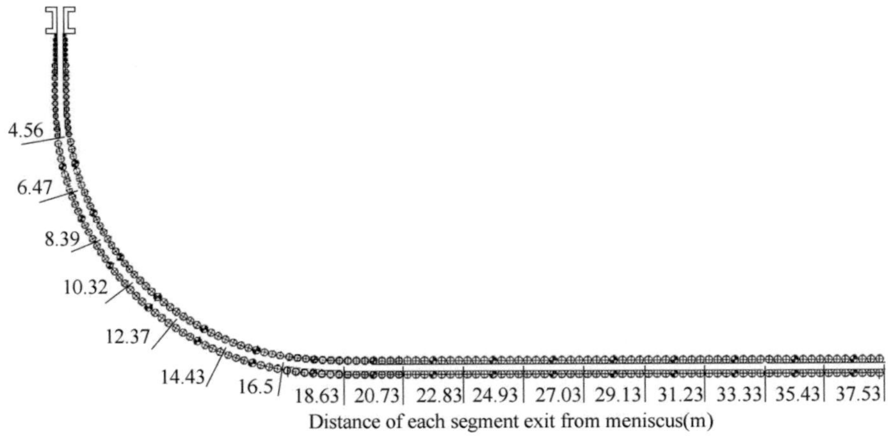

Fig. 5.22 Roller array diagram of the continuous caster

Table 5.4 Water volume of each zone of secondary cooling of wide-thick slab continuous casting

Cooling zone	1N	1I/O	2I/O	3I/O	4I/O	5I	6I	7I	8I	9I	10I	11I
Water volume (L/min)	91	105	410	320	305	104	126	70	37	22	0	0

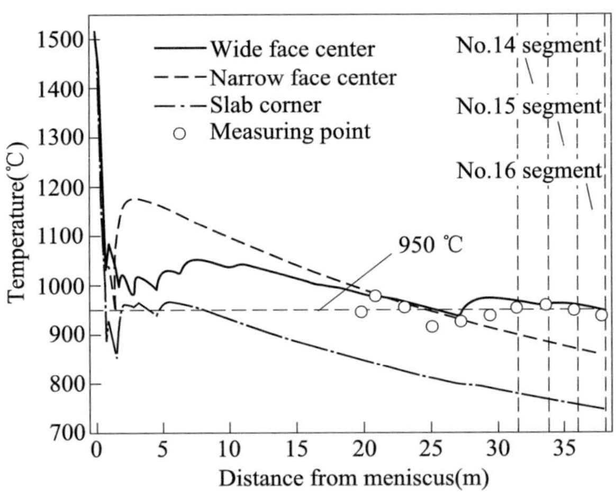

Fig. 5.23 Evolution of the surface temperature fields of a Nb-bearing steel slab

As the quenching time further extends, the slab surface temperature continues to decrease to the minimum about 398 °C after the slab exits the quenching segments, meeting the requirements for the optimal quenching termination temperature of the slab surface. However, in the middle and late stages of quenching, the cooling rate of the slab surface significantly decreases to above 5 °C/s. However, in the early stage of quenching, the slab surface temperature has quickly decreased from about 965 °C to below 600 °C, and the cooling rate in this process is far higher than 5 °C/s, and the microalloy carbonitrides of the slab surface structure will be dispersed.

Figure 5.25 shows the temperature and cooling rate evolution at a depth of 10 mm wide subsurface of the slab during the quenching process. As can be seen from Fig. 5.25, under quenching, the cooling rate at 10 mm below the wide face of the slab also reaches a maximum of about 12.5 °C/s in the early stage of quenching, and the temperature of the slab structure in this process quickly decreases from about 1000 °C to about 700 °C, which can also ensure the dispersed precipitation of Nb (C, N). After the slab exits the quenching segments, the temperature at 10 mm below the surface decreases to the minimum about 550 °C, also meeting the temperature for the transition from austenite to ferrite.

Due to the absence of spray cooling in the No.16 bow segment of a certain steel plant, the cast slab after quenching only undergoes radiative heat transfer and contact

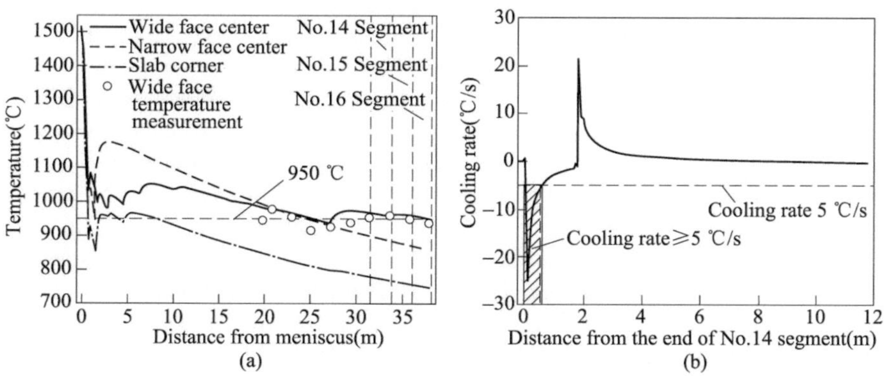

Fig. 5.24 Evolution of temperature and cooling rate during the quenching process in No.15 bow segment. **a** Slab surface temperature; **b** Cooling rate

Fig. 5.25 Temperature evolution at 10 mm below the slab during the quenching process in No.15 segment

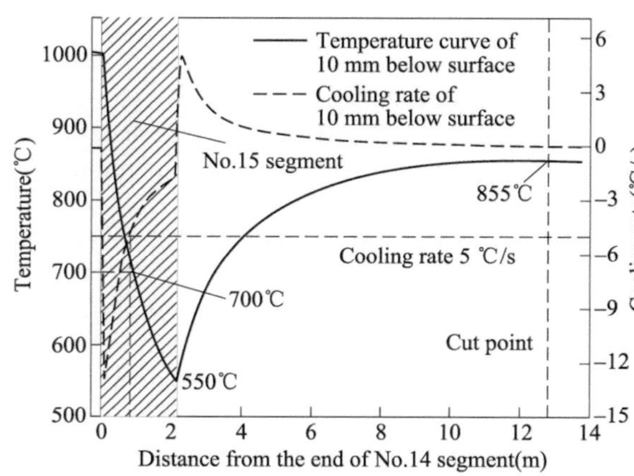

heat transfer with the casting roller (calculated uniformly by the heat transfer coefficient of the slab surface). Affected by the heat transfer from the high temperature core of the slab, the quenching layer gradually reheats from inside to outside, showing a trend of rapid initial heating and gradual slowing down in the middle and later stages. In this reheating process, the maximum reheating rate of the slab surface and 10 mm subsurface reached 22.4 °C/s and 5.1 °C/s, respectively. Due to this high reheating rate, the quenching structure of the slab surface will promote grain refinement during the austenitization process. When the slab moves forward to the flame cutting machine, the temperatures of the slab surface and 10 mm subsurface have returned to 833 and 855 °C, respectively, and the structure is austenitized again, forming a refined surface structure.

Figure 5.26 shows the temperature evolution of the surface and 10 mm subsurface of the slab during the quenching and reheating process under its maximum casting speed of 1.6 m/min when the No.15 segment is used as the quenching segment. As can be seen from Fig. 5.26, during the surface quenching process of the slab, the trends of cooling and reheating on its surface and 10 mm subsurface are basically the same. However, due to the increase of the heat content of the slab cross section under high casting speed conditions, and the rate of heat transfer from the core to the outside is lower than the rate of heat taken away by surface quenching, the surface temperature of the slab decreases to the minimum of about 395 °C after surface quenching, while the temperature of the slab at 10 mm subsurface only decreases to 557 °C. Even so, during the whole quenching process of the slab, the temperature and cooling rate within the 0–10 mm range subsurface all meet the requirements for the dispersion of Nb (C, N) and the phase transition of the structure to ferrite. When the quenched slab runs to the cutting machine, the reheating temperature of the slab surface and subsurface have increased to 861 and 879 °C, respectively. However, under high casting speed conditions, the water volume in the slab quenching process increases significantly. Under wide-thick slab quenching conditions, when the casting speed is increased from 1.4 to 1.6 m/min, the water volume needs to increase by about 22.4%, and the greatly increased water volume poses more stringent requirements for the water supply capacity of the quenching pipeline.

From the surface temperature evolution of Nb-bearing steel wide-thick slab under the main continuous casting speed of a certain steel plant in Fig. 5.23, it can be known that both the No.15 and No.16 segments meet the requirements for the optimal starting quenching temperature. However, the temperature of the slab at the No.15 segment is generally higher, and the required quenching water volume is also larger. Therefore, this book investigates the temperature evolution of implementing surface quenching of the slab at the No.16 segment.

Figure 5.27 shows the temperature evolution at the surface and 10 mm subsurface of the slab during the quenching and reheating process under the main casting speed of 1.4 m/min, with No.16 bow segment as the quenching segment. As can be seen from Fig. 5.27, when the slab surface is quenched in No.16 bow segment, under the condition of satisfying the lowest surface temperature similar to the quenching in No.15 segment, the temperatures at the surface and 10 mm subsurface of the slab after the quenching segment are reduced to the lowest 392 and 544 °C, respectively,

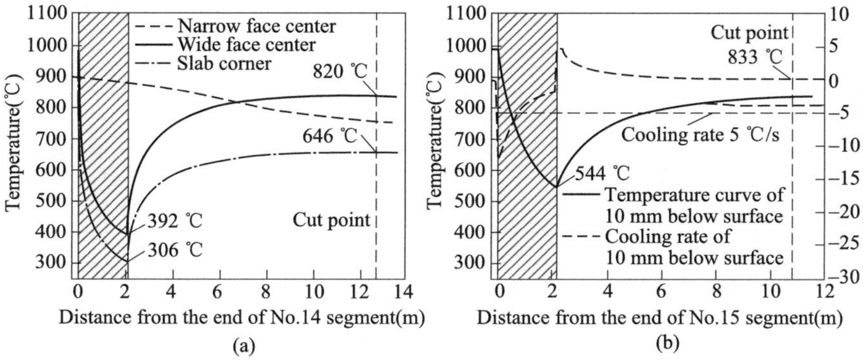

Fig. 5.26 Evolution of temperature and cooling rate at 10 mm below the slab during the quenching process at the casting speed of 1.6 m/min in No.15 segment. **a** Temperature of slab surface; **b** Temperature and cooling rate at 10 mm below the slab

and the maximum cooling rate also exceeds 10 °C/s. The temperatures at the surface and 10 mm subsurface of the slab when it reaches the cutting machine position also reheat to 820 and 833 °C, both meeting the optimal quenching process requirements (under the high casting speed condition of 1.6 m/min, the trend of the slab temperature evolution is similar, also meeting the optimal quenching process requirements, which will not be repeated in the following text). However, the quenching water volume under the casting speed conditions of 1.4 and 1.6 m/min can be decreased by 9.6 and 11.4%, respectively, compared to the water volume for quenching in No.15 segment, and the quenching water volume under the highest casting speed condition of 1.6 m/min is only about 8.5% higher than the quenching water volume under the main casting speed of 1.4 m/min in No.15 segment, which can greatly alleviate the pressure of excessive water supply capacity requirements.

5.3.5 Development of New Continuous Quenching System at the End of Continuous Casting Caster

Based on the research of key process parameters of continuous casting slab surface quenching and the optimal quenching casting flow position, the method of canceling the last segment of the continuous caster can be adopted according to the actual production of the continuous caster, and the last segment is designed as the slab surface quenching device as shown in Fig. 5.28. The last or penultimate segment of the corresponding continuous caster can also be designed as the ultra-strong spray quenching segment with dense nozzles structure as shown in Fig. 5.29, and independent, dynamic ultra-strong quenching can be implemented on the inner and outer arc of the slab.

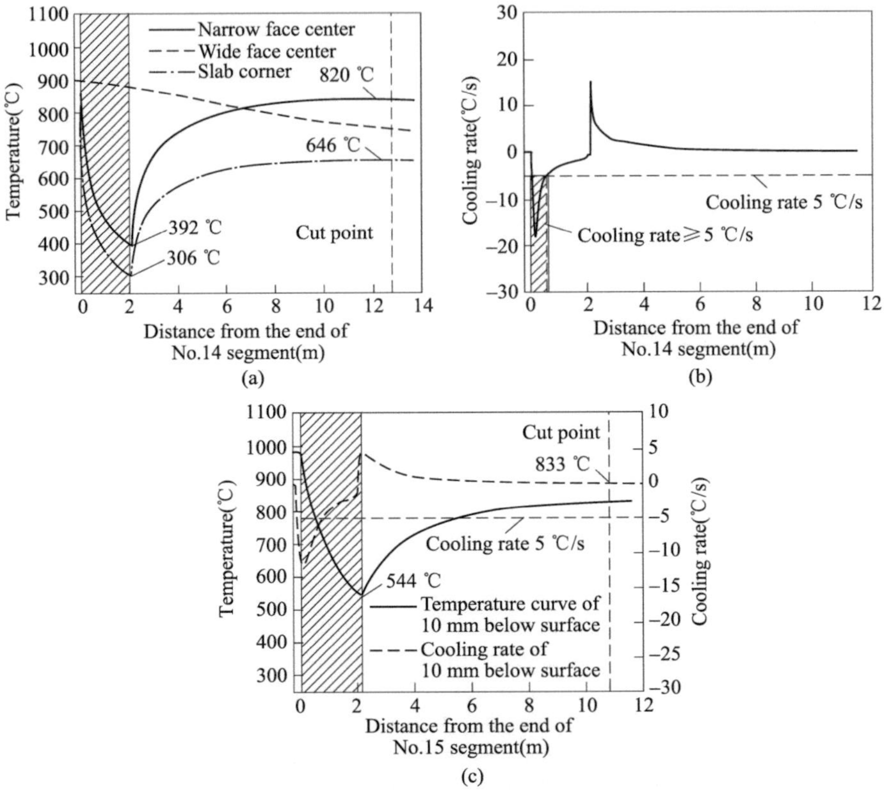

Fig. 5.27 Evolution of temperature and cooling rate of slab surface and 10 mm below the slab during the quenching process in No.16 segment (1.4 m/min). **a** Slab surface temperature; **b** Slab surface cooling rate; **c** Temperature and cooling rate 10 mm below the slab surface

5.3.6 *Application of the New Slab Quenching and the Effect*

In the actual production under the slab surface quenching process, in addition to the mentioned optimal quenching casting flow position and design of the water supply capacity, for slab continuous casting, due to the different heat transfer conditions of its inner and outer arcs, the cooling water of the inner arc after spraying onto the surface will stay for a long time, and the slab will continue to undergo heat transfer. However, the cooling water of the slab outer arc is separated from the slab immediately after spraying onto the surface. If the water volume ratio of the slab inner and outer arcs is irrational, the slab will form upward or downward warping deformation as shown in Fig. 5.30, which will affect continuous casting production. Therefore, in the actual implementation of this technology, it is necessary to control the large difference in water volume between the inner and outer arcs of the slab.

Figure 5.31 shows the evolution of the slab surface temperature at the end of the wide-thick slab continuous caster of a certain steel plant during the quenching

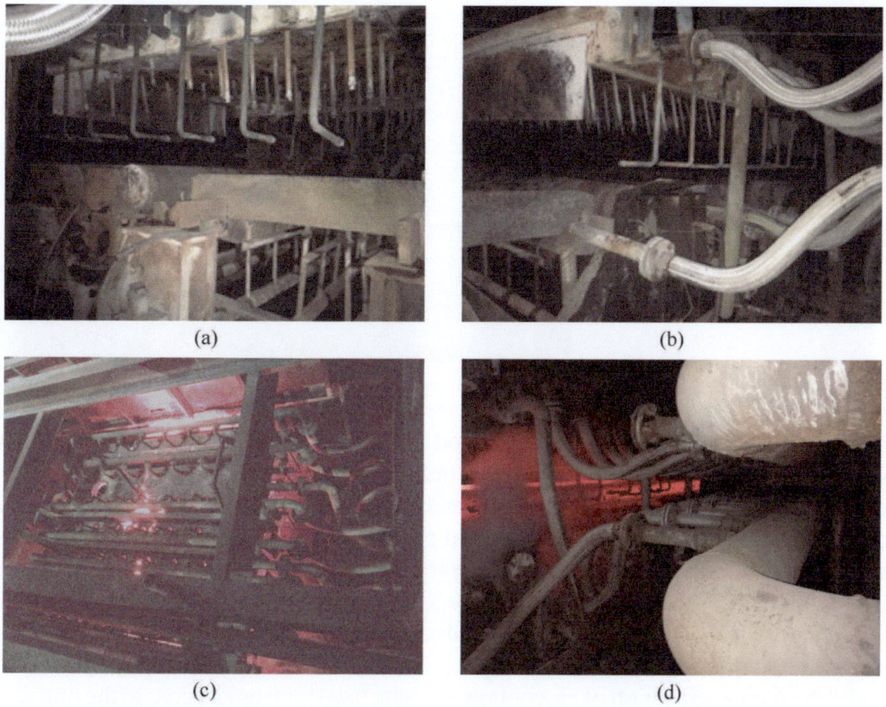

Fig. 5.28 Quenching device at the end of continuous caster. **a** Non-water supply side of the quenching device; **b** Water supply side of the quenching device; **c** Quenching spray rack; **d** Water supply pipeline of spray rack

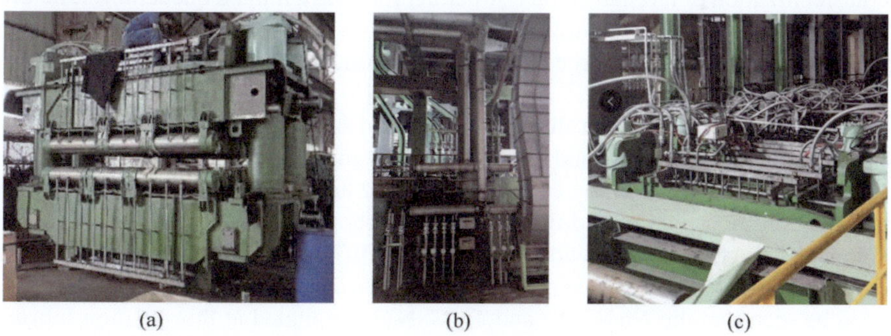

Fig. 5.29 Quenching segment and its water supply structure of the continuous caster. **a** Quenching segment; **b** Water supply circuit; **c** Line on the quenching segment

Fig. 5.30 Warping deformation of the slab after quenching

process, quenching end, flame cutting and cutting end position. It can be seen that in the actual quenching process of a certain steel plant, the surface of the slab is cooled rapidly in the first half of the quenching device (under the casting speed and quenching water volume of a certain steel plant, the time of the slab surface cooling to black is generally 50–60 s), completing the quenching structure transition. Then in the second half of the quenching, the quenching depth of the slab surface continues to increase toward its core. When the slab comes out of the quenching machine, its surface temperature generally reduces to 380–410 °C. Then the slab quickly reheats under the combined action of air cooling and heat transfer from its core to the outside. When the slab moves to the cutting point, its surface temperature rises back to 620–670 °C. When the cutting of the slab is completed, the slab surface temperature increases back to the highest temperature of about 780–820 °C, thus completing the re-austenitization of the slab surface structure, forming a refined quenching structure layer. Then the slab temperature gradually decreases during the roller table charging process. When the slab is charged to the heating furnace, the slab surface temperature is mostly 550–620 °C.

Samples were taken from the slab before and after the application of the new surface quenching process, and the precipitation morphology and structure transition of the surface microalloy carbonitrides were analyzed. Figure 5.32 shows the precipitation morphology and distribution of carbonitrides at the 5 mm below the wide face and corner of the wide-thick slab of Nb-Ti-bearing steel produced under the traditional continuous casting process of a certain steel plant. From Fig. 5.32, it can be seen that the Nb (C, N) and Ti (C, N) precipitated at this position subsurface of the slab is large, mainly in range of 100–200 nm, and often appear in chain distribution. As can be inferred from Sect. 5.1 on the causes of hot charging crack on the surface of micro-alloyed steel continuous casting slab, this large size chain precipitation of carbonitrides will significantly reduce the hot ductility of the slab surface structure during heating, which is prone to form hot charging crack on the slab surface.

Figure 5.33 shows the morphology of carbonitrides precipitated at 5 mm below the wide face and corner of the Nb-Ti-bearing steel wide-thick slab after adopting the

Fig. 5.31 Morphology evolution of the wide-thick slab during quenching process. **a** Quenching process; **b** Quenching device exit; **c** At the flame cutting device; **d** Flame cutting device exit

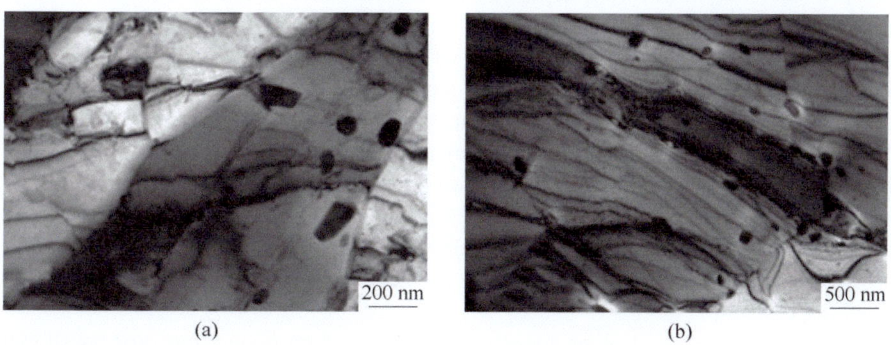

Fig. 5.32 Precipitate morphology at 5 mm below the wide face and corner of the slab during the traditional continuous casting process. **a** Wide face; **b** Corner

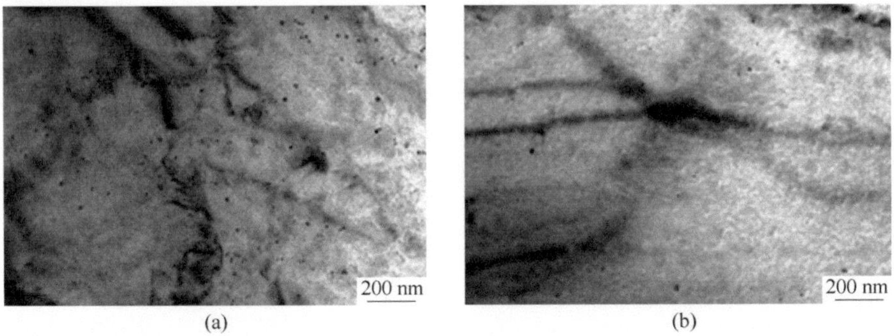

Fig. 5.33 Precipitate morphology at 5 mm below the wide face and corner of the slab during the surface quenching process. **a** Wide face; **b** Corner

slab surface quenching process shown in Fig. 5.31 in a certain steel plant. It can be seen that the precipitation at the same position is dispersed and small, with most of the precipitation being less than 20 nm. As can be known from the mechanism of slab surface high temperature quenching process and the micro-structure control, these small and dispersed precipitation can provide a large number of nucleation sites for ferrite nucleation during the subsequent quenching process of the slab, refining the grains during the transition process from austenite to ferrite. At the same time, it will also significantly refine the grains of the slab surface structure during the reheating austenitization process, improving the crack resistance of the slab surface structure during the heating process in the heating furnace.

Figure 5.34 shows the micro-structure morphology of Nb-Ti-bearing micro-alloyed steel slab at different depths under the wide face of slab produced by traditional continuous casting process in a steel plant. It can be seen that the room temperature micro-structure at different positions in the wide face subsurface of the slab in the traditional continuous casting process is composed of ferrite and pearlite, and the morphology is basically the same, which are characterized by low ductility Widmanstatten and grain boundary proeutectoid ferrite film micro-structure. As can be known from Chap. 2 and Sect. 5.1, the hot ductility of the micro-structure on the slab surface is low. Affected by this, if the surface micro-structure of the hot charging slab is in a dual-phase zone micro-structure when entering the heating furnace, it will easily crack at the grain boundary during the heating process and form hot charging cracking defects.

Figure 5.35 is a steel plant adopting the slab surface quenching process shown in Fig. 5.31 to produce Nb-Ti-bearing micro-alloyed steel wide-thick slab, and the micro-structure of the slab at different depths in the wide face subsurface of the slab. As can be seen that the micro-structure at different depths of the quenched slab is also composed of ferrite and pearlite, the morphology of the ferrite differs significantly from that of the traditional continuous casting process. After quenching, the ferrite grain size of the surface micro-structure of the slab is fine and uniform, and there is no Widmanstatten micro-structure and proeutectoid ferrite film micro-structure.

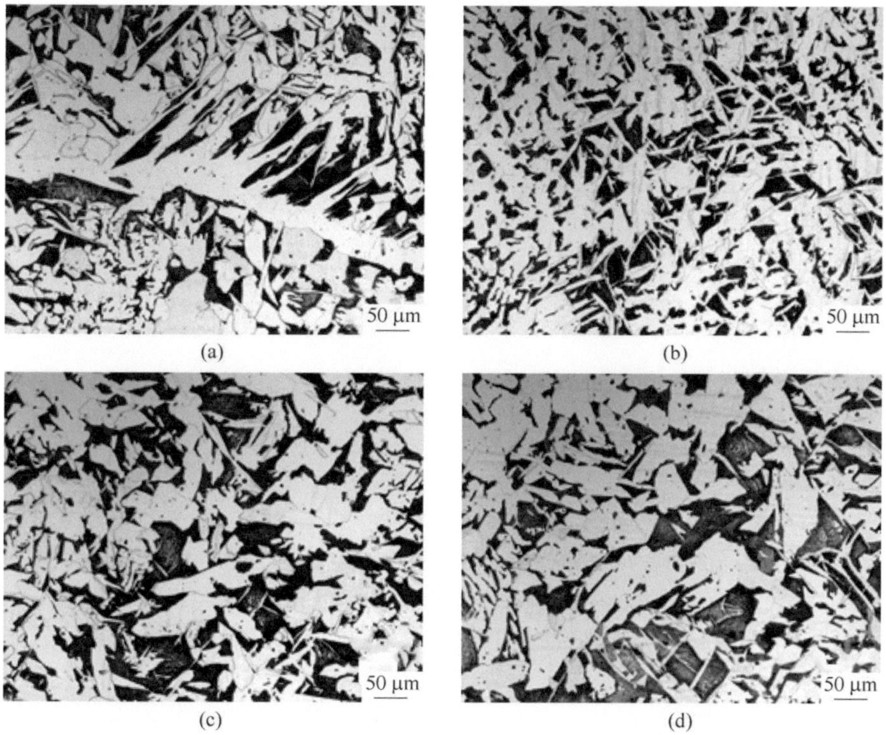

Fig. 5.34 Micro-structure morphology at different depths below the wide face of the slab during the traditional continuous casting process. **a** 1 mm; **b** 10 mm; **c** 20 mm; **d** 40 mm

The main reason for the significant transition in the surface micro-structure of the quenched slab is that during the strong quenching process, the rapid cooling causes the microalloy carbonitrides to diffuse and precipitate in the steel matrix, thus providing a large number of nucleation particles for the subsequent cooling process, forming a substantially refined ferrite grain. Then, the surface quenched slab is subjected to rapid heat transfer and reheating in its core, and the quenched micro-structure is austenitized again, thus forming an austenite micro-structure with fine grain. When this austenite micro-structure is cooled to room temperature during the subsequent air cooling process, the micro-structure is transformed again to form extremely fine, uniform equiaxed ferrite and pearlite micro-structure.

However, as can be seen from the fig., as the distance from the slab surface increases, the size of the surface ferrite grain of the quenched slab gradually increases. When the depth from the slab surface reaches 20 mm, the refined slab micro-structure begins to become unclear. When the depth from the slab surface reaches 40 mm, its ferrite morphology is close to that of the traditional continuous casting process. This is mainly because the surface quenching process of a steel plant is to ensure the dispersion of carbonitrides in the surface micro-structure within 10 mm from the

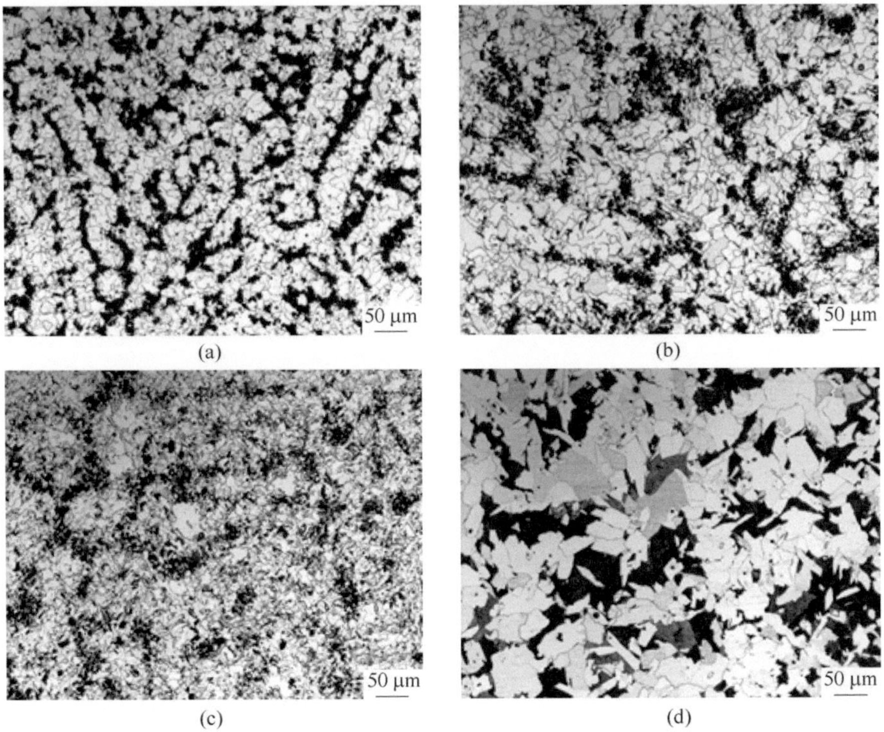

Fig. 5.35 Micro-structure at different depths below the wide face of the slab during the surface quenching process. **a** 1 mm; **b** 10 mm; **c** 20 mm; **d** 40 mm

slab surface and the significant refinement of the micro-structure grain within 15 mm.

Figure 5.36 shows the micro-structure at different depths under the corner of a slab subsurface produced by a steel plant adopting the traditional continuous casting process for Nb-Ti-bearing micro-alloyed steel wide-thick slab. As can be seen, the corner micro-structure of the slab under the traditional continuous casting process is also basically the same as that of the wide face at different depth. It is also composed of large ferrite and pearlite, and there is also a certain proportion of Widmanstatten micro-structure and proeutectoid ferrite film.

As can be seen from Fig. 5.37, the micro-structure of the corner of the slab subsurface quenching process is mainly composed of ferrite, pearlite and bainite, the morphology of its ferrite is quite different from the traditional continuous casting process, which is mainly acicular micro-structure. The acicular ferrite has the characteristics of large angle grain boundary and high dislocation density, so it can effectively refine grain and has good strength and toughness matching. At the same time, as can be seen, the corner micro-structure of the slab after quenching has basically eliminated the Widmanstatten micro-structure and the proeutectoid ferrite film. The main reason is that during the surface quenching process of the high temperature

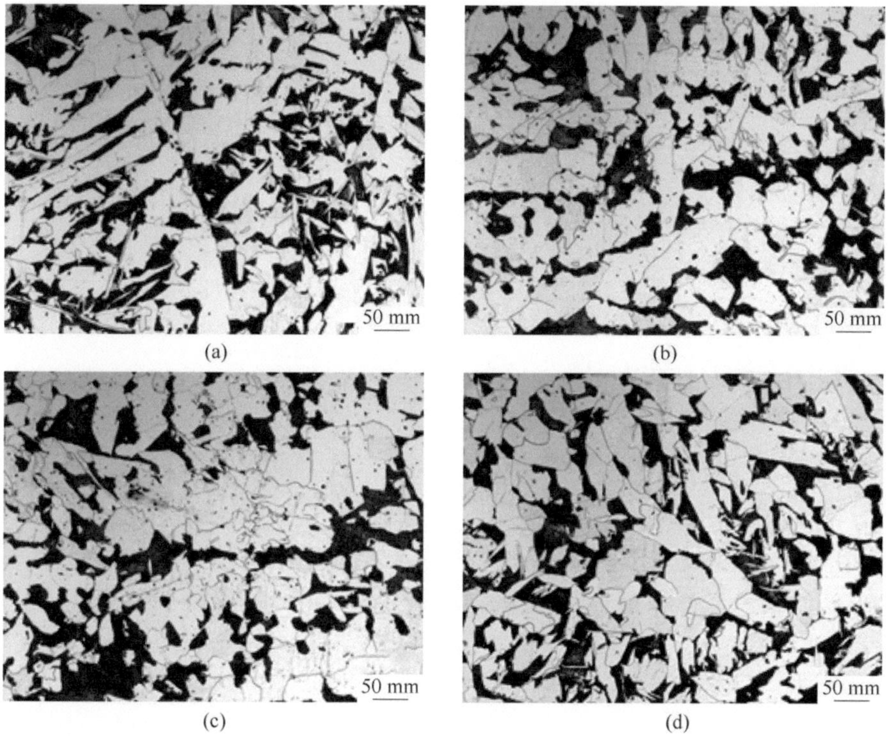

Fig. 5.36 Micro-structure at different depths below the corner of the slab during the traditional continuous casting process. **a** 1 mm; **b** 10 mm; **c** 20 mm; **d** 40 mm

slab, when the corner micro-structure of the slab is subjected to rapid cooling, its cooling rate is rapid and the temperature is relatively low, forming a large amount of intragranular acicular ferrite. After quenching, due to the low reheating ability of the corner of the slab, its reheating maximum temperature is far lower than A_{c1}, which cannot achieve the reheating austenization process and retain the rapid cooling quenching micro-structure to a certain extent.

In addition, with the increase of depth from the slab surface, the morphology of the corner structure is also quite different. In the range of 1–10 mm below the surface, the ferrite is distributed in a fine needle-like and interlaced with each other. However, when the depth increases to 20 mm below the surface, the growth trend of acicular interlaced ferrite is significantly weakened, and when the depth increases to 40 mm below the corner surface, although the structure has changed to a certain extent, it is gradually not obvious. This shows that the surface quenching process adopted by a steel plant, the quenching depth of the corner is nearly 40 mm.

It can be seen from Fig. 5.28 that the surface quenching device of the wide-thick slab in a steel plant also quenched and cooled the narrow face of the slab during the implementation process. Figure 5.38 shows the structure morphology at different depths below the narrow face of the Nb-Ti-bearing micro-alloyed steel wide-thick

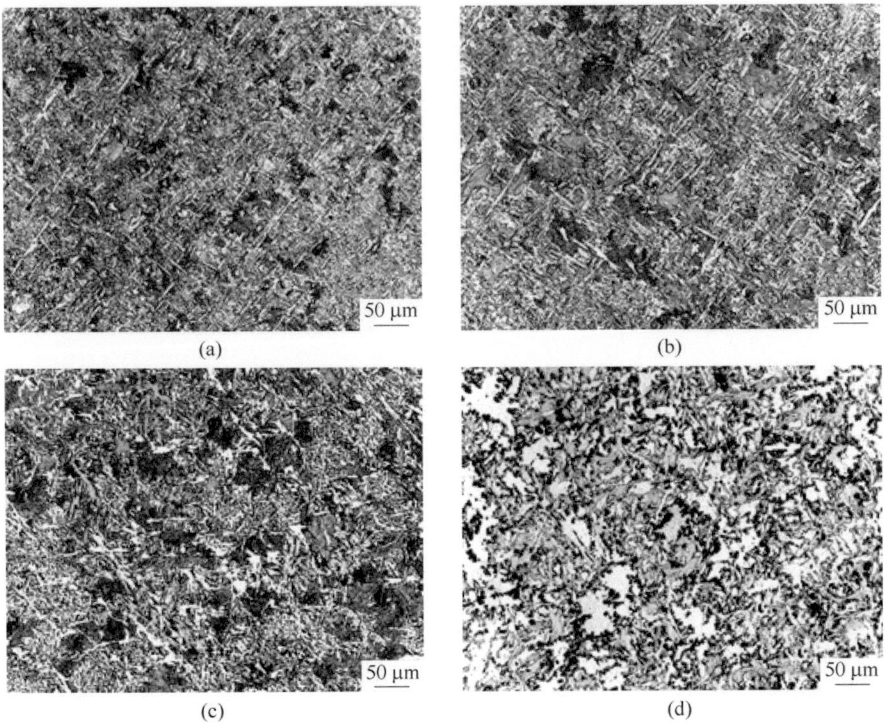

Fig. 5.37 Micro-structure at different depths of the corner of the slab subsurface quenching process

slab produced by the traditional continuous casting process. It can be seen that the narrow face structure of the slab under the traditional continuous casting process is similar to the wide face, which is also mainly composed of ferrite and pearlite, the structure morphology at different depths is basically the same, and there is also Widmanstatten micro-structure and proeutectoid ferrite film.

As can be seen from Fig. 5.39, the structure of the slab narrow face under the surface quenching process is significantly different from the wide face, showing the same structural distribution as the corner, which is mainly composed of ferrite, pearlite, and bainite. Moreover, the ferrite in the depth range of 0–10 mm in the surface is mainly fine needle-like, and there is no obvious Widmanstatten micro-structure and proeutectoid ferrite film. However, when the depth from the narrow face of slab increases to 20 mm, its structure is similar to the quenching structure 40 mm below the corner. The cause is similar to the transition of the quenching structure at the corner of the slab. When the depth from the narrow face of the slab reaches 40 mm below the surface, the structure is close to the traditional continuous casting process.

The surface quenching process of slab developed in this book can be changed in time according to the steel grade and continuous casting speed, the dynamic automatic adjustment of the inner and outer arc water volume (including opening

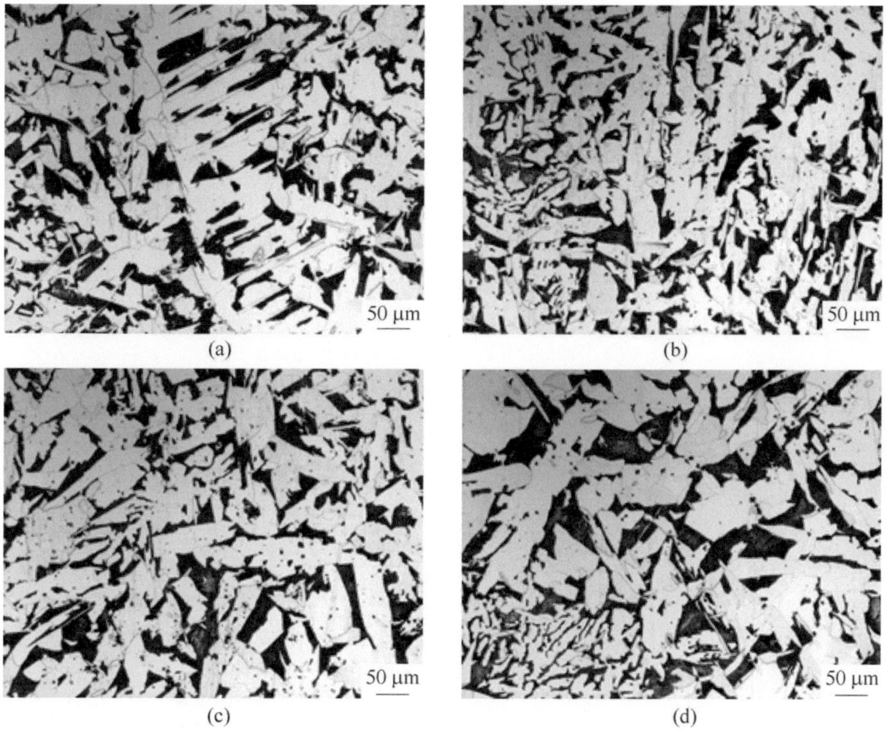

Fig. 5.38 Micro-structure at different depths below the narrow face of the slab during the traditional continuous casting process. **a** 1 mm; **b** 10 mm; **c** 20 mm; **d** 40 mm

and closing) can meet the slab stable quenching implementation of steel grades with different high temperature solidification characteristic. After the implementation of quenching process in a steel plant, the hot charging rate of the micro-alloyed steel wide-thick slab increased from less than 42% to about 98%, which ensured the high quality, high efficiency and green production of the wide-thick slab of micro-alloyed steel.

Therefore, based on carbonitride precipitation and structure evolution of the slab surface under the high temperature continuous quenching technology in this book, it can be seen that the implementation of the quenching process can significantly improve the structure morphology of the slab surface, eliminate proeutectoid ferrite film and Widmanstatten micro-structure, and form equiaxed or acicular ferrite fine grain layer with about 10–20 mm deep on the slab surface. At the same time, it also realizes the dispersion precipitation of the carbonitrides in a certain depth range below the slab surface, thereby significantly improve the hot ductility of the slab surface structure, fundamentally avoid the generation of hot charging cracks on the slab surface. It is an advanced and efficient (without adding special quenching steps, without affecting the production process) slab hot charging process.

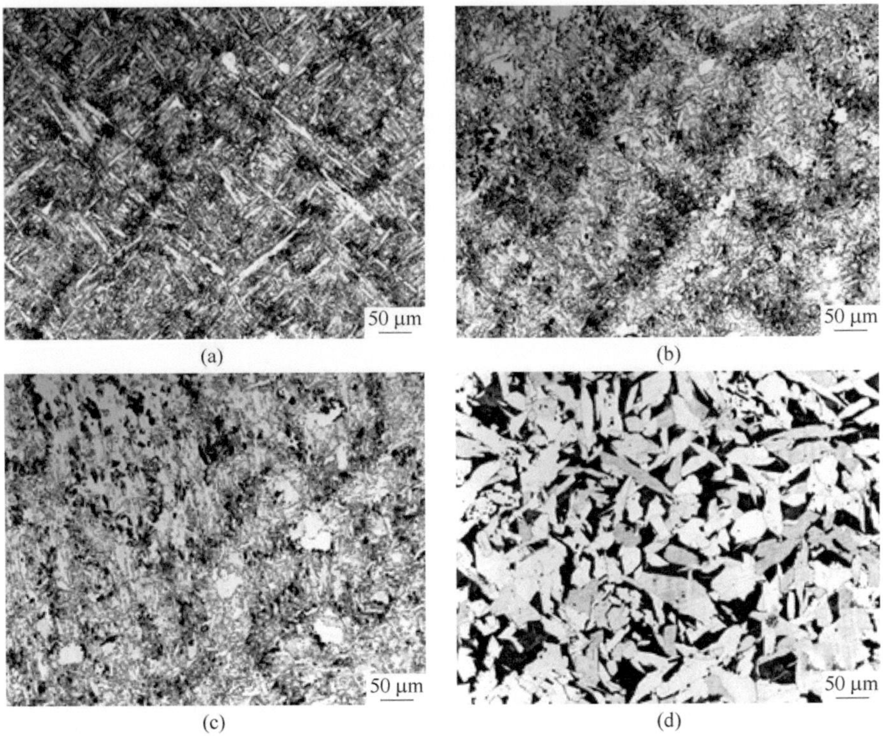

Fig. 5.39 Micro-structure at different depths below the narrow face of the slab during the surface quenching process. **a** 1 mm; **b** 10 mm; **c** 20 mm; **d** 40 mm

5.4 Summary

In this chapter, by analyzing the temperature field and surface structure evolution of the typical micro-alloyed steel wide-thick slab during continuous casting and hot charging, the cause of hot charging cracks are explored, and proposed a hot charging crack control technology based on slab surface high temperature quenching control. On this basis, taking the continuous casting process of wide-thick slab in a domestic steel plant as the research object, the optimal quenching process parameters and quenching casting flow position were analyzed and obtained, and the ultra-strong continuous quenching process and equipment technology at the end of the continuous caster was developed and applied. The main conclusions are as follows:

(1) The dual-phase structure of the surface of micro-alloyed steel slab with a hot charging surface temperature of 550–650 °C, which has a large difference in size and presents a mixed grain structure during the heating process in the heating furnace. The carbonitrides precipitated along the grain boundary during the continuous casting process, hinder the grain boundary movement of the slab during the heating expansion process, resulting in a significant decrease in its

hot ductility and stress concentration. They are the main cause of the surface hot charging cracks in the micro-alloyed steel continuous casting slab. Reducing the temperature of the slab surface micro-structure or significantly refined and diffuse the austenite micro-structure of the slab surface and its carbonitride precipitation before entering the heating furnace is the key to solving the problem of hot charging cracks in the micro-alloyed steel continuous casting slab.

(2) In the surface quenching process of continuous casting slab, the optimal starting quenching temperature should be higher than the "nose point" temperature of its microalloy carbonitride precipitation and the transition temperature from austenite to ferrite, the optimal cooling rate should not be less than 5 °C/s. The optimal quenching termination temperature of slab surface under the composition of main micro-alloyed steel at home and abroad is 400–450 °C.

(3) The optimal quenching casting flow position of the typical slab continuous caster is the appropriate area after the reduction segment of solidification end. The implementation of the quenching process based on the end of the continuous caster or the rapid cooling segment can significantly diffuse the carbonitride precipitation of its slab surface structure and improve its structure morphology, generating an equiaxed or acicular ferrite fine grain about 10–20 mm, which significantly improves the ductility of the slab structure, thus fundamentally avoiding the occurrence of hot charging cracks on the slab surface. It is an advanced and efficient slab hot charging process.

References

1. Yu ZX (2002) Hot delivery and hot charging technology of continuous casting slab. Metallurgical Industry Press, Beijing
2. Gao ZL, Wen Z, Liu ML (1997) Current situation and prospects of energy-Saving technology of heating furnace for steel rolling. Rolling 6:48–52
3. Zhao LC, Zhou JZ, Sun BR (1990) Analysis of connection between continuous casting and rolling steel. Steelmaking 1:29–37
4. Zhuchkov SM, Kulakov LV, Lokhmatov AP et al (2004) Ways of reducing energy costs in the continuous rolling of sections. Metallurgist 48(3):174–180
5. Zhang ST (1999) Systenatic optinizing technology of C C-D R & C C-H C R. Contin Cast 3(1):12–14
6. Wang DW (2004) Development and prospect of hot delivery and hot charging technology of continuous casting slab. Metallurg Manage 12:50–52
7. Tian JL (2007) An evaluation method of energy saving effect for hot delivery and hot charging of continuous casting billet. Metallurg Energy 26(4):12–14
8. Chen C, Ding CJ (2016) Research on surface cracks of sensitive materials during C C-D H C R process. Ind Heat 45(3):58–60
9. Wan YT (2008) Mechanism analysis of surface crack in medium plate during hot slab sending and charging. Iron Steel Res 3:14–16
10. Wada T, Tsukamoto H, Suga M (1988) Effect of hot charge rolling conditions from austenite region on micro-structure and mechanical properties of Nb and Ti bearing steel plates. ISIJ Int 74(7):1438–1445
11. Gao Y, Sun JL (2011) Effect of micro-structure characteristic of Q460C steel on surface cracks. Mater Sci Technol 19(5):139–143

12. Herman JC, Donnay B, Leroy V (1992) Precipitation kinetics of microalloying additions during hot rolling of HSLA steels. ISIJ Int 32(6):779–785
13. He D, Yu X, Chang J et al (2003) Temperature holding hood for hot charging of continuous casting slab in Tangshan iron and steel company. ISIJ Int 43(11):1742–1750
14. Wu XY, Tang GB, Pei YH et al (2009) Cause analysis of surface cracking on Ti-microalloyed steel slab during rough rolling. Phys Test 27(4):5–10
15. Xiao JG, Lin WF (1998) Effect of hot-charging process of C C slab on properties of steel plates. Steel 33(11):19–22
16. Shuai XY (2006) Effect of C C-H C R technology on properties of pipeline steel. Wugang Technol 44(4):14–18
17. Zhang PC, Wang LB, Tang D et al (2008) Effect of hot charging temperature on micro-structure and precipitation behavior of X80 pipeline steel. Metal Heat Treatment 33(10):99–102
18. Xia WY, Zhu ZH, Gan Y (2011) Experimental study on crack formation of micro-alloyed steel during hot charging process. Steel 46(12):29–32
19. Wang JG, Wang H, Li RJ (2017) The mechanism and control measures research of hot charging cracks. Contin Cast 42(2):22–26
20. Wang XH, Liu XY, Lv WJ (1998) Carbide and nitride precipitation and hot ductility of continuous cast steel slabs containing Nb, V, Ti. J Iron Steel Res 10(6):32–36
21. Maehara Y, Ohmori Y (1984) The precipitation of AlN and NbC and the hot ductility of low carbon steels. Mater Sci Eng 62:109–119
22. Wang XH, Zhang L et al (1997) The reduction of ductility and the precipitation of Nb(C, N) in Nb-bearing steel slab at 700–1000 °C. Acta Metall Sin 33(5):485–491
23. Mintz B, Yue S (1991) Hot ductility of steels and its relationship to the problem of transverse cracking during continuous casting. Int Mater Rev 36(5):187–271
24. Suzuki HG, Nishimura S, Yamaguchi S (1982) Characteristic of hot ductility in steels subjected to the melting and solidification. ISIJ Int 22(1):48–56
25. Li WG, Wang Q, Zhang HL et al (2016) Tertiary cooling of some continuous caster's products. Contin Cast 41(6):31–36
26. Zhang HM (2014) Cause analysis and preventive measures of surface cracks on steel plate rolled from hot delivered slab. Wide-thick Plate 20(1):21–23
27. Tercelli C (1995) Surface quenching: an effective tool for hot charging of special steels. Metall Plant Technol Int 18(1):69–71
28. Wang Q, Li YG, Lu YJ et al. A method to realize direct charging of high strength low alloy steel continuous casting billet: China, CN102228968B. 1 Jan 2014
29. Lu YJ (2013) Study on the formation and prevention mechanism of hot delivery cracks in medium-thick slab of low alloy steel. Chongqing University, Chongqing
30. Teshima T, Kitagawa T, Miyahara S et al (1988) The secondary cooling technology of continuous casting for the production of high temperature and high-quality slab. Tetsu-to-Hagane 74(7):1282–1289
31. Cai CS (2007) Slab surface hardening technology for hot delivery. In: Proceedings of the 8th national continuous casting conference, continuous casting branch of China Metal Society, 6, pp 71–76
32. Carboni A, Ruzza DW, Feldbauer SL (1999) Quenching for improved direct hot charging quality. Iron Steelmak 26(8):39–42
33. Maehara Y, Nakai K, Yasumoto K et al (1987) Hot cracking of low alloy steels in simulated continuous casting-direct rolling process. ISIJ Int 73(7):876–883
34. Kamada Y, Hashimoto T, Watanabe S (1990) Effect of hot charge rolling condition on mechanical properties of Nb bearing steel plate. ISIJ Int 30(3):241–247
35. Luo YZ, Zhang JM, Xiao C et al (2011) Research status of surface cracks in medium and heavy plate of microalloyed steel during hot charging. Steelmaking 27(4):74–78
36. Zhang Q (2004) Experimental study on the hot delivery and hot charging process of continuous casting slab. Master's Thesis of Beijing University of Science and Technology, Beijing
37. Li YC, Li BX, Guo MY (2016) Development and application of casting blank surface quenching technology. Hebei Metall 3:44–47

38. Kato T, Ito Y, Kawamoto M et al (2003) Prevention of slab surface transverse cracking by micro-structure control. ISIJ Int 43(11):1742–1750
39. Ma F, Wen G, Tang P et al (2011) Effect of cooling rate on the precipitation behavior of carbonitride in micro-alloyed steel slab. Metall Mater Trans B 42(1):81–86
40. Cai ZZ, An JZ, Liu ZY et al (2019) Research on development and application of micro-alloyed steel slab corner transversal crack control technology. J Iron Steel Res 31(2):117–124
41. Xu G (2009) Determination and drawing of CCT Curve of Metal Materials. Chemical Industry Press, Beijing
42. Wang X, Li HY, Tang W et al (2021) Determination and analysis of CCT curve of a high strength Steel. J Central South Univ (Nat Sci Edn) 52(4):1090–1098
43. Zheng DS, Jian HG, Wang SC et al (2018) Study on transformation behavior of hot-rolled microalloyed TRIP Steel. J Iron Steel Res 30(12):1013–1020
44. Bo XT (2017) Tensile strength of common steel at high temperature. Heat Treatment 32(6):62
45. Mintz B, Crowther DN (2010) Hot ductility of steels and its relationship to the problem of transverse cracking in continuous casting. Int Mater Rev 55(3):168–196

List of Publications

Book	Authors
Innovative Technology and Equipment for Efficient Utilization of Iron Ore	Yuexin Han, Yongsheng Sun, Peng Gao, Yanjun Li, Wenbo Li, Shuai Yuan
Theory and Technology for Profound Reduction of Refractory Iron Ore	Yongsheng Sun, Yuexin Han, Peng Gao, Qiang Zhang
Interaction Theory of Flotation Reagents for Non-sulphide Ores	Yimin Zhu, Jie Liu, Yanjun Li
Beneficiation of Boron-Bearing Iron Ore	Peng Gao, Jianwen Yu, Yanjun Li, Zhihang Li
Experiment, Modelling and Engineering Application of Non-linear Flow in Mining-Disturbed Rock Mass	Tianhong Yang, Bin Yang, Zenghe Xu, Wenhao Shi, Xin Yang
Low-Carbon Ironmaking Technology	Jue Tang, Mansheng Chu
Ladle Refining Technology of Bottom Powder Injection	Miaoyong Zhu, Wentao Lou, Zhongfu Cheng
Electroslag Remelting Technology for High Quality Special Steels	Zhouhua Jiang, Yanwu Dong, Xin Geng, Fubin Liu
The Control Technology for Slab Surface Crack During Continuous Casting of Micro-alloyed Steel	Zhaozhen Cai, Miaoyong Zhu
Solidification End Reduction of Continuous Casting Process	Cheng Ji, Miaoyong Zhu, Rui Guan
Technology and Application of Heat Treatment Equipment for Medium and Heavy Steel Plate	Zhaodong Wang, Jiadong Li, Tianliang Fu, Yong Li
New Generation TMCP Technology for Medium and Heavy Steel Plates	Zhaodong Wang, Bingxing Wang, Bin Wang, Yong Tian, Yanmei Li
Microstructure Control Theory and Performance of Low Carbon Medium Manganese Steel Plates	Linxiu Du, Xiuhua Gao, Hongyan Wu, Xiangyu Qi

(continued)

© Metallurgical Industry Press 2025

Z. Cai and M. Zhu, *The Control Technology for Slab Surface Crack During Continuous Casting of Micro-alloyed Steel*, Key Steel Manufacturing Technologies by Collaborative Innovation, https://doi.org/10.1007/978-981-96-7171-7

(continued)

Book	Authors
New Generation TMCP Technology for Hot Strip	Guo Yuan, Zhenlei Li, Jian Kang
Studies on Oxidation Behavior of Steels During Hot Rolling and Development and Application of the Scale Control Technologies	Zhenyu Liu, Guangming Cao
Advanced Technology of Processing and Equipment for Cold Rolled Strips	Jianping Li, Fuan Huan
Regulation Mechanism of Microstructure and Properties of High Performance Electrical Steel Produced by Twin-Roll Strip Casting	Yuanxiang Zhang, Yang Wang, Feng Fang, Guo Yuan
Ultra-High Strength Structural Steel	Zhaodong Wang, Xiangtao Deng
Fabrication of Bulk Nano/Submicron—Grained Steels	Linxiu Du, Hongyan Wu, Huifang Lan, Guosheng Sun
Technology and Process of Vacuum Rolling Cladding	Zongan Luo, Guangming Xie, Yingying Feng
High Performance Green Steels	Zhenyu Liu et al.
Key Intelligent Technology of Steel Strip Production Process	Dianhua Zhang, Hongru Li
Research and Application of Fine and Intelligent Control Technology in Plate Rolling Process	Jun Wang, Chunyu He, Zhijie Jiao, ZhongZhao, Zhiqiang Wu, Jianzhong Xu